ADVANCES IN
Chromatography

VOLUME 42

ADVANCES IN
Chromatography

VOLUME 42

Edited by

Phyllis R. Brown
UNIVERSITY OF RHODE ISLAND
KINGSTON, RHODE ISLAND

Eli Grushka
HEBREW UNIVERSITY OF JERUSALEM
JERUSALEM, ISRAEL

CRC Press
Taylor & Francis Group
Boca Raton London New York

CRC Press is an imprint of the
Taylor & Francis Group, an **informa** business

First published 2003 by Marcel Dekker, Inc.

Published 2019 by CRC Press
Taylor & Francis Group
6000 Broken Sound Parkway NW, Suite 300
Boca Raton, FL 33487-2742

© 2003 by Taylor & Francis Group, LLC
CRC Press is an imprint of Taylor & Francis Group, an Informa business

First issued in paperback 2019

No claim to original U.S. Government works

ISBN 13: 978-0-367-44685-7 (pbk)
ISBN 13: 978-0-8247-0950-1 (hbk)

**Visit the Taylor & Francis Web site at
http://www.taylorandfrancis.com**

**and the CRC Press Web site at
http://www.crcpress.com**

Library of Congress Cataloging-in-Publication Data
A catalog record for this book is available from the Library of Congress.

ADVANCES IN
Chromatography

VOLUME 42

Edited by

Phyllis R. Brown
UNIVERSITY OF RHODE ISLAND
KINGSTON, RHODE ISLAND

Eli Grushka
HEBREW UNIVERSITY OF JERUSALEM
JERUSALEM, ISRAEL

CRC Press
Taylor & Francis Group
Boca Raton London New York

CRC Press is an imprint of the
Taylor & Francis Group, an **informa** business

First published 2003 by Marcel Dekker, Inc.

Published 2019 by CRC Press
Taylor & Francis Group
6000 Broken Sound Parkway NW, Suite 300
Boca Raton, FL 33487-2742

© 2003 by Taylor & Francis Group, LLC
CRC Press is an imprint of Taylor & Francis Group, an Informa business

First issued in paperback 2019

No claim to original U.S. Government works

ISBN 13: 978-0-367-44685-7 (pbk)
ISBN 13: 978-0-8247-0950-1 (hbk)

**Visit the Taylor & Francis Web site at
http://www.taylorandfrancis.com**

**and the CRC Press Web site at
http://www.crcpress.com**

Library of Congress Cataloging-in-Publication Data
A catalog record for this book is available from the Library of Congress.

Contributors

Jason Anspach, Ph.D. Department of Chemistry, Natural Sciences Complex, University of Buffalo, State University of New York, Buffalo, New York, U.S.A.

Brice Bouyssiere CNRS–Université de Pau et des Pays de l'Adour, Pau, France

Phyllis R. Brown, Ph.D. Chemistry Department, University of Rhode Island, Kingston, Rhode Island, U.S.A.

Carsten A. Bruckner Center for Process Analytical Chemistry (CPAC), Department of Chemistry, University of Washington, Seattle, Washington, U.S.A.

Annalisa Castagna Department of Agricultural and Industrial Biotechnologies, University of Verona, Verona, Italy

Héctor Colón Department of Chemistry, Natural Sciences Complex, University of Buffalo, State University of New York, Buffalo, New York, U.S.A.

Luis A. Colón   Department of Chemistry, Natural Sciences Complex, University of Buffalo, State University of New York, Buffalo, New York, U.S.A.

Carlos G. Fraga   Center for Process Analytical Chemistry (CPAC), Department of Chemistry, University of Washington, Seattle, Washington, U.S.A.

David S. Hage, Ph.D.   Department of Chemistry, University of Nebraska, Lincoln, Nebraska, U.S.A.

Mahmoud Hamdan   Computational, Analytical and Structural Sciences, GlaxoSmithKline Group, Verona, Italy

Mark A. Hayes, Ph.D.   Department of Chemistry and Biochemistry, Arizona State University, Tempe, Arizona, U.S.A.

Kevin J. Johnson   Center for Process Analytical Chemistry (CPAC), Department of Chemistry, University of Washington, Seattle, Washington, U.S.A.

Gaëtane Lespes   CNRS–Université de Pau et des Pays de l'Adour, Pau, France

Ryszard Lobinski   CNRS–Université de Pau et des Pays de l'Adour, Pau, France

Michela Maione   Institute of Chemical Sciences, University of Urbino, Urbino, Italy

Todd D. Maloney   Department of Chemistry, Natural Sciences Complex, University of Buffalo, State University of New York, Buffalo, New York, U.S.A.

Filippo Mangani   Institute of Chemical Sciences, University of Urbino, Urbino, Italy

Patrick D. McDonald, Ph.D.   Waters Corporation, Milford, Massachusetts, U.S.A.

Pierangela Palma   Institute of Chemical Sciences, University of Urbino, Urbino, Italy

Bryan J. Prazen   Center for Process Analytical Chemistry (CPAC), Department of Chemistry, University of Washington, Seattle, Washington, U.S.A.

Pier Giorgio Righetti   Department of Agricultural and Industrial Biotechnologies, University of Verona, Verona, Italy

Robert E. Synovec   Center for Process Analytical Chemistry (CPAC), Department of Chemistry, University of Washington, Seattle, Washington, U.S.A.

Joanna Szpunar   CNRS–Université de Pau et des Pays de l'Adour, Pau, France

Todd O. Windman   Department of Chemistry and Biochemistry, Arizona State University, Tempe, Arizona, U.S.A.

Barb J. Wyatt   Department of Chemistry and Biochemistry, Arizona State University, Tempe, Arizona, U.S.A.

Contents

Contents of Other Volumes

Volume 9

Volume 10   out of print

Volume 11

Volume 12

Volume 23

Volume 24

Volume 25

Volume 26

Volume 27

Volume 28

Volume 29

Volume 30

Volume 37

Volume 38

Volume 39

Volume 40

Volume 41

1

Chemometric Analysis of Comprehensive Two-Dimensional Separations

Robert E. Synovec, Bryan J. Prazen, Kevin J. Johnson, Carlos G. Fraga, and Carsten A. Bruckner   *University of Washington, Seattle, Washington, U.S.A.*

I. INTRODUCTION

The combination of chemometrics and chromatography is a broad area of research. Analytical chemists are continually developing and applying better chemical separation methods combined with more powerful tools for extracting useful information from the data, as is the role of chemometrics. Within the wide array of published chemometrics research, there have been numerous investigations into the use of chemometric techniques for quantitative or qualitative analysis of chromatographic and electrophoretic data. Although there are many examples of chemometric techniques applied to one-dimensional (1-D) separations, this chapter will focus on applications of chemometrics with comprehensive two-dimensional (2-D) separations. Chemometrics and 2-D separations can, independently, offer vast selectivity enhancements over conventional 1-D analytical separations and traditional peak identification and quantification methods. These methods can also be used to reduce sample run times. Thus, the combination of comprehensive 2-D separations with chemometrics represents the development of a new and powerful tool in the analysis of complex mixtures. Applications of chemometrics to 1-D separations will be reviewed briefly in Section III in order to put this exciting area of research into context with the issues and advantages provided by 2-D separations.

The principles and advantages of comprehensive 2-D separations have been well described in recent reviews [1–4]. Fundamental aspects of chemometric algorithms and applications of those algorithms have been the subject of several books [5–8], and there are two journals centered on the development of chemometric algorithms: *Journal of Chemometrics* from John Wiley & Sons, Inc., and *Chemometrics and Intelligent Laboratory Systems* from Elsevier Science. Many of the journals related to analytical chemistry feature articles about chemometric techniques as well. The journal *Analytical Chemistry* hosts a fundamental review of chemometric applications biannually.

This chapter is organized in the following fashion. First, we discuss the impetus and methodology behind both comprehensive 2-D separations and chemometric analysis of data arising from these techniques. Next, two applications of chemometric analyses to 2-D gas chromatographic data will be presented. The first application uses the chemometric technique of generalized rank annihilation method (GRAM) coupled to objective run-to-run retention time alignment. The second application combines pattern recognition and feature selection techniques. Following this will be a discussion of the scope and future directions of chemometric analysis of 2-D separation data. Here, a theoretical study investigates the utility of applying a chemometric method such as GRAM to the analysis of comprehensive 2-D separations such as GC × GC, LC × CE, and other comprehensive 2-D separation methods.

II. COMPREHENSIVE 2-D SEPARATIONS

Giddings has phrased the goal of analytical chemical separations as "to multiply the quantity of detailed information available about complex mixtures and to enhance the quality of that information [9]." In this regard, the developments of 1-D capillary chromatography and capillary electrophoresis are particularly important because of the enormous separating power each possesses. More recently, comprehensive 2-D separation methods have been emerging due to the pioneering efforts of scientists such as Giddings, Jorgenson, Phillips, and Lee [1,3,9–16]. The first comprehensive 2-D separations were published in 1990 by Jorgenson and Bushey [12,13,17]. These separations were comprehensive 2-D liquid chromatographic separations (LC × LC) and comprehensive 2-D liquid chromatography–capillary electrophoresis separations (LC × CE). A recent comprehensive 2-D micro-reversed-phase liquid chromatographic–capillary zone electrophoretic separation (LC × CE) from the Jorgenson group is shown in Fig. 1a [18]. In this separation, human urine components tagged with fluorescein 5–isothio-cyanate are analyzed. The high peak capacity of this separation led to over 400 resolved components in this example. The random scatter of peaks indicates that RPLC and CE are very complementary techniques. Comprehensive 2-D gas chromatography (GC × GC) is a multidimensional technique pioneered by Phillips to analyze complicated mixtures that had previously been impractical or impossible to analyze using 1-D GC [15]. Fig. 1b is a GC × GC separation of pesticides published by Phillips et al. [16]. In this example, baseline separation of the 19 components was achieved in 4.5 minutes. Quantification of the components in this sample was by peak volume. This example demonstrates the high peak capacity and selectivity of comprehensive 2-D separations.

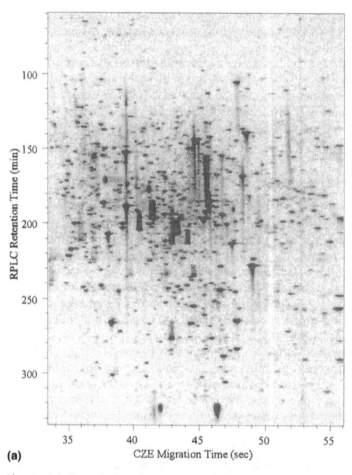

(a)

Fig. 1 (a) Gray-scale image of a comprehensive two-dimensional micro-reversed-phase liquid chromatographic–capillary zone electrophoretic separation of fluorescein 5-isothio-cyanate tagged human urine. (Reprinted with permission from Ref. 18. Copyright 1997 American Chemical Society.) (b) Comprehensive two-dimensional gas chromatogram of a mixture of 17 pesticides and two internal standards. Peak identities: 1, dicamba; 2, trifluralin; 3 dicloran; 4, phorate; 5, pentachlorophenol; 6, atrazine; 7, fonofos; 8, diazinon; 9, chlorothalonil. (Reprinted with permission from Ref. 16. Copyright 1994 American Chemical Society.)

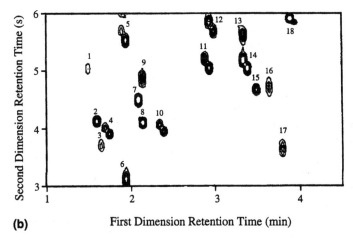

(b) First Dimension Retention Time (min)

Fig. 1 Continued

Chromatographic separations of complex mixtures are made possible by the vast increase in peak capacity afforded by subjecting each component of the mixture to comprehensive separations on two columns with complementary selectivity. Essentially, comprehensive 2-D separations provide a substantially larger peak capacity for a given length of analysis time relative to 1-D separation methods [10,19]. Here, we must be careful in the practice of comprehensive 2-D separations to properly adhere to the appropriate experimental criteria as outlined by Liu and Lee [3]. Of critical importance in 2-D separation instrumentation is the injection device that provides periodic sample introduction onto the second column, as well as the run time on the second dimension column. The frequency of the second column injections governs the range of available second column run times as well as the frequency with which the first column effluent stream is sampled. A comprehensive 2-D separation requires that the narrowest individual analyte peak eluting from the first column be sampled and a separation obtained no less than four times by the sampling interface/second column combination. Meeting this requirement preserves the resolution provided by the first column separation and provides quantitative data [20].

In practice, chemical separations become more difficult with increasingly complex mixtures, such as petroleum distillates, botanical extracts, and biological samples. At some point, complete resolution of all analytes of interest may be impractical, or even impossible, with current analytical separation methods. Even with a large peak capacity, the probability of peak overlap in 2-D separations can become quite severe, especially for complex samples. Giddings and Davis have demonstrated this concept using theoretical peak distributions and Poisson statistics [21,22]. The probability of peak overlap becomes even more likely if one desires to speed up the analysis by designing a given separation to conclude in a reduced run time. This is because a reduction in run time generally goes hand in hand with a reduction in the resolving power along the first dimension separation. Here, the separation scientist must confront the issue of how much resolution is required to answer the analytical question. What is deemed "necessary" and "reasonable" in terms of separation performance and required resolution is highly subjective and depends largely on the number of samples that are to be analyzed and the urgency and context with which knowledge of the sample is needed. While it may be acceptable for one analyst to scrutinize samples for three hours each, a separation taking only one-tenth the time may be completely acceptable to another analyst, even when coupled with the loss in resolution. In general, a separation method should be deemed optimized when it is reduced to the shortest possible run time while still generating data containing the desired information about the sample.

Figure 2 illustrates the relationship between chromatographic resolution of a pair of peaks and the separation run time. Chromatographic theory states that resolution increases with the square root of the column length, whereas run time is linear with column length [9]. If the two components in Fig. 2 are initially analyzed at a resolution of 1.5, using traditional data analysis techniques, the run time could be reduced fourfold if comparable quantitative precision and accuracy could be achieved using a data analysis technique capable of resolving peaks at a resolution of 0.75. Thus, when short analysis times are important, the chemometric resolution of peaks is of interest. A rule of thumb is that, under standard signal-to-noise conditions, many chemometric techniques function at a resolution down to 0.3, but quantitative precision losses can become an issue. If quantitative analysis precision can be maintained within an acceptable range at a resolution of 0.3, the sample run time can be reduced by a factor of 25 from the run time at a resolution of 1.5. This is a sub-

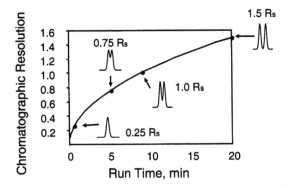

Fig. 2 The relationship between chromatographic resolution and separation run time. Chromatographic resolution increases with the square root of separation run time by changing column length, with the mobile phase linear flow velocity held constant. Thus, small reductions in the resolution requirements for an analysis result in substantial savings in run time and analysis time.

stantial savings in analysis time, but requires the analyst to think differently about the role of separations in the overall analytical procedure. Thus, integration of chemometrics with high-speed 1-D and 2-D separations is essential for optimization of many chemical analysis strategies. This chapter will focus on recent experimental and theoretical results that optimize the amount of information gleaned from high-speed separation processes. The implications of using chemometric analysis as an integral part of any 2-D separation-based chemical analysis strategy will be conveyed.

III. CHEMOMETRIC ANALYSIS OF CHEMICAL SEPARATION DATA

A. Chemometrics Background

Chemometrics can be loosely defined as "the development and use of mathematical techniques to extract useful information out of data acquired through chemical analyses" [5–7]. The fairly broad definition includes both data analysis techniques as well as techniques for optimally designing analytical experiments and instrumentation. Though chemometrics has arguably been around as long as analytical chemistry itself, it has only recently come into its own as a field of study, principally concerned with the development and application of methods to analyze and manipulate multivariate data acquired from chemical analyses. This can be traced directly to the appearance and subsequent wide availability of powerful, off-the-shelf computers. For the purpose of this discussion, the term "chemometrics" will refer to the development and application of multivariate data analysis techniques to data acquired from analytical instrumentation.

Multivariate analysis refers to "the simultaneous investigation of two or more variables over a set of objects [23]." In the context of analytical chemistry, the term "objects" refers to the chemical samples that are analyzed, or to the data sets that result from the analysis of those samples. The term "variables" refers to the set of several measurements of which each data set referring to a single object is composed. In chromatography, the "objects" are chromatograms and the "variables" are the different retention times (elution time interval) at which the detector signal is recorded. In spectroscopy, the objects are spectra and the variables are the wavelengths at which intensities are measured. The general goal in using a multivariate analysis technique is to mathematically transform a series of measured variables into a property of interest. Multivariate calibration techniques such as classical least squares (CLS), multiple linear regression (MLR), principal components regression (PCR), and partial least squares regression (PLS) are used to generate models that predict quantitative, continuous chemical properties of a sample [5]. On the other hand, multivariate pattern recognition techniques such as k-nearest neighbors (KNN), linear discriminate analysis (LDA), hierarchical cluster analysis (HCA), and soft independent modeling by class analogy (SIMCA) are typically used to predict abstract qualities for the purpose of classifying the sample as belonging to one particular subset of a larger population of samples [6]. Though either of these two tasks can also theoretically be accomplished with univariate data (for example, calibration of the concentration of red dye in solution using a linear regression of absorbance measurements), multivariate techniques have several advantages [6]. By making several redundant or highly correlated measurements (for example, the acquisition of a number of data points across a spectroscopic or chromatographic peak), the precision of a model can be improved in a process analogous to signal averaging. The acquisition of multiple variables for each sample also helps to facilitate fault detection through a change in the relative magnitudes between responses in the presence of an interfering factor in the analysis. Finally, multivariate techniques allow for the construction of models to analyze for different properties simultaneously.

Because of these advantages, multivariate analysis techniques allow the analyst more latitude in choice of instrumentation. It becomes possible to obtain selective information without using particularly selective detectors, as evidenced by the success demonstrated by multivariate analyses of NIR data [24–27]. Conversely, when applied to already very selective data, multivariate techniques allow for the elucidation of even more complex analysis problems. This brings us to the application of chemometric techniques to chromatographic data, which is an interesting proposition because selectivity in chromatographic techniques is related directly to resolution, and thus to chromatographic run time. Therefore, chromatographic separations that are designed to be analyzed with chemometric techniques can be run at higher speed, allowing more complicated samples to be analyzed in more reasonable, shorter periods of time, or even possibly allowing analyses that were not possible with standard chromatographic methods.

Currently, chromatograms are normally interpreted by means of a systematic comparison of varying peak areas or heights across two or more chromatograms, relying on the chromatographic separation to provide enough selectivity to solve the analysis problem at hand. By utilizing chemometric techniques on chromatographic data, the analyst can partially shift the burden of resolution (or selectivity, depending on your point of view) from the instrument, which is generally time consuming and expensive to operate, to computational devices. Because over the past decade computational devices have increased in speed and decreased in cost at a phenomenal rate, implementation of chemometrics has become more appealing. As such, applying chemometrics is a paradigm shift in the way that chromatographic analysis is done, in that it loosens the traditional link between "resolution" and "information" in chromatography. When utilizing chemometric techniques to analyze chromatographic data, the emphasis becomes no longer the achievement of maximum chromatographic resolution, but the acquisition of a maximal amount of useful information from the chromatogram. This viewpoint comes full circle again to Giddings' description of the goal of analytical chromatography that was discussed earlier, which is "to multiply the quantity of detailed information available about complex mixtures and to enhance the quality of that information [9]."

Current published applications of chemometric techniques to chromatographic data fall into the two categories of multivariate analysis previously discussed: pattern recognition methods that use high-speed (partially resolved) GC "fingerprints" to provide qualitative information about a sample, and regressive calibration models that provide quantitative information about a sample. There are many examples of successful applications of chemometric algorithms [6,28,29] to chromatographic data for the purpose of sample classification by pattern recognition. The types of samples that are analyzed fall into three broad categories: food and beverages, biological samples, and petrochemical samples. Food and beverage studies generally center on quality control, detection of impurities, or identification of sample origin. Coffee and tea varieties have been characterized by principal components analysis (PCA) of both HPLC and GC data [30–33]. PCA has also been used to identify orange juice samples (based on a GC analysis of volatiles) as either freshly squeezed or reconstituted [34,35]. A separate study used PCA as well as hierarchical cluster analysis (HCA) and soft independent modeling by class analogy (SIMCA) on HPLC data to detect the presence of an adulterant in orange juice samples [36]. Blanco-Gomis used several chemometric methods to characterize differently filtered apple juice samples [37]. PCA and statistical discriminate analysis were used by Lee to characterize fatty acid composition in several vegetable oils and to detect adulteration of one oil with another [38]. Reynes used PCA to classify three varieties of dates [39]. Kim used statistical discriminate analysis to distinguish among four brands of wine based on their GC amino acid profiles [40]. In the area of biological applications, Keller used SIMCA and HCA to monitor microbial growth in fermentation processes [41]. Kim used statistical discriminate analysis to detect uterine myoma or cervical cancer from patients' urinary organic acid GC profiles [42]. Dunlop used PCA, SIMCA, and LDA to identify various species of eucalyptus trees [43]. Baiocchi used PCA, HCA, and KNN to characterize and predict poplar tree resistance to the fungus *Discosporium populeum* [44]. In the area of petrochemical samples, Lavine and coworkers have used SIMCA, KNN, artificial neural networks (ANN), and several statistical discriminate methods to identify the fuel type of both weathered and neat jet fuel samples [45–48].

Aside from a 1-D chromatographic data analysis, there are also examples of the chemometric analysis of data arising from 2-D "hyphenated" systems such as gas chromatography with mass spectral detection (GC-MS) [49], or liquid chromatography with diode array UV-vis absorbance detection (e.g., LC-DAD). Additionally, there have been texts and reviews published that focus on numerical analysis issues in chromatographic data such as statistics of peak overlap, curve resolution, Fourier analysis techniques, and multivariate signal resolution [50,51]. Chemometric methods can be used to understand chromatographic mechanisms [52–54], to classify stationary phases [55], and to optimize chromatographic separations [56–59]. Chemometrics has also been used to evaluate the quality of chromatographic data and the age of chromatographic columns [60,61].

With enhanced selectivity and data structure, 2-D separation methods provide data that are ideal candidates for chemometric analysis. The vast peak capacity enhancement 2-D separations provide over conventional 1-D methods make 2-D separations a natural fit for analyses of complex mixtures. The 2-D data structure created can be analyzed by a host of powerful chemometric techniques. These chemometric techniques, originally designed to take advantage of the two- or three-dimensional data sets provided by other hyphenated analytical techniques, include the generalized rank annihilation method (GRAM) [62–67], parallel factor analysis (PARAFAC) [68–70], and trilinear partial least squares (Tri-PLS) [71–75]. Owing to the extra amount of information that can be contained in a 2-D data set, these techniques demonstrate some surprising benefits. One such benefit of 2-D data is the ability of a GRAM calibration to deconvolute chromatographic and/or spectroscopic profiles and quantify only partially resolved chemical components in a sample data set with the use of a single calibration standard, even in the presence of interference from an unknown mixture component. GRAM calibration has been shown to obtain useful information from GC × GC data without complete chromatographic resolution, thus extending the number of analyzable peaks in a complex separation [49,62,65,66].

B. Objective Run-to-Run Retention Time Alignment

Although there have been many applications to chromatographic data, the bulk of chemometric research to date has centered on the development and application of chemometric algorithms toward spectroscopic rather than chromatographic data. This is because chromatographic data exhibit time axis uncertainty due to unavoidable variations in instrument performance between and during separations. The uncertainty, or limited precision, is observed as shifting in the retention times of chromatographic peaks between chromatograms. Factors such as mobile phase flow fluctuations, temperature gradients or changes, and column degradation over time contribute to this variation, and ensure that no two chromatographic separations ever have exactly the same time axis, even though their overall run times may be identical. For example, in GC, several factors lead to variations in retention time for a single analyte over a series of many chromatographic runs. The reasons for variation include temperature and flow fluctuations between runs or during a run, matrix effects between samples, and stationary phase composition changes between columns (as well as column degradation over time). Other separation methods, such as capillary electrophoresis, raise similar issues to consider. Retention time variation complicates multivariate analyses of the chromatographic profiles, as the same retention time in different chromatograms may not reflect the same chemical information about the sample mixtures represented.

Multivariate chemometric analysis techniques are generally constructed as to interpret chemical variation as changes from sample to sample in only the signal magnitude contained at each elution time interval and, because of this, are inherently sensitive to retention time precision. In order to accurately ascertain differences in chemical composition between samples using chromatograms, there must be a one-to-one correspondence between the elution time intervals of the two chromatograms being compared. That is, the same chemical component should exhibit the same retention time in each chromatogram from sample run to sample run. By applying a multivariate technique to chromatographic data, one is assuming that this is true. If this is not true, two chromatograms offset from each other in retention time, but identical in every other respect, would tend to be incorrectly interpreted by a multivariate analysis technique as being chemically different.

A number of researchers have investigated this area. In particular, Malmquist provides a discussion on the sensitivity of principal components analysis (PCA) to retention time shifting in chromatographic data [76]. Bahowick reports on the effect of retention time precision on quantification of analytes from liquid chromatographic data using classic least squares (CLS) regression [77]. Poe reports on its effect on quantification of analytes from data acquired from a liquid chromatographic separation with diode-array fluorescence detection by generalized rank annihilation method (GRAM) [78]. Prazen and Fraga each discussed the effect of retention time precision on quantitation of analytes in 2-D GC data by GRAM [63,65,66]. Thus, in order to reliably perform a direct, point-by-point comparison of chromatograms that have been acquired at different times (e.g., days, weeks, or months intervening between runs), peak shifting must be corrected prior to analysis by an objective retention time alignment method.

Retention time alignment methods center on the ability to identify chromatographic features that correspond between a "target" chromatogram and a chromatogram that is to be aligned to the target. Using these features, the shift required to align a given chromatogram relative to the target chromatogram can be estimated for every point in the chromatogram, and a retention time correction can be made. In this manner, the retention times of chromatographic features not present in the target chromatogram can be corrected as well. A useful alignment algorithm must preserve chemical selectivity differences between samples of differing composition while minimizing run-to-run retention time differences in the analysis of the samples. The alignment algorithm must also be computationally efficient if it is to align large data sets in a reasonable time. Finally, the algorithm must provide retention time precision that is significantly better than that initially provided by the instrumentation.

Several techniques have been developed for the alignment of 1-D chromatograms, with varying degrees of success. Andersson and Hämäläinen created a simplex-optimized chromatographic profile alignment utility in which the retention time shifting is modeled by a linear equation containing a constant term for offset and a slope reflecting the degree of stretch or shrink relative to the target chromatogram [79]. Optimization of these terms for best alignment was based on correlation between target and sample chromatograms in two retention time windows. Grung and Kvalheim developed a retention time alignment method for hyphenated chromatographic data that was based on Bessel's inequality [80]. Malmquist and Danielsson's alignment algorithm involves four rounds of iterative shifting in order to optimize sample-to-target correlation [76]. First, the entire profile is shifted to correct for a scalar offset in retention times between the sample and target, followed by alignment of a smaller subset of the largest peaks for coarse peak-to-peak retention time shifting. Fine-tuning is then accomplished with iterative shifting of a slightly larger subset containing more peaks, and a fourth step that involves noninteger shifting of a smaller subset of peaks. The noninteger shifts required for each peak in the subset are retained and used to construct a time displacement function that will correct the sample chromatogram's retention times.

Nielsen and coworkers developed a method known as correlation optimized warping [81]. In this method, the chromatographic profiles to be aligned are divided up into a series of retention time "windows" of equal length. Alignment occurs through the systematic, iterative stretching and shrinking of the sample chromatogram's retention time windows so as to find a combination of stretches and shrinks that optimizes sample to target correlation. An extension of this method to 2-D data for which a liquid chromatographic separation was combined with a spectrophotometric detector was presented as well. Unfortunately, this iterative global optimization approach leads to long execution times for programs modeled on this algorithm. Bro and coworkers demonstrated that PARAFAC2 could be used as an approach for handling retention time shifts [82]. The PARAFAC2 model loosens the restrictions on retention time precision present in the original PARAFAC algorithm.

With comprehensive 2-D separation methods such as GC × GC, the data contain two chromatographic time axes, each of which can exhibit retention time uncertainties. Prazen has demonstrated a rank-based algorithm for the alignment of the primary chromatographic axes of subsets of 2-D data prior to analysis by GRAM [63]. The method requires the analyst to estimate the rank (number of independent chemical components present) of the submatrix, and calculates the magnitude of the residual variance not accounted for by these factors in an augmented matrix constructed from the submatrix and the corresponding region of a target chromatogram. This variance is monitored as the submatrix is shifted relative to the target, and the shift at which a minimum variance is found is taken as the retention time correction required. Fraga extended this technique to include both dimensions as well as interpolative subinteger retention time shifts [65,83].

IV. COMPREHENSIVE 2-D SEPARATIONS/ GRAM ANALYSIS

A. Instrument Performance and Bilinear Data Structure

A bilinear data structure is important for the utilization of chemometrics to analyze comprehensive 2-D separation data. The ability to obtain bilinear data is related to the performance of the 2-D separation instrument, the principles of which will now be considered. The general schematic for an instrument that performs a comprehensive 2-D separation is shown in Fig. 3a. Key to the instrument design is a high-speed sample injector between column 1 and column 2. Column 2 is designed to provide a separation that is considerably faster, more efficient, and yet complementary to column 1. In the case of GC × GC, the detector is often a flame ionization detector (FID) and is located after column 2.

Consider a comprehensive 2-D separation in which the separation conditions provided by column 1 result in two analytes coeluting from column 1, while those of column 2 allow the analytes to be readily separated at the end of that column. If a detector was placed directly after column 1, only a single peak would be observed by it, similar to what is shown in the upper trace of Fig. 3b. In the GC × GC experiment, an injection is made onto column 2 from column 1 at each dot in the First Column Separation in Fig. 3b. Following the second stage of the 2-D separation, what appears to be a single peak eluting from column 1 is actually composed of two components that are readily separated on column 2. Each data set acquired from the 2-D separation in Fig. 3b reflects the detector signal over the entire time of a particular separation run. When this vector is plotted as a function of time, it appears as a series of consecutive column 2 separations, the outline of which follows the trace of the column 1 elution profile.

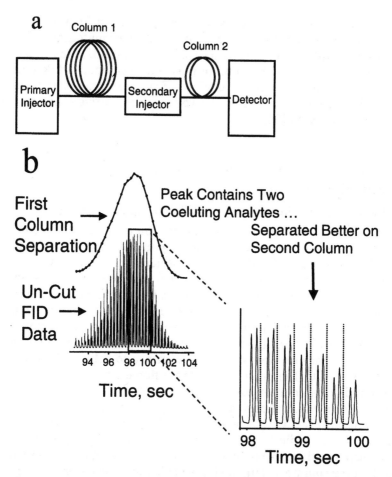

Fig. 3 (a) Diagram of comprehensive two-dimensional separation instrument. The unique portion of the instrument is the secondary injector, which should sample the effluent of column 1 at a rate such that each column 1 single-analyte peak is sampled at least four times. Thus, column 2 separations must be relatively fast. (b) Raw data from a GC × GC separation. This data contains the signals of two components. The peaks of the two components are not separated on column 1 but are fully separated on column 2. The outline of the uncut flame ionization detector (FID) data shows the column 1 separation that appeared to contain only one peak. The close-up view contains dotted lines to mark the injection onto the second column. The column 2 separations each contain two analyte peaks.

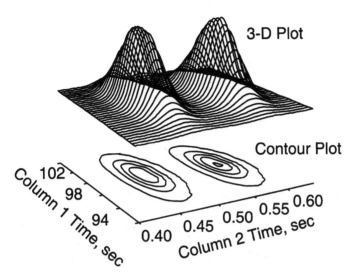

Fig. 4 Three-dimensional plot and contour plot of the data presented in Fig. 3a. The data vector from the flame ionization detector (FID) is stacked into a two-dimensional matrix such that one dimension depicts the separation on column 1 and the other dimension depicts the separation on column 2.

Once acquired, the data collected from the 2-D separation can be plotted for visualization in two forms, either a three-dimensional (3-D) plot or a contour plot as shown in Fig. 4. The 2-D separation data from Fig. 3 has been reshaped such that each column 2 separation is stacked side by side, forming a matrix of data in which the columns reflect the column 1 time and the rows reflect the column 2 time. This matrix can be visualized as a 3-D surface or as a topographical contour map with the x and y axes set as first and second column time axes and the z axis set as the signal. The data presented in Fig. 3 and Fig. 4 adhere to a bilinear data structure. We will now turn our attention to what it means mathematically to have a bilinear data structure, and how it relates to chemometric data analysis performance. In this regard we will focus our attention on the chemometric method referred to as generalized rank annihilation method (GRAM) [84–89].

The pure 2-D signal from a single component has a bilinear structure if it can be described as the product of two vectors: one

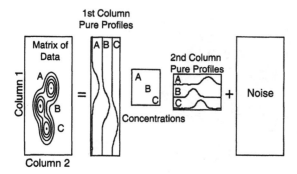

Fig. 5 Decomposition of bilinear data is demonstrated. Data from most comprehensive 2-D separation systems, for example the data in Fig. 4, can be decomposed into three signal matrices and an error matrix. Columns of the first decomposition matrix contain the pure column 1 separation profiles of each component. The second decomposition matrix is a diagonal matrix containing the concentration information. Rows of the third matrix contain the column 2 separation profiles. The last matrix contains noise that is filtered from the data by the bilinear decomposition. (Adapted and modified from Ref. 65.)

representing the component's column 1 chromatographic profile and the other representing its column 2 chromatographic profile. Figure 5 illustrates the concept of bilinear structure as it pertains to a mixture of overlapped comprehensive 2-D chromatographic signals. In this case, the data matrix obtained by the comprehensive 2-D analyzer is described as the product of two matrices containing the pure chromatographic profiles of each analyte and a diagonal matrix of concentration values, plus a noise matrix.

B. Rank Alignment

GRAM decomposes the data matrix into the useful information illustrated in Fig. 5. Essentially, GRAM is a chemometric technique that utilizes bilinear data structure to resolve and quantify the overlapped comprehensive 2-D peaks, while also separating each analyte signal profile from unwanted noise. GRAM compares comprehensive 2-D data obtained from a sample and a calibration standard to determine the pure signals for analytes present in both the sample and standard. In addition, GRAM determines the ratio of the sample and calibration standard comprehensive 2-D signals. With standard addition, the GRAM ratio is the concentration of the analyte in the sample divided by the concentration of the analyte plus standard in the spiked sample. Assuming a linear response to the analyte, the analyte concentration in the sample is calculated from the GRAM ratio and the known quantity of analyte added or spiked into the original solution [90].

In order to use chemometric methods such as GRAM to resolve and quantify peaks that are not fully resolved, the shifting of a 2-D peak's retention time(s) between sample and standard runs must be objectively corrected. In the past, run-to-run retention time shifting has been a severe impediment to the use of chemometric methods on chromatographic data [76–78]. A recent advance has been a rank-based alignment method to correct retention time shifts along one time axis [49,63,65]. When used in conjunction with GRAM, this objective retention time alignment algorithm resulted in excellent peak identification and quantification of severely overlapped peak profiles for GC × GC separations.

In the following explanation of retention time alignment and GRAM analysis, a typical comprehensive 2-D data set will be analyzed. In this example, a 46-compound mixture is separated with a high-speed, diaphragm valve–based GC × GC [62,65,67] in just over 2 min. A representative 2-D separation of the sample is shown in Fig. 6. In this separation most compounds were resolved and could be quantified using peak height or 2-D peak volume. Two compounds

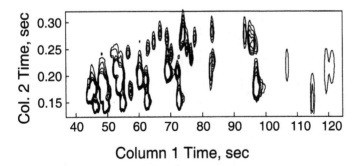

Fig. 6 GC × GC chromatogram of a 46-component mixture containing alkanes, alkyl benzenes, ketones, alcohols, esters, pyrazines, and thiazoles. Most of the components in this mixture were resolved in only 2 min.

that eluted around 75 s on column 1 required GRAM analysis to obtain resolution between the chromatographic peaks. Examples for the sample, **M**, and standard, **N**, data matrix of these samples are shown in Fig. 7. In most cases the data matrices **M** and **N** that are being analyzed by a chemometric method are small regions of separation data matrices containing overlapped peaks.

The rank-based retention time alignment method relies on the data matrix formed by augmenting the sample and standard data

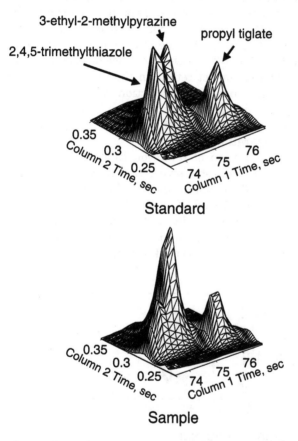

Fig. 7 Three-dimensional plots of the region of the sample and calibration standard analyzed using generalized rank annihilation method (GRAM) for peak resolution and quantitation. This data is from a portion of the separation shown in Fig. 6.

sets having a minimum pseudorank when the peaks in sample and standard are perfectly aligned. The pseudorank is defined as the rank of a data matrix in the absence of noise [91–94]. For a comprehensive 2-D separation, the pseudorank of the data matrix ideally equals the number of analyte peaks or chemical components. Thus for the example in Fig. 7, the pseudorank is three, including a relatively resolved peak and the two overlapped peaks. The alignment method augments sample and standard data sets as illustrated in Fig. 8. In Fig. 8, the augmented matrix is represented by two stacked contour plots, one representing the sample and the other the standard. The sample and standard data sets in Fig. 8 are the same data shown in Fig. 7 in 3-D mesh plots. These two data sets contain the peaks for the same three components but at different concentrations. Sample data and standard data are joined such that the augmented matrix has twice the number of data points on the column 2 axis than either the sample or standard data set. When the peaks between the sample and the standard are aligned on the column 1 axis, the pseudorank of the augmented matrix is at a minimum value. However, when the peaks between sample and standard are shifted on the column 1 axis, then the pseudorank of the augmented matrix will be artificially high. Hence, for shifted peaks, the alignment method essentially shifts the peaks in the sample matrix until the pseudorank of the augmented matrix reaches a minimum value. In Fig. 8 the double arrows represent the shifting of the sample relative to the standard.

In practice, the alignment method does not measure the pseudorank of the augmented matrix as a function of peak shift. Instead, the percent residual variance is measured as a function of peak shift. The percent residual variance is defined as the sum of the secondary eigenvalues from the singular value decomposition (SVD) of the augmented matrix divided by the sum of all the singular values and multiplied by 100 and a term related to the degrees of freedom [63]. The secondary eigenvalues are those eigenvalues associated with the data variance of measurement error or noise. An eigenvalue plot is shown in Fig. 9. The first three principal components contain signal and are called the primary eigenvalues. In a plot as shown in Fig. 9, the primary eigenvalues are the principal components where the eigenvalues are significantly above zero. The principal components with eigenvalues approximately on a straight line just above or essentially at zero are secondary eigenvalues. Notice in Fig. 9, for

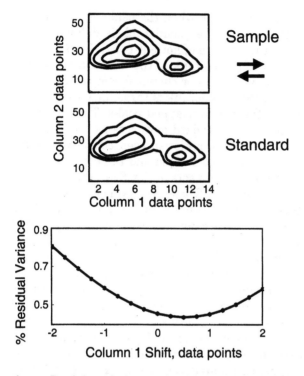

Fig. 8 Rank-based alignment is demonstrated. By connecting the sample data matrix, shown in Fig. 7, to the standard matrix, also shown in Fig. 7, to form an augmented matrix, retention time alignment along the first dimension (column 1) of the 2-D separation can be achieved by shifting the sample data relative to the standard data until a minimum pseudorank is achieved. The correct alignment for sample matrix to the standard matrix on the column 1 time axis is illustrated by the minimum in the % Residual Variance plot. Data interpolation was used to find the best time shift and minimum % Residual Variance. The effect this alignment procedure has on the principal components can be seen in Fig. 9. (Adapted and modified from Refs. 63 and 65.)

a given principal component, that the difference in each eigenvalue after alignment minus the eigenvalue before alignment are positive for the primary eigenvalues [3] and negative for the secondary eigenvalues (all the others). The percent residual variance plot captures this trend observed in the eigenvalue difference plot. In the align-

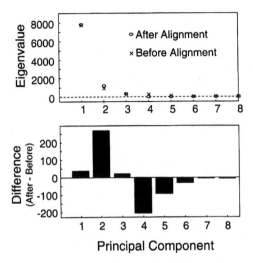

Fig. 9 Eigenvalues from the singular value decomposition (SVD) of the addition of the sample and calibration standard data matrices shown in Figs. 7 and 8. The first three principal components contain signal; the remaining components have eigenvalues just above or nearly at zero and contain noise. An eigenvalue of zero is marked by a dotted line. Eigenvalues before and after using the rank-based alignment are shown. The eigenvalue of each principal component containing signal increased as a result of alignment, whereas the eigenvalues containing noise decreased after alignment.

ment method, the percent residual variance reaches a minimum when the pseudorank of the augmented matrix reaches a minimum. Fig. 8 depicts the percent residual variance plot for the augmented matrix. The column 1 peak shift required for the alignment is the peak shift producing a minimum in the percent residual variance plot. In Fig. 8, the percent residual variance reaches a minimum value at a shift of 0.5 data points. In other words, the three peaks in the sample (see Fig. 7) elute 0.5 data points earlier than the same peaks in the standard. Note that each data point represents a chromatographic run on column 2. Thus, a data interpolation algorithm is applied in order to find the optimum shift correction of 0.5. A maximum expected peak shift of two was input into the alignment algorithm to generate the percent residual variance plot in Fig. 8.

Recently, an extension of this retention time alignment method was presented that objectively corrects for run-to-run retention time variation on both separation dimensions of comprehensive 2-D separations prior to application of chemometric data analysis algorithms [83]. The alignment algorithm is easy to apply and robust. Thus, data from 2-D separation techniques such as GC × GC, LC × LC, and LC × CE that, in many instances, would benefit from time correction on both axes, can be readily analyzed by various chemometric methods to increase chemical analysis capabilities. By applying the original alignment method on the augmented matrix [M|N] and then subsequently on the augmented matrix [M/N], peak shifts occurring along both time axes can be independently and successively corrected. This method was demonstrated and evaluated using simulated comprehensive LC × CE data. The data was simulated in the sense that real data collected from a GC × GC was subsequently inflicted by retention time shifting consistent with reported precision from recent reports for LC × CE [13,18]. The improved alignment algorithm corrects both dimensions in an independent, stepwise fashion. Correction of retention time shifting on both separation dimensions broadens the scope considerably for the use of chemometric methods, because most 2-D separation techniques produce run-to-run shifting on both dimensions that may be significant and detrimental to chemometric applications, if not corrected.

C. GRAM Analysis

A few requirements must be met for the successful application of GRAM [62,67]. First, detector response must be linear with concentration. Second, peak profile shape for a given analyte must not change. In comprehensive 2-D separations, this means that the retention times of each slice of a compound taken from the effluent of the first column must have a consistent second column retention time, and all the retention times of the compounds in the analytical sample and the standard must be the same. These requirements are consistent with the need to provide a bilinear data structure. GRAM analysis also requires that overlapped peaks have some resolution on both separation dimensions. The final requirement for appropriate use of GRAM is that no two compounds within the window of data analyzed can perfectly covary in concentration from the standard to the sample. This covariance possibility is minimized by performing GRAM on subsections of the entire 2-D separation data set, which reduces the number of peaks analyzed at a time.

In the first step for GRAM algorithm, SVD is performed on the addition matrix,

$$(\mathbf{M} + \mathbf{N}) = \mathbf{U} \, \Theta \, \mathbf{V}^T \tag{1}$$

which is the sum of the sample data $\mathbf{M}$ and the standard data $\mathbf{N}$ (matrices are denoted in boldface type). In this decomposition, Θ is a diagonal matrix, $\mathbf{U}$ and $\mathbf{V}$ are orthogonal matrices, and superscript T denotes a transpose. SVD is a well-known method of decomposing a matrix [6,95]. If there are different components contained in the sample and standard data, performing SVD on $(\mathbf{M} + \mathbf{N})$ ensures all the components are included in the model.

In the next step, the decomposed components of the addition matrix, $\mathbf{U}$, Θ, and $\mathbf{V}$, are truncated according to the number of chemical components. For the ideal situation, the data does not contain noise, and each chemical component has a unique retention time in both separation dimensions. In this case, the number of significant factors, or rank, of the addition matrix $(\mathbf{M} + \mathbf{N})$ is equal to the number of components. An illustration of this process is demonstrated in Fig. 9. As in the alignment algorithm, the rank of the addition matrix $(\mathbf{M} + \mathbf{N})$ is determined to be three for the current example (Figs. 6–10). In this case only the first three columns of the $\mathbf{U}$ and $\mathbf{V}$ matrices, and only the first three diagonal components of Θ, are retained. Malinowski has shown that a data matrix can be accurately reconstructed using only the primary principal components found using SVD [94]. The number of primary principal components used in the GRAM analyses of the data presented here were easily determined through inspection of the eigenvalues like those shown in Fig. 8 [91,92]. Many other statistical methods for estimating rank have been developed, any one of which could be applied to comprehensive 2-D separation data [93,94,96–98]. In the third step of GRAM, an eigenvalue problem that contains the truncated SVD components from the addition matrix $(\mathbf{M} + \mathbf{N})$ and the standard matrix $\mathbf{N}$ is solved.

$$(\overline{\Theta}^{-1}\overline{\mathbf{U}}^T\mathbf{N}\overline{\mathbf{V}})\mathbf{T} = \mathbf{T}\Pi \tag{2}$$

In Eq. (2), $\mathbf{T}$ is the resulting matrix of the eigenvectors, and Π is the associated diagonal matrix containing the eigenvalues. The overbars denote the truncated versions of the associated matrices, superscript

T denotes the transpose of a matrix, and superscript -1 denotes an inverse.

A vector containing the ratio of each analyte concentration in the standard, C_N, relative to that in the sample matrix, C_M, is determined from the eigenvalues.

$$\text{diagonal}(\mathbf{\Pi}) = \frac{C_N}{C_M + C_N} \tag{3}$$

The pure second column elution profiles, **X**, and the pure first column elution profiles, **Y**, of all the components common to both matrices (**M** and **N**) are determined with the eigenvectors and the decomposed components from SVD.

$$\mathbf{X} = \mathbf{\overline{U}\overline{\Theta}T} \tag{4}$$

$$\mathbf{Y} = \mathbf{\overline{V}(T^{-1})^T} \tag{5}$$

The eigenvectors from this step can be thought of as a rotation of the ambiguous SVD vectors of the addition matrix (**M** + **N**) into physically meaningful information. Although the signals may not be resolved in the sample or the standard, this is achievable because both the addition matrix and the standard matrix contain signals for the analytes of interest. The comparison of these two data sets allows an otherwise ambiguous decomposition to be rotated into the first and second dimension chromatographic profiles. For example, the multiplication of X_1 and Y_1 gives the comprehensive 2-D signal created by the first analyte in the sample. Fig. 10a and Fig. 10b are the GRAM resolved signals of the three compounds in Fig. 7 on the column 1 axis and the column 2 axis.

V. FEATURE SELECTION AND PATTERN RECOGNITION

The fundamental goal of pattern recognition is to classify a sample of a population as belonging to a specific subset of that population. In terms of analytical chemistry, pattern recognition is an attempt to classify (or identify) chemical samples according to particular criteria through chemometric analysis of measurements that are indirectly related to the criteria. The important distinction between pattern recognition and regressive calibration techniques is that the former is designed to provide qualitative information while the latter is designed to provide quantitative information.

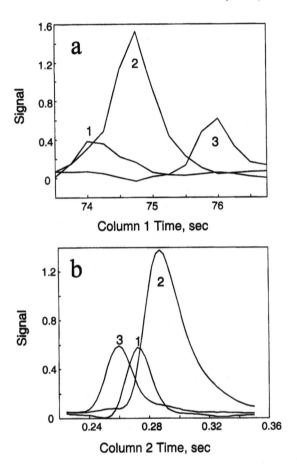

Fig. 10 Results of GRAM resolution of (1) 2,4,5-trimethylthiazole, (2) 3-ethyl-2-methylpyrazine, and (3) propyl tiglate (a) column 1 chromatographic profiles (b) column 2 chromatographic profiles. The resolved peaks are shown as the summation of the 2-D signals onto the column 1 and column 2 axes. Previous steps of the analysis are given in Figs. 6 through 9.

Pattern recognition techniques are often most useful in situations where the data to be examined is sufficiently complex to preclude alternative methods of analysis (e.g., petrochemicals [45] and biochemistry/pharmaceuticals analysis [42]), as well as when the analysis itself is of a qualitative nature (e.g., flavor/fragrance analysis [31,35]). Invariably, these are cases where mixtures of analytes and interfering compounds were simply too complex to be approached by most standard means of analysis. In terms of chemical separations, pattern recognition methods are most useful in the analysis of complex chromatograms (or electropherograms) with many analyte peaks and/or less than complete resolution of analyte peaks. Such chromatograms are treated as chemical "fingerprints" in a pattern recognition analysis and are exactly the kind of data that can be generated quickly by a 2-D separation method when applied to complex mixtures. As stated earlier, the vast enhancement in peak capacity that 2-D separations provide over 1-D separations makes 2-D separations a natural replacement of 1-D methods in the analysis of complex mixtures. Because of this, it is warranted to hypothesize, as current GC × GC literature has suggested [99,100], that 2-D chromatographic data of complex mixtures should be ideally suited to analysis by pattern recognition techniques.

Recently, GC × GC was applied to a pattern recognition problem involving classification of jet fuel mixtures [101]. A feature selection method based on analysis of variance (ANOVA) calculations was used to identify and select chromatographic features relevant to a given classification in two studies. Principal component analysis (PCA) was then used for classification of samples according to mixture composition or fuel type. In the first of these two studies, the effective combination of GC × GC, ANOVA-based feature selection and PCA was developed and evaluated as a chemical analysis tool. In the second study, this analysis tool was used to screen and classify a set of jet fuel samples consisting of three different jet fuel types, each from three different geographic origins, in order to rapidly screen and classify the samples according to jet fuel type.

A high-speed GC × GC chromatogram of a jet fuel sample is shown as a contour plot in Fig. 11a. Here, the GC × GC was purposely run in a high-speed mode to save time, as well as to determine whether such high-speed runs preserve enough selectivity in the data to achieve the desired classification via PCA. Initial attempts at classification using the entire GC × GC chromatograms were not successful, indicating much of the information contained in the data was not pertinent in the classification of interest. While the GC × GC had spread out the analytes in a 2-D space, there was still too much superfluous data to achieve a successful classification by PCA.

A feature selection procedure was developed to objectively select the appropriate chromatographic features for a desired classifica-

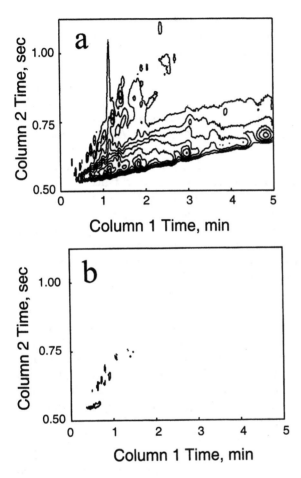

Fig. 11 (a) Contour plot representation of a GC × GC chromatogram of a sample mixture containing JP-7 and JP-5 type jet fuels. (b) Contour plot of the same chromatogram displaying only the selected features generated from an f-ratio threshold of 500 (all nonselected regions of the chromatogram are set to zero).

tion. This was accomplished by means of a point-by-point ANOVA calculation at each retention time in the chromatograms of a suitable training set with samples of known classification. The ANOVA calculation establishes whether the values of two different sample populations are statistically different [23]. The ANOVA calculation provides a value (known as an f-ratio) that is an estimation of the ratio of sample-to-sample variance to within-sample variance for two or more sample populations. The f-ratio, calculated at each retention time, can be used to gauge the amount of separation among different classes of jet fuel mixtures provided at each retention time. Retention times that gave an ANOVA f-ratio greater than a selected threshold were extracted from subsequently acquired chromatograms and analyzed by PCA for fuel type classification. A typical plot of the GC × GC jet fuel data following the ANOVA procedure is shown in Figure 11b. In this first study, a one percent volumetric composition change in mixtures of JP-5 and JP-7 type jet fuels was readily distinguished by PCA of ANOVA-selected features. It is important to note that, in this case, only about 0.3% of the total area of the GC × GC chromatogram was selected as pertinent to the desired fuel-type classification (compare Fig. 11a to 11b). Furthermore, many of these small chromatographic features would have been completely overwhelmed by larger, less selective chromatographic features in a 1-D separation of the same sample and the same chromatographic run time.

The benefits of applying the ANOVA feature selection procedure were also demonstrated by a second study involving the analysis of three samples each of three different jet fuel types, JP-5, JP-8, and JP-TS. For each fuel type, the three samples originated from different locations in the United States. These samples were analyzed to determine if a classification based on fuel type was possible in the presence of sample variability (due to geographic origin) with GC × GC/pattern recognition analysis. Using the ANOVA procedure, chromatographic features were selected that are adept at classification of jet fuel type and are insensitive to geographic origin. Data similar to Fig. 11a and b were generated, but not shown for brevity. Figure 12a depicts the scores plot generated from PCA of complete GC × GC chromatograms of samples of three different jet fuel types from nine different geographic locations, with each sample run in triplicate. Without feature selection, it is not possible to distinguish fuel type due to considerable confusion brought about by *inconsistent* features in the chromatograms within groups of samples of a given fuel type (due to geographical location variation). Following ANOVA-based feature selection, the inconsistent features from within a given fuel type are removed while leaving features that are useful for distinguishing the fuels of different fuel type. Fig. 12b depicts a scores plot of feature-selected data from the same chromatograms.

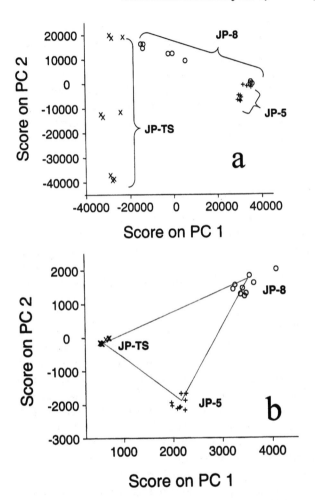

Fig. 12 Scores plots from the JP-5/JP-8/JP-TS jet fuel mixture study. (a) Scores plot generated from principal components analysis (PCA) of the mean-centered chromatograms before feature selection. (b) Scores plot generated from PCA of the selected features obtained with the f-ratio threshold set to 150. (Adapted and modified from Ref. 101.)

The feature selected 2-D data was considerably more adept at providing a useful classification according to fuel type. Thus, the combination of GC × GC with ANOVA-based feature selection was found to be a useful tool to enhance the chemical selectivity, and thus the classification power, of the analytical procedure. It is likely that this analytical approach of "smart" pattern recognition may have applications to data from other separation methods and research areas as well, such as with LC × CE data acquired in proteomics research [102–104].

VI. SCOPE AND FUTURE DIRECTIONS

It is important to obtain a better understanding of the overall scope and the future directions for applying chemometrics to 2-D separation data. Simulations have been useful to ascertain the advantage a chemometric method such as GRAM provides in the analysis of 2-D separation data [66]. Results from these recent studies will now be reviewed. Implications of the studies should have a positive impact on a variety of comprehensive 2-D separation methods [3] such as GC × GC [1,15,49,62,65,67,105–109], LG × GC [110], LC × LC [12,102], and LC × CE [13,17,18]. The theoretical enhancement provided by the GRAM analysis of unresolved peaks in comprehensive 2-D separations was carefully modeled and critically evaluated. Simulations were used to determine the conditions where the use of GRAM results in the successful analysis of unresolved peaks. The methods and conditions applied are outlined in more detail in a recent report [66]. A wide range of experimental conditions and instrument performance criteria were simulated, typical to many available 2-D separation methods. First, the quantitative precision (%RSD) was calculated for a simulated pair of 2-D peaks (analyte and interference) for realistic 2-D experimental separation conditions with resolution ranging from 0.1 to 1.0 for column 1 (R_{s1}) and column 2 (R_{s2}). Quantitative bias was also evaluated but not found to be the limiting factor, so the bias results are not included here for brevity. The %RSD plot as a function of R_{s1} and R_{s2} then served as a guide and target to determine the minimum level of combined resolution needed to achieve an acceptable level of quantitative precision. A %RSD of 4% or better was chosen as acceptable for the discussions here. Second, the increase in the analyzable peaks provided by GRAM relative to the analyzable peaks provided by traditional methods of data analysis, such as peak height and peak area measurements for simulated 2-D separations over a wide range of peaks per peak capacity was evaluated. Simulated comprehensive separations were produced with randomly distributed peaks over a space 10 peaks wide along dimension one and 5 peaks wide along dimension two for a total peak capacity of 50 [111].

A typical plot of the %RSD for a one-to-one peak height ratio of the analyte to interference is shown in Fig. 13. The conditions for generating Fig. 13 are identical to those used to generate Fig. 4 (A) of reference 66, while using interpolation with the retention time alignment along the column 1 time axis. Note that the %RSD determined from the simulations is consistent with experiment [62,65, 67], providing confidence in the use of these simulations to draw further conclusions. A simulated 2-D separation of 15 analytes is shown in Fig. 14a. In Fig. 14a there are nine singlet peaks and three doublets, a reasonable simulation of a real separation while practicing good separation skill with a mixture of 15 analytes in a space that could hold 50 resolved analyte peaks. Here, we define a singlet peak as containing a single analyte and a doublet as con-

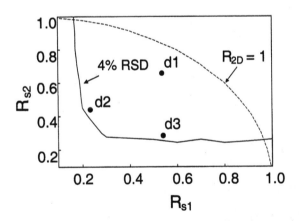

Fig. 13 Typical plot of the percent relative standard deviation (%RSD) for a one-to-one peak height ratio of an analyte to an interferent. The conditions for generating this plot included using interpolation with the retention time alignment along the column 1 time axis. The coordinates of the three doublets from Fig. 14a are indicated in the plot. R_{2D} is the net 2-D resolution for R_{s1} and R_{s2}. (Adapted and modified from Ref. 66.)

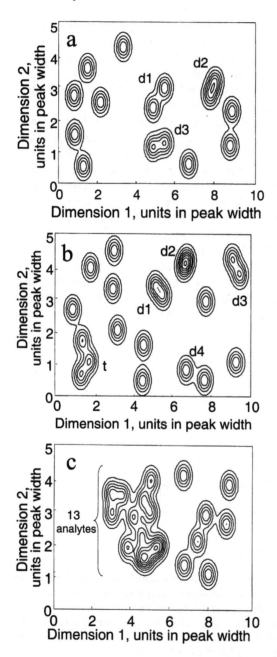

taining two analytes, and so on. The three doublets are at a resolution that would make quantitative analysis difficult using traditional methods of peak height or peak volume (area) that rely upon a minimum resolution of one along the diagonal connecting the center of adjacent peaks, but readily quantifiable using a chemometrics method such as GRAM. The resolution coordinates of the six analytes from the three doublets are indicated in Fig. 13. Indeed, the %RSD achieved for these overlapped analytes is quite acceptable. In the example separation of Fig. 14a, GRAM increased the number of quantifiable analyte peaks by 67%, going from nine analyte peaks quantifiable using traditional quantification methods, to 15 analyte peaks using the sum of traditional methods and GRAM. Thus, the total %Quantifiable analyte peaks relative to using traditional methods alone is 167% in this example.

In Fig. 14b is shown a typical 2-D separation one might obtain for a real mixture of 20 analytes, again, practicing good separation skills, thus utilizing most of the 2-D separation space. Here, we see nine singlet peaks, four doublets, and one triplet. The benefit of GRAM in the analysis of data in Fig. 14b is also evident. With the exclusion of the doublet d2, the analyte signals from the other three doublets and the triplet are all readily quantifiable using GRAM. The %Quantifiable, relative, was 200% with this example.

During the course of this study, one would randomly produce rather poor 2-D separations since a random number generator is not a very good separation scientist. A typical example of such a separation is shown in Fig. 14c. Here, very little of the separation space has been used and, in practice, one would tend to further optimize the separation in Fig. 14c prior to attempting quantitative analysis.

Using simulated 2-D separations with the 4% RSD criteria as in Fig. 13, the number of quantifiable analyte peaks when using GRAM was determined and compared to the number of quantifiable

---

Fig. 14 (a) Simulated 2-D separation of 15 analytes with a peak capacity of 50. Three doublets and nine singlets are observed. The %RSD for the analytes from the doublets is indicated in Fig. 13. (b) Simulated 2-D separation of 20 analytes with a peak capacity of 50. Four doublets, one triplet, and nine singlets are observed. (c) Simulated 2-D separation of 20 analytes with a peak capacity of 50 in which the separation does not satisfactorily utilize a sufficient fraction of the available peak capacity.

analyte peaks using only traditional quantitative methods such as peak integration or height. In this comparison, the %Quantifiable, relative, was determined as previously outlined in Fig. 13 and Fig. 14a. A summary of these results is presented in Fig. 15, in which the number of analyte peaks was varied from zero to 50 for the 2-D space with a peak capacity of 50. For the sake of objectivity, all separations, even those of the caliber in Fig. 14c were included in Fig. 15 results. The conditions for generating Fig. 15 are identical to those used to generate Fig. 8 of Ref. 66. GRAM clearly provides a benefit in improving the utility of 2-D separations by increasing the %Quantifiable analyte peaks. Although separations as in Fig. 14c weighed down the benefit, separations as in Fig. 14a and Fig. 14b necessarily showed a larger enhancement than average. In general, the application of GRAM should increase the number of analyzable peaks for all forms of comprehensive 2-D separations and should be seriously considered in the modern data analysis strategy.

There is certainly much more to be done in this interdisciplinary area of research, combining state-of-the-art separations with novel chemometrics. Considerably more can be done in the area of 2-D retention time alignment algorithms, especially when one is inter-

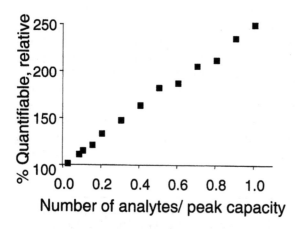

Fig. 15 Percent (%)Quantifiable, relative, of analytes as a function of the number of analytes per peak capacity. The %Quantifiable, relative, is the sum of GRAM quantified plus traditional method quantified divided by traditional method quantified analytes, then expressed as a percentage. (Adapted and modified from Ref. 66.)

ested in analyzing the entire 2-D space with a given chemometric algorithm. There are many new and exciting developments on the horizon for investigating various chemometric methods such as tri-PLS [71–75], PARAFAC and PARAFAC2 [68–70,82], PCA [30–34,38,39,43,44,94,101], and GRAM [62–67,84–89]. All of these developments will be critical for applications dealing with high throughput chemical analysis such as in proteomics and drug discovery. For example, an analyst may be interested in acquiring quantitative information on the composition of complex mixtures. Or, rather than desiring quantification of single analytes, an analyst may wish to obtain quantitative information about a group or class of chemical analytes, or even information about an abstract property related in some fashion to the chemical composition of the mixture. For this type of problem, the tri-PLS algorithm coupled to a high-speed comprehensive 2-D separation method may be the right choice. In some ways, this approach could be thought of as quantitative pattern recognition, as the goal is to quantify a pattern of analyte peaks, rather than a single analyte peak. Of course, quantitative regression-based calibration techniques are used rather than pattern recognition techniques. Further work may focus on characterizing the quality of such calibrations and assessment of the method's potential to provide quantitative information in analysis problems currently treated with pattern recognition techniques. Judging by the body of current work and the potential of research to come, the future for the combined technique of comprehensive 2-D separations analyzed by chemometrics is indeed bright.

REFERENCES

1. J. B. Phillips and J. Xu, *J. Chromatogr. A, 703*: 327 (1995).
2. H. J. deGeus, J. deBoer, and U. A. T. Brinkman, *Trends Anal. Chem., 15*: 398 (1996).
3. Z. Liu and M. L. Lee, *J. Microcol. Sep., 12*: 241 (2000).
4. J. B. Phillips and J. Beens, *J. Chromatogr. A, 856*: 331 (1999).
5. R. Kramer, *Chemometric Techniques for Quantitative Analysis*, Marcel Dekker, New York, 1998.
6. K. R. Beebe, R. J. Pell, and M. B. Seasholtz, *Chemometrics: A Practical Guide*, Wiley-Interscience, New York, 1998.
7. M. A. Sharaf, D. L. Illman, and B. R. Kowalski, *Chemometrics*, Wiley, New York, 1986.

8. H. Martens and T. Naes, *Multivariate Calibration*, John Wiley & Sons, New York, 1989.
9. J. C. Giddings, *Unified Separation Science*, Wiley, New York, 1991.
10. J. C. Giddings, *Anal. Chem.*, *56*: 1258A (1984).
11. J. C. Giddings, *J. High Resolut. Chromatogr.*, *10*: 319 (1987).
12. M. M. Bushey and J. W. Jorgenson, *Anal. Chem.*, *62*: 161 (1990).
13. M. M. Bushey and J. W. Jorgenson, *Anal. Chem.*, *62*: 978 (1990).
14. G. J. Opiteck, J. W. Jorgenson, M. A. Moseley, and R. J. Anderegg, *J. Microcol. Sep.*, *10*: 365 (1998).
15. Z. Liu and J. B. Phillips, *J. Chromatogr. Sci.*, *29*: 227 (1991).
16. Z. Liu, S. R. Sirimanne, D. G. Patterson Jr., L. L. Needham, and J. B. Phillips, *Anal. Chem.*, *66*: 3086 (1994).
17. M. M. Bushey and J. W. Jorgenson, *J. Microcol. Sep.*, *2*: 293 (1990).
18. T. F. Hooker and J. W. Jorgenson, *Anal. Chem.*, *69*: 4134 (1997).
19. J. C. Giddings, *J. Chromatogr. A*, *703*: 3 (1995).
20. R. E. Murphy, M. R. Schure, and J. P. Foley, *Anal. Chem.*, *70*: 1585 (1998).
21. J. M. Davis and J. C. Giddings, *Anal. Chem.*, *55*: 418 (1983).
22. J. M. Davis and J. C. Giddings, *Anal. Chem.*, *57*: 2178 (1995).
23. S. K. Kachigan, *Statistical Analysis—An Interdisciplinary Introduction to Univariate & Multivariate Methods*, Radius Press, New York, 1986.
24. M. K. Alam and J. B. Callis, *Anal. Chem.*, *66*: 2293 (1994).
25. I. A. Cowe and J. W. McNicol, *Appl. Spectrosc.*, *39*: 257 (1985).
26. C. C. Wu, J. D. S. Danielsen, J. B. Callis, M. Eaton, and N. L. Ricker, *Process Contr. Qual.*, *8*: 1 (1996).
27. D. L. Wetzel, *Anal. Chem.*, *55*: 1165A (1983).
28. B. K. Lavine, in *Practical Guide to Chemometrics* (S. J. Haswell, Ed.), Marcel Dekker, New York, 1992, p. 211.
29. R. G. Bereton, *Chemometric: Applications of Mathematics and Statistics to Laboratory Systems*, Ellis Horwood, New York, 1990.
30. C. P. Bicchi, O. M. Panero, G. M. Pellegrino, and A. C. Vanni, *J. Agr. Food Chem.*, *45*: 4680 (1997).
31. F. Carrera, M. Leon-Camacho, F. Pablos, and A. G. Gonzalez, *Anal. Chim. Acta*, *370*: 131 (1998).

32. P. Valera, F. Pablos, and A. G. Gonzalez, *Talanta*, *43*: 415 (1996).
33. N. Togari, A. Kobayashi, and T. Aishima, *Food Res. Int.*, *28*: 495 (1995).
34. P. E. Shaw, B. S. Buslig, and M. G. Moshonas, *J. Agr. Food Chem.*, *41*: 809 (1993).
35. P. E. Shaw, M. G. Moshonas, and B. S. Buslig, in *Fruit Flavors* (R. L. Rouseff and M. M. Leahy, Eds.), ACS Symposium Series 596, American Chemical Society, Washington, D.C., 1995, p. 33.
36. K. Robards, X. Li, M. Antolovich, and S. Boyd, *J. Sci. Food Agr.*, *75*: 87 (1997).
37. D. Blanco-Gomis, P. Fernandez-Rubio, M. D. Gutierrez-Alvarez, and J. J. Mangas-Alonso, *Analyst*, *123*: 125 (1998).
38. D. S. Lee, B. S. Noh, S. Y. Bae, and K. Kim, *Anal. Chim. Acta*, *358*: 163 (1998).
39. M. Reyens, M. LeBrun, and P. E. Shaw, *J. Food Quality*, *19*: 505 (1996).
40. K. R. Kim, J. H. Kim, E. J. Cheong, and C. M. Jeong, *J. Chromatogr. A*, *722*: 303 (1996).
41. S. E. Keller, D. S. Stewart, and S. M. Gendel, *J. Ind. Microbiol.*, *13*: 382 (1994).
42. K. R. Kim, H. G. Park, M. J. Paik, H. S. Ryu, K. S. Oh, S. W. Myung, and H. M. Liebich, *J. Chromatogr. B*, *712*: 11 (1998).
43. P. J. Dunlop, C. M. Bignell, J. F. Jackson, and D. B. Hibbert, *Chemom. Intell. Lab. Syst.*, *30*: 59 (1995).
44. C. Baiocchi, E. Marengo, M. A. Roggero, D. Giacosa, A. Giorcelli, and S. Toccori, *J. Chromatogr. A*, *715*: 95 (1995).
45. B. K. Lavine, H. Mayfield, P. R. Kromann, and A. Faruque, *Anal. Chem.*, *67*: 3846 (1995).
46. B. K. Lavine, A. J. Moores, H. T. Mayfield, and A. Faruque, *Anal. Lett.*, *31*: 2805 (1998).
47. B. K. Lavine, A. J. Moores, H. Mayfield, and A. Faruque, *Microchem. J.*, *61*: 69 (1999).
48. B. K. Lavine, J. Ritter, A. J. Moores, M. Wilson, A. Faruque, and H. T. Mayfield, *Anal. Chem.*, *72*: 423 (2000).
49. B. J. Prazen, C. A. Bruckner, R. E. Synovec, and B. R. Kowalski, *J. Microcol. Sep.*, *11*: 97 (1999).
50. A. Felinger, in *Advances in Chromatography* (P. R. Brown and E. Grushka, Eds.), 39, Marcel Dekker, Inc., New York, 1998, p. 201.

51. A. Felinger, *Data Analysis and Signal Processing in Chromatography*, Elsevier, Amsterdam, 1998.
52. M. Reta, P. W. Carr, P. C. Sadek, and S. C. Rutan, *Anal. Chem.*, *71*: 3484 (1999).
53. K. Heberger, M. Gorgenyi, and M. Sjostrom, *Chromatographia*, *51*: 595 (2000).
54. M. Thorsteinsdottir, C. Ringbom, D. Westerlund, G. Andersson, and P. Kaufmann, *J. Chromatogr. A*, *831*: 293 (1999).
55. M. H. Abraham, C. F. Poole, and S. K. Poole, *J. Chromatogr. A*, *842*: 79 (1999).
56. D. Bylund, A. Bergens, and S. P. Jacobsson, *Chromatographia*, *44*: 74 (1997).
57. P. H. Lukulay and V. L. McGuffin, *Anal. Chem.*, *69*: 2963 (1997).
58. G. Vivo-Truyols, J. Torres-Lapasio, and M. Garcia-Alvarez-Coque, *J. Chromatogr. A*, *876*: 17 (2000).
59. V. M. Morris, C. Hargreaves, K. Overall, P. J. Marriott, and J. G. Hughes, *J. Chromatogr. A*, *766*: 245 (1997).
60. M. Fransson, A. Sparén, B. Lagerholm, and L. Karlsson, *Anal. Chem.*, *73*: 1502 (2001).
61. A. Block, A. K. Smilde, and C. H. P. Bruins, *Chemom. Intell. Lab. Syst.*, *46*: 1 (1999).
62. C. A. Bruckner, B. J. Prazen, and R. E. Synovec, *Anal. Chem.*, *70*: 2796 (1998).
63. B. J. Prazen, R. E. Synovec, and B. R. Kowalski, *Anal. Chem.*, *70*: 218 (1998).
64. B. J. Prazen, C. A. Bruckner, R. E. Synovec, and B. R. Kowalski, *Anal. Chem.*, *71*: 1093 (1999).
65. C. G. Fraga, B. J. Prazen, and R. E. Synovec, *Anal. Chem.*, *72*: 4154 (2000).
66. C. G. Fraga, C. A. Bruckner, and R. E. Synovec, *Anal. Chem.*, *73*: 675 (2001).
67. C. G. Fraga, B. J. Prazen, and R. E. Synovec, *J. High Resolut. Chromatogr.*, *3*: 215 (2000).
68. K. Dahl, M. Piovoso, and K. Kosanovich, *Chemom. Intell. Lab. Syst.*, *46*: 161 (1999).
69. G. Andersson, B. Dable, and K. Booksh, *Chemom. Intell. Lab. Syst.*, *49*: 195 (1999).
70. J. L. Beltran, J. Guiteras, and R. Ferrer, *J. Chromatogr. A*, *802*: 263 (1998).

71. S. de Jong, *J. Chemom.*, *12*: 77 (1998).
72. A. K. Smilde, *J. Chemom.*, *11*: 367 (1997).
73. R. Bro, *J. Chemom.*, *10*: 47 (1996).
74. K. Hasegawa, M. Arakawa, and K. Funatsu, *Chemom. Intell. Lab. Syst.*, *47*: 33 (1999).
75. K. E. Miller, E. Bramanti, B. J. Prazen, M. Prezhdo, K. J. Skogerboe, and R. E. Synovec, *Anal. Chem.*, *72*: 4372 (2000).
76. G. Malmquist and R. Danielsson, *J. Chromatogr. A*, *687*: 71 (1994).
77. T. J. Bahowick and R. E. Synovec, *Anal. Chem.*, *67*: 631 (1995).
78. R. B. Poe and S. C. Rutan, *Anal. Chim. Acta*, *283*: 845 (1993).
79. R. Andersson and M. D. Hämäläinen, *Chemom. Intell. Lab. Syst.*, *22*: 49 (1994).
80. B. Grung and O. M. Kvalheim, *Anal. Chim. Acta*, *304*: 57 (1995).
81. N. P. V. Nielsen, J. M. Carstensen, and J. Smedsgaard, *J. Chromatogr. A*, *805*: 17 (1998).
82. R. Bro, C. Andersson, and H. Kiers, *J. Chemom.*, *13*: 295 (1999).
83. C. G. Fraga, B. J. Prazen, and R. E. Synovec, 24th International Symposium on Capillary Chromatography and Electrophoresis, Las Vegas, NV, USA, May 22, 2001, Abstract 64.
84. K. Faber, A. Lorber, and B. R. Kowalski, *J. Chemom.*, *11*: 95 (1997).
85. E. Sánchez and B. R. Kowalski, *Anal. Chem.*, *58*: 496 (1986).
86. L. S. Ramos, E. Sánchez, and B. R. Kowalski, *J. Chromatogr.*, *385*: 165 (1987).
87. B. Antalek and W. Windig, *Journal of the American Chemical Society*, *118*: 10331 (1996).
88. E. Sánchez, L. S. Ramos, and B. R. Kowalski, *J. Chromatogr.*, *385*: 152 (1987).
89. B. E. Wilson, E. Sánchez, and B. R. Kowalski, *J. Chemom.*, *3*: 493 (1989).
90. D. C. Harris, *Quantitative Chemical Analysis*, W. H. Freeman and Company, New York, 1995.
91. N. M. Faber, L. M. C. Buydens, and G. Kateman, *Anal. Chim. Acta*, *296*: 1 (1994).
92. N. M. Faber, L. M. C. Buydens, and G. Kateman, *Chemom. Intell. Lab. Syst.*, *25*: 203 (1994).

93. K. Faber and B. R. Kowalski, *Anal. Chim. Acta, 337*: 57 (1997).
94. E. R. Malinowski and D. G. Howery, *Factor Analysis in Chemistry*, Wiley, New York, 1991.
95. J. J. Dongarra, C. B. Molar, J. R. Bunch, and G. W. Stewart, *Linpack Users' Guide*, Siam, Philadelphia, 1979.
96. K. Faber and B. R. Kowalski, *J. Chemom., 11*: 53 (1997).
97. E. R. Malinowski, *Anal. Chem., 49*: 612 (1977).
98. E. R. Malinowski, *J. Chemom., 3*: 49 (1988).
99. J.-M. D. Dimandja, S. B. Stanfill, J. Grainger, and J. Donald G. Patterson, *J. High Resolut. Chromatogr., 23*: 208 (2000).
100. H.-J. d. Geus, I. Aidos, J. d. Boer, and J. B. Luten, *J. Chromatogr. A, 910*: 95 (2001).
101. K. J. Johnson and R. E. Synovec, *Chemom. Intell. Lab. Sys., 60*: 225 (2002).
102. G. J. Opiteck, S. M. Ramirez, J. W. Jorgenson, and M. A. Moseley, *Anal. Biochem., 258*: 349 (1998).
103. J. P. Larmann, Jr., A. V. Lemmo, A. W. Moore, Jr., and J. W. Jorgenson, *Electrophoresis, 14*: 439 (1993).
104. L. A. Holland and J. W. Jorgenson, *J. Microcol. Sep., 12*: 371 (2000).
105. J. B. Phillips, R. B. Gaines, J. Blomberg, F. W. M. van der Wielen, J. M. Dimandja, V. Green, J. Granger, D. Patterson, L. Racovalis, H. J. de Geus, J. de Boer, P. Haglund, J. Lipsky, V. Sinha, and E. B. Ledford Jr., *J. High Resolut. Chromatogr., 22*: 3 (1999).
106. G. S. Frysinger, R. B. Gaines, and E. B. Ledford Jr., *J. High Res. Chromatog., 22*: 195 (1999).
107. C. J. Venkatramani and J. B. Phillips, *J. Microcol. Sep., 5*: 511 (1993).
108. A. L. Lee, A. C. Lewis, K. D. Bartle, J. B. M. Quaid, and P. J. Marriott, *J. Microcol. Sep., 12*: 187 (2000).
109. J. V. Seeley, F. Kramp, and C. J. Hicks, *Anal. Chem., 72*: 4346 (2000).
110. W. W. C. Quigley, C. G. Fraga, and R. E. Synovec, *J. Microcol. Sep., 12*: 160 (2000).
111. J. M. Davis, *Anal. Chem., 63*: 2141 (1991).

2

Column Technology for Capillary Electrochromatography

Luis A. Colón, Todd D. Maloney, Jason Anspach,
and Héctor Colón *University at Buffalo, The State University*
of New York, Buffalo, New York

I. INTRODUCTION

Capillary electrochromatography (CEC) is a relatively new analytical separation technique, which is considered a hybrid between capillary electrophoresis (CE) and high-performance liquid chromatography (HPLC). This combination allows the use of an electrically driven flow to transport solvent and solutes through a chromatographic column, making use of a pluglike flow profile to provide high efficiency, with the possibility of retaining a variety of retention mechanisms and selectivity afforded by HPLC. CEC is capable of separating neutral solutes by differential interactions between the mobile and stationary phases. Additionally, separation of charged species is influenced by differential electromigration. Interest in CEC has been stimulated by the possibility of increasing the limited peak capacity of traditional liquid chromatography. The fundamental aspects of CEC have been discussed in a recent publication of this series [1], as well as others [2–5].

One very active research area in CEC has been column technology. Columns used in CEC include open tubes, packed columns, monoliths, and microfabricated structures. In open tubular CEC, the stationary phase is affixed to the inner wall of a capillary column [3,4,6–17]. It is recommended to use columns with an internal diameter of less than 20 µm for best performance [4]. The other type of columns, conversely, consists of capillaries containing chromatographic material through the entire diameter of the column. These columns can be classified into three different groups: (1) columns packed with particles [18–46], (2) monolithic or continuous bed columns [47–55], and (3) columns containing entrapped or embedded particulate material, which are a combination of the first two groups [56–63]. The microfabricated structures are of the open channel format [64–66], similar to those originally developed for CE [67–72]; most recently, chromatographic material has also been incorporated inside the microfabricated devices [73–77]. Herein, we present an overview (not exhaustive or inclusive) of the state-of-the-art on column technology for CEC. However, since the vast majority of the CEC has been performed using packed capillary columns, we include details on the column fabrication process for such columns.

II. OPEN TUBULAR COLUMNS

The use of open tubes is the simplest way to implement CEC. In open tubular CEC (OT-CEC), the inner walls of a capillary column are modified with the stationary phase. As it is the case in CE, the negatively charged silanol groups at the inner surface provide for the generation of the electroosmotic flow (EOF), which is responsible for the solvent flow through the capillary. Alternatively, charged species can migrate *via* electromigration while the EOF is suppressed. As the species migrate through the column, they have differential interaction with the stationary phase at the capillary walls, as well as differential migration due to the applied electric field. Theoretical aspects of OT-CEC and the incorporation of various types of stationary phases have been reported [3,4,6,8].

In OT-CEC, plate height is minimized by decreasing the column inner diameter. Figure 1 shows plate height versus linear velocity for open tubes operated in the CEC and LC mode [3,4]. One can notice that column efficiency in OT-CEC is about twice that of OTLC for a given column diameter. OT-CEC can also afford higher linear velocities than OTLC with minimal loss in separation efficiency,

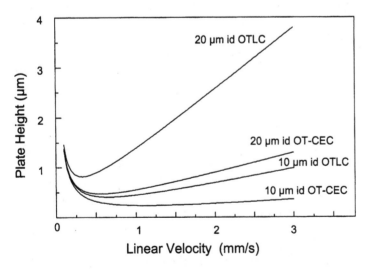

Fig. 1 Plate height vs. linear velocity plots for two different columns operated in the OTLC and OT-CEC mode.

translating to faster analysis times. Figure 1 also indicates that columns for OT-CEC can have twice the diameter of columns in OTLC with similar separation efficiencies. The larger column diameter in CEC is beneficial for on-column detection by spectroscopic means, in addition to facilitating column preparation.

One of the earliest reports on OT-CEC was presented by Tsuda et al. [78]. They fabricated columns by drawing large-bore soda-lime tubes, to which octadecylsilane was attached at the inner surface. They separated a series of hydrocarbons using acetonitrile/water as the mobile phase and applying 13 kV as the separation voltage. In general, OT-CEC fused silica columns have been fabricated using procedures previously utilized in open tubular liquid chromatography (OTLC) [79]. The most typical approach includes the derivatization of the inner walls of fused silica capillaries with a silane reagent, which serves as the stationary phase. This approach, however, provides low surface coverage, which leads to low phase ratios, hence low sample capacity.

The problem of sample capacity has been addressed by several means. One approach to enhance solute sorption is to affix polymeric films at the capillary surface [16,80–86]. Most notably is the incorporation of polyacrylate films by in situ photo- or thermal-polymerization. Thick polymeric films, however, suffer from poor solute diffusion in the stationary phase. Recent work in the use of polymeric material for OT-CEC includes the anchoring of thick polymethacrylate films to capillary walls [16,86] and polymerization of N-terbutylacrylamide and 2-acrylamido-2-methyl-1-propanesulfonic acid, providing for hydrophobicity and charged groups for the generation of EOF [15]. Another approach to OT-CEC columns is to increase the surface area of the inner surface of the capillary by etching procedures. Most notable is the work by Pesek and coworkers [6,8–13,87–92]. In their method, a capillary column is filled with concentrated HCl and heated at 80°C overnight (with the ends sealed). The column is then washed with deionized water, acetone, and diethyl ether. After drying at room temperature with a stream of nitrogen, the capillary is filled with a solution of 5% ammonium hydrogen difluoride in methanol, and left to stand for an hour. The methanol is removed with a stream of nitrogen, the column's ends are sealed, and the capillary is heated to temperatures between 300° and 400°C. The morphology of the inner surface depends on the time and temperature used for this heat treatment [9,87]. Figure 2 shows scan-

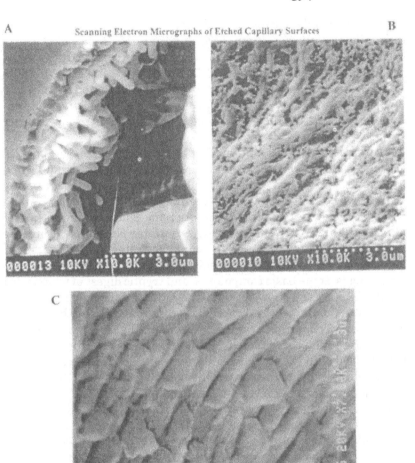

Fig. 2 Scanning electron micrographs (SEM) showing the morphology of the inner surface of fused silica capillaries after the etching process. (A) 3 h at 300°C, (B) 2 h at 300°C followed by 2 h at 400°C, and (C) 2 h at 300°C followed by 1 h at 400°C. (Reprinted from Ref. 92 with permission. Copyright Elsevier 2000.)

ning electron micrographs (SEM) of the inner surface for etched fused silica capillary columns. Etching times of 4 h at 400°C seems to produce somewhat more uniform and regular surface. Reportedly, the etching procedure increases the surface area by 100- to 1000-fold. Because an anodic EOF at a pH < 4.5 is observed in such columns, Pesek and coworkers have suggested that nitrogen-containing species from the etching agent are incorporated in the inner surface of the capillary.

After etching, the stationary phase is affixed at the surface through silanization/hydrosilation reactions. This is accomplished by reacting the etched capillary walls with triethoxysilane, forming a silica hydride surface, which is then reacted with a terminal olefin (e.g., 1-octadecene) to form the stationary phase. This method of affixing the stationary phase yields very stable phases at the surface due to Si–C linkage. OT-CEC columns fabricated by this procedure have been used to separate tetracyclines, cytochrome C, lysozymes, amino acids, some basic compounds, and tryptic digest of transferrin [9–12,90]. Figure 3 shows OT-CEC separations of cytochrome c's on 20 μm inner diameter (i.d.) capillaries containing C_{18} as the stationary phase. The etched capillaries render higher resolution and greater retention than unetched columns. Chiral selectors have also been affixed onto the etched capillaries (e.g., lactone, β-cyclodextrin, and naphthylethylamine) as well as liquid crystals (i.e., cholesterol-10-undeceneoate and 4-cyano-4'-pentoxybiphenyl) [13], and their potential as a stationary phase for OT-CEC was evaluated.

Another approach to fabricate OT-CEC columns is using the sol-gel method introduced by our laboratory [93,94], also explored by Malik's [95,96] and Freitag's [97] groups. Silica-based materials fabricated by the sol-gel process have a characteristic high surface area, which can provide an alternative to etching. For example, Fig. 4 shows SEMs of a capillary surface coated with octadecylsilane (ODS) using the sol-gel process [95]. Typically, the OT-CEC capillaries are fabricated by the acid catalyzed hydrolysis and condensation of a precursor mixture containing alcoxysilanes (e.g., tetraethoxysilane and octyltriethoxysilane). It must be noted that this is a one step in which all the precursors are mixed in a single "pot" reaction, which simplifies the column preparation procedure. After the reaction has proceeded, the solution is introduced inside the capillary columns to form a thin film on the capillary inner surface. The organosilica film thus formed provides a surface-bonded stationary phase containing

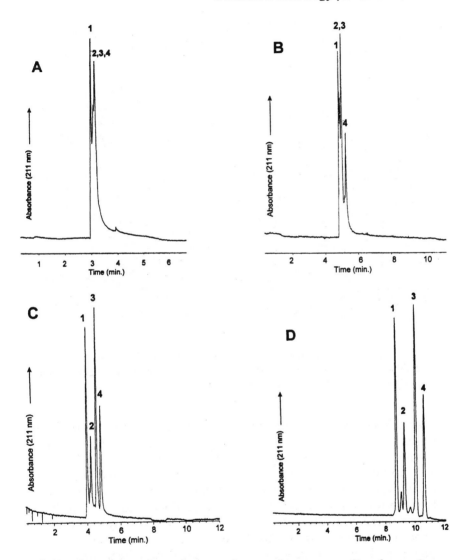

Fig. 3 Separation of cytochrome c from various sources (1 = horse, 2 = bovine, 3 = chicken, 4 = tuna) on different 20 μm i.d. capillaries: (A) bare silica, (B) unetched C_{18} modified, (C) and (D) etched C_{18} modified columns. Separations conducted at 30kV, except (D), which was conducted at 15kV. (Reprinted from Ref. 9 with permission. Copyright Elsevier 1999.)

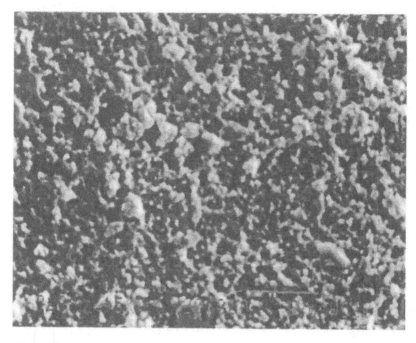

Fig. 4 Scanning electron micrograph (SEM) of a capillary surface coated with ODS using sol-gel processing. (Reprinted from Ref. 95 with permission. Copyright American Chemical Society 2001.)

desirable chromatographic surface groups on the inner wall of the column. It has been shown that coating materials prepared by this method produce capillary columns that are stable under acidic (pH < 1) and basic (pH ~ 11.5) conditions and are capable of achieving 500,000 plates/m [93]. By controlling the coating procedure, column-to-column retention was shown to be reproducible (8% RSD or less) [98]. Selectivity of the stationary phase can be tuned by selecting an appropriate precursor for the sol-gel reaction process. For example, phenyl and octadecyl groups have been introduced by using phenyl-dimethylsilane and n-octadecyldimethyl[3-(trimethoxysilyl)propyl] ammonium chloride as precursors for the reaction [95]. Another example is the incorporation of a fluorinated precursor to prepare a fluorinated stationary phase for OT-CEC [99], used to separate fluorobenzene isomers (see Fig. 5) and other halogenated compounds.

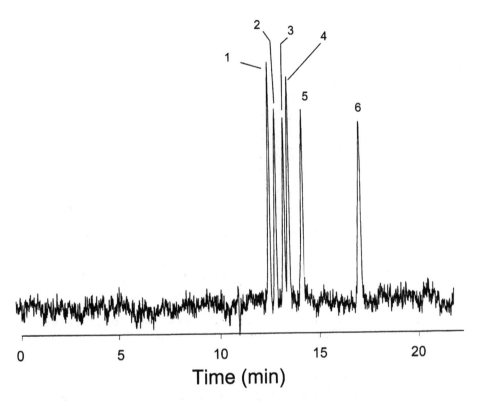

Fig. 5 Separation of five fluorinated compounds by OT-CEC using a capillary column coated with TEOS:F_{13}-TEOS via sol-gel processing. The test mixture contained (1) fluorobenzene, (2) 1,4-difluorobenzene, (3) 1,3-difluorobenzene, (4) 1,2-difluorobenzene, (5) 1,2,4-trifluorobenzene, and (6) 1,2,3,5-tetrafluorobenzene. (Reprinted from Ref. 99 with permission. Copyright Elsevier 1997.)

Porous-layer open-tubular (PLOT) columns have been used by Horváth and coworkers in OT-CEC to separate basic proteins and peptides [14]; these columns also provide high surface areas, leading to high permeability and high loading capacity (see Fig. 6). The polymeric PLOT columns have been prepared by first silanizing the inner wall of fused silica capillaries using 3-(trimethoxysilyl)propyl methacrylate. This facilitates anchoring of the polymeric layer by covalent bonds via the vinyl functionality. Vinylbenzyl chloride and

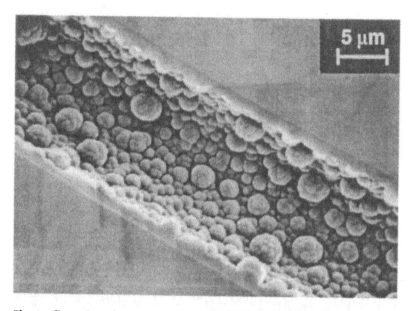

Fig. 6 Scanning electron micrograph (SEM) of a polymeric porous layer in a capillary column. (Reprinted from Ref. 14. Copyright Elsevier 1999.)

divinylbenzene are polymerized in situ, and 2-octanol is used as a porogenic agent. A chromatographic surface with fixed C_{12} alkyl chains is obtained by reacting N,N-dimethyldodecylamine with the chloromethyl functionality at the surface of the porous polymeric support layer. This also leads to a positively charged surface due to the quaternary ammonium groups, which is responsible for the generation of EOF upon application of the electric field. It was reported that this type of column provided an increase in retention with a concomitant increase in the concentration of the organic modifier in the mobile phase. The PLOT columns were successfully used in the separation of basic proteins [14].

Chiral separations using OT-CEC columns have mainly used cyclodextrin-type of stationary phases attached to the inner walls of fused silica columns [100,101]. Examples include the separation of a racemic mixture of mephobarbitol [102] using permethylated allyl-substituted β-cyclodextrin; and the separation of epinephrine enantiomers [103]. In a comparison study using a single Chirasil-Dex coated capillary column, Schurig et al. performed enantiomeric separations in the GC, SFC, CEC, and LC modes, using hexobarbital enantiomers as model solutes [104]. The separations performed by CEC were shown to be superior in all aspects significant to chiral separations, although the analysis time was longer for the CEC separation than for the other modes (see Fig. 7). CE separations with cyclodextrins additive for racemic mixtures of carprofen and hexobarbitol have also been compared to OT-CEC separations using β- or γ-Chirasil-Dex stationary phase coatings [105]. The chiral selectors affixed to the capillary inner surface provided increased resolution of enantiomers when compared to the use of cyclodextrin derivatives as a pseudo-stationary phase in the CE running buffer. In contrast to buffer additives, chiral selectors affixed to the capillary walls as stationary phase are not wasted and the column can be used for multiple separations. This becomes important when using exotic chiral selectors and cost is an issue to consider.

A recent development in OT-CEC is the use of DNA aptamers as stationary phase [106,107]. DNA aptamers are short, usually single-stranded, oligonucleotides that have been identified via combinatorial selection processes for high affinity and specific binding to the

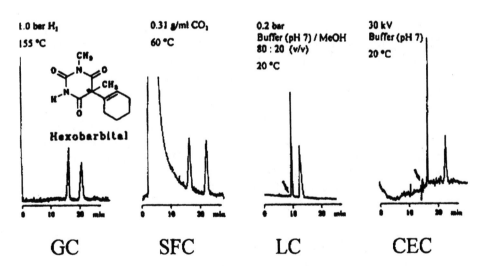

Fig. 7 Separations of hexobarbitol enantiomers by gas chromatography (GC), supercritical fluid chromatography (SFC), liquid chromatography (LC), and capillary electrochromatography (CEC). (Reprinted from Ref. 104 with permission. Copyright Elsevier 1995.)

A   B

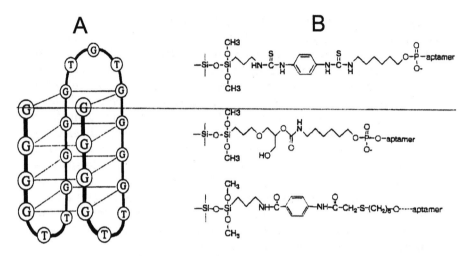

Fig. 8 (A) Structure of 4-plane aptamer. (Reprinted from Ref. 107 with permission. Copyright Wiley-VCH 2001.) (B) Structures of possible linkers. (Reprinted from Ref. 106 with permission. Copyright American Chemical Society 2000.)

molecule of interest. OT-CEC capillaries are prepared by covalently attaching an organic linker molecule to the surface (e.g., sulphosuccinimidyl 4-(N-maleicmidomethyl) cyclohexane-1-carboxylate) and reacting with the aptamer after reducing the 5′ disulfide bond to a free thiol. A schematic representation of a DNA aptamer is illustrated in Fig. 8, as well as possible linkers to attach the aptamer to the capillary inner surface. The potential application of OT-CEC columns using aptamer stationary phases has been demonstrated in the separation of amino acids (D-trp and D-tyr), enantiomers (D-trp and L-trp), and bovine β-lactoglobulin variants A and B. These types of phases also have the potential to be used in the study of solute-aptamer specific interactions.

III. PACKED COLUMNS

Packed capillaries are by far the most used columns in CEC. Most commonly, capillaries have been packed using HPLC type of chromatographic packing material. New alternatives, however, are emerging to fabricate chromatographic materials with greater CEC functionality. With the new advances in column technology, the preference for using columns packed with particles may change, particularly with the emerging technologies in monolithic columns and microfabricated devices.

Fused silica tubes with inner diameters of 100 μm or less are typically used to fabricate packed CEC columns. The typical architecture of a packed CEC column consists of two sections: one contains chromatographic material and the other is the tube without packing. Figure 9 depicts a CEC column with a packed and an open segment. The packing material most often used consists of reverse-phase HPLC spherical particles of 1.5–5 μm diameter. Retaining frits are employed to hold the chromatographic material in place. It has been found that the electroosmotic flow (EOF) velocity in the open and packed segment of the CEC column is different [108], and the overall EOF velocity depends on the fraction of the packed segment [108,109]. For the purpose of spectroscopic, on-column detection, the polyimide coating must be removed close to the outlet retaining frit extending toward the open segment. Most commonly, an on-column optical window is made by burning off the polyimide coating of the fused silica capillary; however, this must be done carefully so as not to expose the packing material to excessive heat, which could result in additional sintering of the packing material. Optical detection through the packed bed has been performed [34,40,43], but it is not a common practice due to the reduced sensitivity caused by the light scattering from the particles.

It is possible to completely pack capillary columns with the desired packing material leaving no open segment; however, if a detection scheme other than through the packed bed is desired, connection to the appropriate detection system is necessary. This can be accomplished by using a segment of an open capillary butted against the packed column and securing the connection with PTFE shrinking [110]. Care must be taken because the connection of two different

Fig. 9 Illustration of a typical CEC packed capillary column.

pieces of tubing always has the potential of introducing band broadening to the system. An optimized connection, however, has an insignificant contribution to band spreading [111]. In many situations, capillary butting is necessary to facilitate connections to detection schemes that offer higher detectabilities, such as with the use of a Z-cell for UV detection, or to couple with information-rich detection schemes, such as mass spectrometry (MS) and NMR detection. In such cases, the gain in sensitivity and/or structural information is far more important than the possible loss in efficiency.

A. Packing Materials

Silica-Based Packing

The most common packing material used in CEC is, by far, the silica-based C_{18} reverse-phase packing, with differing particle diameters. Other packing materials with C_8 [112–114], C_{30} [115,116], and phenyl [113,114] functionalities have also been utilized in the reversed-phase mode. Packing materials for CEC must have the ability to support EOF. The EOF is responsible for the bulk transport of the mobile phase and most of the solutes through the CEC column [23,32]. Only species that contain the appropriate charge will be able to migrate toward the detector if there is no EOF present. It is, therefore, desirable to choose packing materials that have favorable characteristics for EOF generation for use in CEC. The primary source of EOF upon application of the electric field on silica-based materials is the result of the negatively charged surface silanol groups of the silica packing support. The main contribution to EOF is from the particulate material; contributions from the fused silica capillary walls appear to be negligible [23,32,117]. Not all C_{18} HPLC materials are suitable for use in CEC, however, since the material must provide for EOF generation. In general, materials with low hydrophobic selectivity and those prepared with octadecyltrichlorosilane, instead of octadecyldimethylchlorosilane, yield high EOF [118]. Electrophoretic mobilities observed on various reversed-phase HPLC supports are summarized in Table 1. The CEC Hypersil and Spherisorb ODS-1 materials appear to support the fastest EOF. The high EOF generated by these materials have been attributed to the fact that they have not been end-capped, and therefore, have a relatively large quantity of surface silanol groups that are capable of generating EOF. As the alkyl substitution at the packing surface increases, the

Table 1 Electroosmotic Mobilities of Various Chromatographic Materials Utilized for CEC

Stationary phase material	Electroosmotic mobility ($\times 10^{-4}$ cm^2/Vs)
BDS-ODS Hypersil[a]	0.99
CEC Hypersil C$_{18}$[a]	2.26
Hypersil ODS[b]	0.14
LiChrospher RP-18[b]	1.45
Nucleosil 5 C$_{18}$[b]	1.56
ODS Hypersil[a]	1.47
Partisil 5 ODS3[b]	<0.01
Prontosil polymeric C30[c]	1.54
Purospher RP-18[b]	<0.01
Rainin polymeric C30[c]	1.54
Spherisorb Dial[b]	0.80
Spherisorb ODS I[a]	2.26
Spherisorb ODS II[a]	1.79
Spherisorb S5 ODS2[b]	0.50
Zorbax BP-ODS[b]	0.68

[a] Adapted from [23]; separation conditions: (80/20) acetonitrile/Tris-HCl 50 mM, pH = 8, 20°C, EOF marker used: Thiourea.
[b] Adapted from [32]; separation conditions: (70/30) acetonitrile/3-cyclohexylamino-2-hydroxyl-1-propanesulfonic acid 25 mM, pH = 9.53, EOF marker used: Thiourea.
[c] Adapted from [115]; separation conditions: (95/5) acetone/1 mM borate buffer, EOF marker used: Acetone.
Reprinted from Ref. 205 with permission. Copyright Elsevier 2001.

residual silanol groups decrease, resulting in a concomitant decrease in EOF [117,119]. Packed columns containing 3 μm particles with C$_{18}$ phases have generated separation efficiencies of 300,000 plates/m [33,46], while separation efficiencies of 500,000 theoretical plates/m have been reported with 1.5 μm nonporous reverse-phase silica materials [120–124]. Such efficiencies were realized in the separation of 14 explosive compounds, which is shown in Fig. 10. In this separation, SDS was added to the mobile phase to stabilize the EOF through dynamic modification of the alkylated surface [122].

Many enantiomeric CEC separations have been performed on silica-base stationary phases [22,101,125–135]. Packing materials that have been used for chiral CEC include cyclodextrins

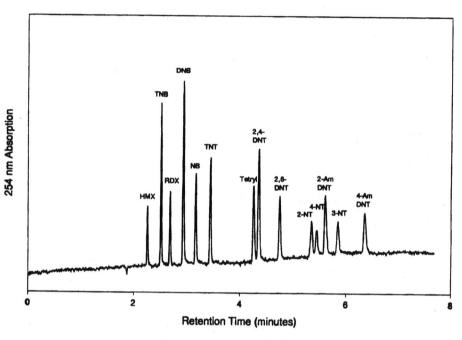

Fig. 10 CEC separation of explosives in a 21 cm (75 μm I.D., total length of 34 cm) capillary column packed with 1.5 μm nonporous reverse-phase silica particles. The mobile phase consisted of 20% methanol, 80% 10 mM MES, and 5 mM SDS; separation voltage of 12 kV. (Reprinted from Ref. 122 with permission. Copyright American Chemical Society 1998.)

[22,126,133] and protein-type selectors [125,127,132], such as α_1-acid glycoprotein [125] and human serum albumin [127]. Figure 11 shows a representation of permethyl-β-cyclodextrin stationary phase linked to silica and its application to the separation of several enantiomeric mixtures [133]. Packed columns operated in the CEC mode have shown higher separation efficiencies than those operated in the LC mode. Other enantiomeric selectors used in CEC include cyclodextrin-based polymer coated silicas [101,130], the silica-linked or silica-coated macrocyclic antibiotics teicoplanin [136,137] and vancomycin [138,139], and weak anion-exchange type chiral phases [140]. Columns packed with naproxen-derived and Whelk-O chiral stationary phases affixed to silica particles have also offered high separation efficiency and excellent resolution for a variety of com-

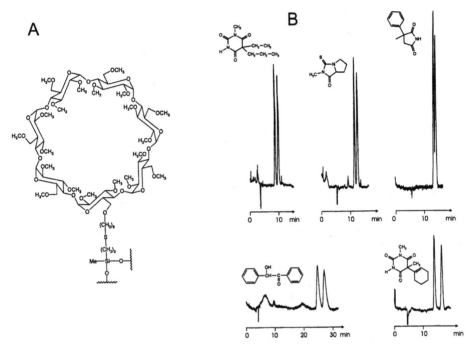

Fig. 11 (A) Structure of permethyl-β-cyclodextrin linked to silica support by means of a sulfido spacer and (B) enantiomeric separation of 5-ethyl-1-methyl-5-(n-propyl)barbituric acid, MTH-proline, α-methyl-α-phenylsuccin-imide, benzoin and hexobarbital on a 23.5 cm (total length, 40 cm) × 100 µm capillary packed with silica containing the phase shown in (A), using a mobile phase of phosphate buffer (5 mM, pH 7.0)-methanol (4:1, v/v) and 15 kV (benzoin, 20 kV). (Adapted from Ref. 133 with permission. Copyright Elsevier 1998.)

pounds [140]. CEC applications have been discussed in detail by Dermaux et al. [141] and Robson et al. [142].

Submicron Packing. Decreasing particle diameter in CEC improves separation efficiency, as can be seen in Fig. 12 down to particles of about 1 µm in diameter. It is also appreciable that the use of EOF in CEC allows one to operate at relatively high linear velocities of the mobile phase without a detrimental effect on efficiency, contrary to pressure-driven LC. The mobile phase linear velocity (u) due to the EOF in CEC is given by Eq. (1):

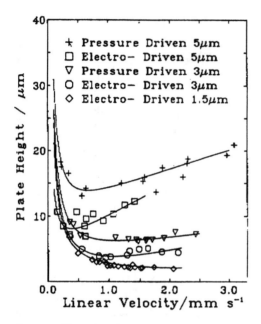

Fig. 12 Plate height vs. mobile-phase linear velocity for columns packed with different particle diameters using pressure and electrically driven flows. (Reprinted from Ref. 27 with permission. Copyright Friedr. Vieweg & Sohn 1991.)

$$u = \frac{\varepsilon_o \varepsilon_r \zeta E}{\eta} = \mu_{eo} E \tag{1}$$

where ε_o, ε_r, ζ, E, η, and μ_{eo} are, respectively, the permitivity in vacuum, the relative permitivity of the medium, the zeta potential, the applied electric field strength, the viscosity of the medium, and the electroosmotic mobility. Eq. (1) shows that the EOF velocity does not depend on particle diameter. This eliminates the pressure limitations found in HPLC and opens the possibility of exploring the performance of columns packed with particles in the submicron regime. In principle, the particle diameter in CEC is limited by the overlapping of the electrical double layer in the flow channels between the particles.

In a packed bed, there are many interconnected channels between particles and the EOF depends on the column packing structure and pore size of the packing material [108,143–148]. The porosity of the packed bed dictates the permeability through the column. The average size of the flow channels between particles in a CEC column can be estimated by assuming that the packed bed is a collection of capillary tubes with an average diameter corresponding to the channel between particles. A relationship between the mean channel diameter (d_c) and particle size (d_p) is given in Eq. (2) [145]:

$$d_c = \frac{0.42 d_p \varepsilon}{1 - \varepsilon} \tag{2}$$

where ε is the interparticle porosity. For a fairly well packed column $\varepsilon = 0.4$ [149], leading to a channel diameter given by:

$$d_c = 0.28 \, d_p \tag{3}$$

A similar value for the channel diameter (i.e., $0.25 \, d_p$) was obtained using the flow resistance parameter of a packed column [150]. Applying classical double-layer theory to thin capillaries, the flow channel must be at least 20 times the thickness of the electrical double layer ($d_c \geq 20\delta$) in order to retain at least 80% of the volume transport due to electroosmosis, minimizing parabolic flow [151]. The EOF would decrease considerably with the electrical double layer overlapping, which becomes significant with dimensions of less than 20δ. The double layer thickness (δ) is given by Eq. (4):

$$\delta = \sqrt{\frac{RT\varepsilon_o \varepsilon_r}{2IF^2}} \tag{4}$$

where F, I, R, and T are, respectively, the Faraday constant, the ionic strength, the gas constant, and the temperature in K. The other variables are defined in Eq. (1). Figure 13 shows calculated electrical double-layer thickness for different aqueous-acetonitrile solutions at different ionic strengths for univalent ions. The flow channel dimensions must be 20–200 nm to satisfy the flow channel dimensions of $d_c \geq 20\delta$ with double-layer thickness in the range of 1 to 10 nm. This corresponds to a minimum in particle diameter of about 0.071–0.71 μm.

It has been shown that EOF is independent of particle size down to 0.2 μm (see Fig. 14) [119], as predicted by Eq. (1). It follows from

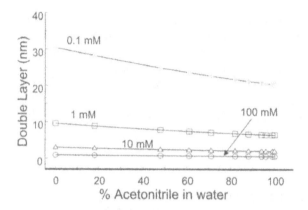

Fig. 13 Thickness of the electrical double layer as a function of the percent acetonitrile in an aqueous solution. Constants for the mixed solvents were obtained from Refs. 186 and 201.

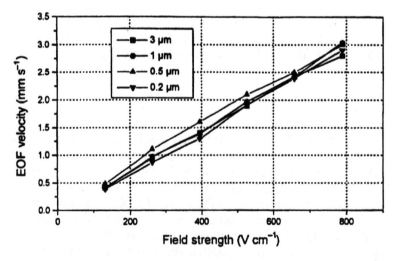

Fig. 14 EOF velocity as a function of applied electric field for capillary columns packed with different particle diameters. (Reprinted from Ref. 119 with permission. Copyright Elsevier 2000.)

Eq. (3) that the average interparticle channel diameter for 0.2 μm particles is about 0.06 μm, and the condition of $d_c \geq 20\ \delta$ is still satisfied. Columns packed with submicron particles have achieved EOF linear velocities above 2 mm/s [119,152,153]. The reported separation efficiencies of columns packed with submicron particles have not offered a significant improvement in performance over those obtained with particle diameters in the 1 μm range [119,152–154]. Figure 15 depicts the separation of a test mixture obtained in a column packed in our laboratory using particles of about 0.5 μm, rendering theoretical plates of about 600,000 plates/m. Plate heights of about three times the particle diameter $(H = 3d_p)$ are achieved with the 0.5 μm [154]. This performance has been attributed to band dispersion due to temperature effects and instrumental limitations [154].

Large Pore Silica Packings. The commercially available porous silica-based packing materials have pore sizes close to 100 Å. Under normal CEC conditions, it is unlikely that EOF can be generated through such pores. Intraparticle EOF, however, has been shown to occur in materials with relatively large pores [61,148,155–157]. Solute flow through the pores results in an enhanced mass transfer, providing improved separation efficiencies. Pore accessibility in pressure-driven systems is by means of diffusion, since there is no flow through the particle pores. Intraparticle EOF has been demonstrated in particles having nominal pore sizes between 500 and 4000 Å [61,155,156,158–160]. Electrical conductivity studies have also shown perfusive currents indicative of EOF through the pores, with the highest for particles having pore size of 1000 Å and above [159]. The concentration of the electrolyte in the mobile phase dictates the EOF within the pores. EOF through the pores is observed at high concentrations of electrolyte. At low concentrations of electrolyte, on the other hand, double layer interaction occurs and transport through the pores is stopped [155,156,159]. Figure 16 shows the effect of pore size and electrolyte concentration on separation efficiency in columns packed with 7 μm (C_{18}) particles [155]. Separation efficiencies of 430,000 theoretical plates/m have been achieved with perfusive transport through such columns. The use of highly porous materials has been proposed as an alternative to the use of very small particles in CEC [155,156,161]. Wide-pore large particles have the advantage over submicron material in that they are relatively easy to pack because particle aggregation is minimal. In addi-

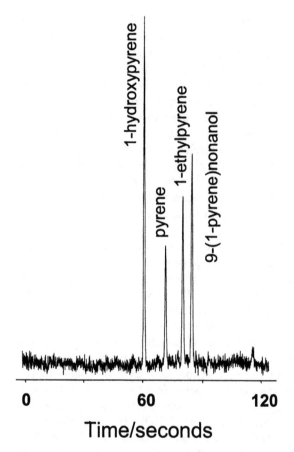

Fig. 15 Separation of a test mixture containing pyrene derivatives in capillary column packed with 0.5 μm (C₈) particles. Column: packed bed of 13 cm (35 cm total length); mobile phase: 80/20 acetonitrile/50 mM Tris buffer at pH = 8; separation voltage of 30 kV; injection, 3 s at 300 V; UV detection (220 nm). (Reprinted from Ref. 2 with permission. Copyright Wiley-VCH 2000.)

tion, problems associated with bubble formation and current breakdown are significantly fewer with the wide-pore large particles. Highly porous materials, however, are fragile and care must be exercised during packing to avoid damage and/or fracturing of the particles. Further, the enhanced particle porosity will decrease sample

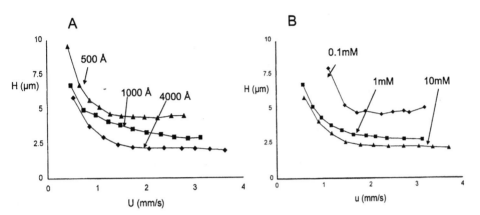

Fig. 16 Effect of the (A) nominal particle pore size and (B) electrolyte concentration in the mobile phase on separation efficiency for columns packed with C_{18}, 7 μm highly porous particles. (Reprinted from Ref. 155 with permission. Copyright Elsevier 1999.)

capacity and column heating can result from the increase in ionic strength to maintain a thin double-layer inside the pores.

Phases with Ion Exchangers

Solvent flow in CEC depends on the surface charge of the packing material. For most commonly used silica-based packing materials, the EOF is generated because of the deprotonation of the surface silanol groups on the silica support, and it is dependent on the pH of the mobile phase. The majority of the silanol groups are deprotonated at a pH of 8, which leads to a strong EOF. Below pH 3, however, there is a significant lack of silanol deprotonation, resulting in a reduction and/or elimination of the EOF. The dependence of the EOF on the pH of the mobile phase can be reduced by using packing materials with a fixed surface charge throughout a wide pH range. A variety of stationary phases, as those shown schematically in Figure 17, containing ion-exchangers or materials specifically tailored for CEC, can provide high surface charge to generate EOF.

Columns packed with ion exchangers (i.e., SAX, SCX) have been used to separate charged inorganic and organic solutes [162–166]. Anionic [62,140,167–169] and cationic [17,117,164,165,169–182] exchange packing materials have been studied for use in CEC. In prin-

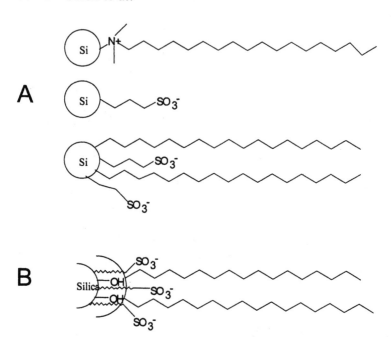

Fig. 17 (A) Generic representation of stationary phases that can provide high surface charge for electroosmotic flow (EOF) generation. (B) Mixed-mode phase specifically designed for CEC (Adapted from Ref. 172 with permission. Copyright Wiley-VCH 1998.)

ciple, a strong and stable EOF over a wide pH range should be achieved by using SCX materials to impart a permanent negative charge at the surface through their sulfonic groups (e.g., C_3-SCX, C_6-SCX, phenyl-SCX), providing a cathodic EOF. An anodic EOF can be provided by means of the positively charged quaternary amines typical of SAX phases.

Some of the materials illustrated in Fig. 17 have been called "mixed-mode" phases in CEC [117,119,169–174,176,178,179,183]. This is because of the possibility of having "mixed" separation mechanisms involving ion exchange and hydrophobic interactions, in addition to differential electromigration. The EOF in these materials has been shown to be higher than in typical reverse-phase materials [117,172,174,178]; the presence of residual silanols on the silica sup-

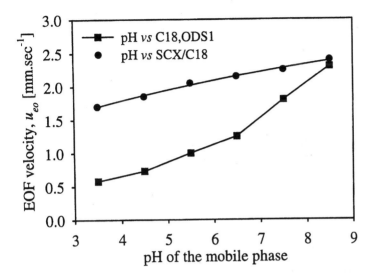

Fig. 18 Dependence of electroosmotic flow (EOF) velocity on the pH of the mobile phase a C_{18} bonded silica and a mixed-mode bonded silica (SCX/C_{18}). (Reprinted from Ref. 178 with permission. Copyright Elsevier 2000.)

port, however, results in a small dependence on the pH of the mobile phase, as shown in Fig. 18 [178]. One type of silica-based mixed-mode material specifically designed for CEC is shown in Fig. 17B, which incorporates a γ-glycidoxypropyltrimethoxysilane sublayer attached to the silica support; then a sulfonated layer is covalently affixed between the sublayer and an octadecyl top layer [170–172, 179]. It has also been reported that lower EOF velocities than expected have been observed with the SCX phases [164]. The behavior of the SCX phases is not completely understood, and more studies on these are needed to completely elucidate their role in CEC.

Packed beds containing anion-exchanger groups have been less popular in CEC. Using a SAX mixed-mode phase, excellent separations of peptides have been reported [167]. Because of the ion exchange-type of interactions contributing to the separation [117, 167–169,171,174,179], the selectivity of the mixed-mode and ion exchange stationary phases in CEC have been shown to be significantly different from the selectivity of conventional reverse-phase silica materials. Changing the pH of the mobile phase, for example,

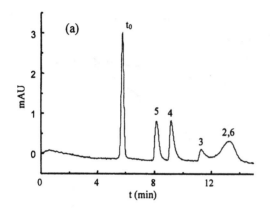

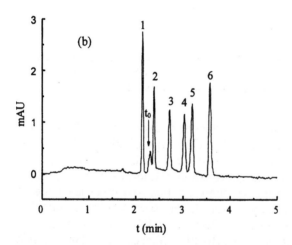

Fig. 19 Separation of acidic compounds in columns packed with (a) 5 μm Spherisorb-ODS and (b) 5 μm Spherisorb-SAX materials. The mobile phase was composed of 60% acetonitrile in 2 mM phosphate buffer (pH = 2.2) for (a) and 50% acetonitrile in 20 mM phosphate buffer (pH = 2.2) for (b); t_o identifies the electroosmotic flow (EOF). The compounds were (1) 3,5-dinitrobenzoic acid, (2) p-nitrobenzoic acid, (3) p-bromobenzoic acid, (4) o-toluic acid, (5) benzoic acid, (6) o-bromobenzoic acid. (Reprinted from Ref. 169 with permission. Copyright Elsevier 2000.)

one can tune the selectivity of the phase by influencing ion exchanging, as illustrated in Figs. 14 and 15 for the separation of peptides and acidic compounds.

B. Retaining Frits

In a packed CEC column, the chromatographic packing material is maintained in place by means of retaining frits. Fabrication of these frits is one of the most critical aspects in column fabrication, influencing column performance [21,184,185]. The bed-retaining frits must be highly permeable to solvent flow and mechanically strong to resist the pressures used to pack and/or rinse the column, as well as to retain the material under the CEC conditions. Several studies have addressed the process of frit fabrication [43,111,157,186–188] in CEC. The most common approach to fabricate the retaining frits is by sintering silica-base packing material using heating. Precautions, however, must be taken to minimize degradation of the alkylated silica [40]. To prevent degradation, sintering time must be optimized for column i.d., particle size, and type of stationary phase material [189]. If excessive heating is applied, there is also the potential of damaging the outer side of the capillary column, particularly at the column entrance, which can create undesirable adsorption sites. Figure 20, for example, depicts the entrance of a column that has been exposed to excessive heat during frit fabrication. Pressure resistance of a frit and its influence on the EOF has been reported to be likely insignificant [187].

One must also be aware that the CEC column becomes fragile at the frits, because during frit fabrication the heat removes the protective polyimide coating. In addition, heating during the frit fabrication can create a substantial number of silanol groups, thereby changing the surface charge of the material at the frit [187]. This changes the characteristics of the material, which can result in different electrical resistivity at the frits, as compared to the open and pack segments of the column. Different electrical properties of the frit are contributors to non-uniformities in EOF, leading to bubble formation at the interface between the frit and the unpacked/open segment of the CEC column [26,108,185]. Discontinuities in the column structure and non-uniformities in EOF can be minimized by manufacturing frits with resistivities similar to either the packed or the open segment. It is possible to resilianize the material at the frit in order to create a more uniform structure, reducing the likelihood

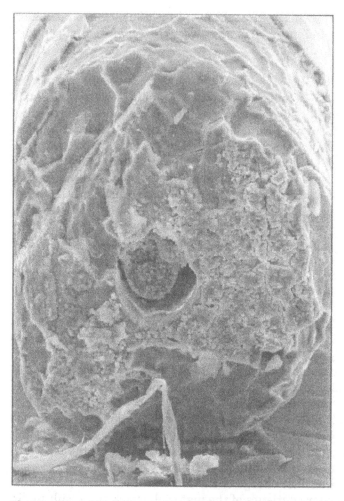

Fig. 20 Scanning electron micrograph (SEM) of a column end that has been exposed to excessive heating during frit fabrication.

of bubble formation [111]. Silanization deactivates undesirable adsorption sites at the frit with which the solutes can interact [190].

Bubble formation has also been attributed to heat generation upon application of the electric field and it can be minimized by pressurization of the mobile phase at both inlet and outlet column reservoirs [3,26,30,33,43]. Additional practices to reduce bubble formation include the use of well-degassed solvents, low concentrations of electrolytes, low conductivity electrolytes (i.e., zwitterionic buffers), working at reduced temperatures (e.g., 15°C), and the use of narrow column inner diameters. Bubble formation has also been addressed by the addition of sodium dodecyl sulfate (SDS) into the mobile phase at low concentrations [120].

The functionality of the packing material has also been reported to affect the final properties of the frits. In a study where frits formed in open tubes using different silica with different functionalities (bare silica, SCX and SAX silicas, and Hypersil ODS silica) the EOF velocity varied with each frit [187]. EOF appeared to depend on the residual groups of the different packings after sintering. EOF velocities of frits formed with SCX and bare silica were higher than an open tube and frits of ODS material. Residual groups at the surface of the frit depended on the time and magnitude of heating used during the formation of the frit, contributing differently to the overall flow through the frit. The same study also showed reversal of EOF in an open tube containing a frit formed with SAX material. A filament at 430 °C was applied for between 12 and 15 seconds to create the frit. If the heating time exceeded 15 seconds, the reversed EOF was reduced, attributed to the reduction of the positive charges responsible for the EOF reversal [187].

Despite the difficulties associated with the sintering of packing material, it continues to be the most common approach to frit fabrication. Other methodologies, however, have been developed to avoid heating. In one method, frits have been fabricated by UV photopolymerization of trimethylpropane trimethacrylate and glycidyl methacrylate solution (UV radiation, 365 nm for 1 hour) [191]. Since heat is not applied, alteration of the stationary phase at the frit and/or weakening of the capillary column by removal of the polyimide coating are avoided. Frits fabricated via this method have shown to be reproducible when the appropriate porogenic solvent is selected. The method has been used on larger bore capillary columns up to (~550 μm I.D.) where sintering of frits has been difficult [192]. Another approach to eliminate heating to form frits involves the polymerization of a potassium silicate solution containing formamide, which was originally developed by Cortes et al. [193] and carefully optimized by Cassidy and coworkers [188]. In this approach, the column is filled with a sodium silicate solution and heated for a few seconds where the outlet frit is desired. The polyimide coating is not affected during this heating step and, therefore, no weak spot on the column is created. The excess silicate solution is pushed out of the column and the frit is then silanized with a solution of 0.02 M trimethylchlorosilane in DMF, using imidazole as an acid acceptor. After the column has been packed, the column entrance is dipped into a less concentrated silicate solution and heat is applied. Sodium silicate solutions of 10.8% and 5.4% (w/v) were found to be optimal for outlet and inlet frits, respectively. Cassidy et al. found that by using this method, relatively short frits (~75 μm) could be created using short heating times, 5–6 s for the outlet frit and about 1 s for the inlet one. Columns with these frits were mechanically stable without problems associated with bubble formation or electrical current stability, even under relatively high current conditions (~27 μA); the columns were also run without pressurization over a wide range of acetonitrile/ water mixtures without any problems [188].

Another alternative to frits is the use of tapered capillaries to retain the packing material within the column. Tapered capillaries have been shown to be useful in coupling CEC with mass spectrometry and NMR [194,195]. Tapered capillaries are schematically depicted in Fig. 21; they are prepared in one of two fashions: internally [194,195] or externally [189,196] tapered. The internally tapered columns are prepared using a high temperature flame to seal the end of the capillary. The sealed end is then carefully ground to produce a small opening suitable for flow; the packing material is maintained in place because of the keystone effect. The externally tapered columns are fabricated by use of a laser-based micropipette puller. Externally tapered columns are inherently fragile; each taper is a weak point on the column since the outer diameter of the column is reduced during their formation. The internally tapered columns are not as fragile, since the outer diameter is not reduced. As a result, internally tapered columns are easily connected to other tubing via shrink tubing. It is important to realize, however, that the end of

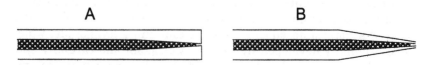

Fig. 21 Schematic representation of (A) internally and (B) externally tapered capillaries.

the connecting capillaries must be carefully prepared by grinding the capillary ends to avoid dead volume at the coupling region. Figure 22 shows the capillary ends prepared by three different procedures. One can clearly see that the most popular method of cutting the end with a ceramic cutter does not necessary lead to acceptable capillary ends for coupling.

C. Column Fabrication

A general procedure to fabricate a CEC packed column is illustrated in Fig. 23 [197]. Slight variations can be found between laboratories, although the general approach is the same. A piece of fused silica capillary with a desired inner diameter and length is selected. It is our practice to rinse the capillary column with a 0.1 M sodium hydroxide solution followed by water. One end of the capillary tube is tapped into a pile of wet silica material to fabricate a temporary frit. The material is sintered at the end of the column by applying heat. The heat needed to form the frits depends on the column and particle diameters [189] and the functionality of the material to be sintered as well as the heating element used. Frits can be fabricated using heating elements such as microtorchs [34], burners [157], thermal wire strippers [44], optical splicers [44,156], and heating elements incorporated into more complicated assemblies [30]. Our laboratory uses a relatively inexpensive setup in which a soldering gun is fitted with a Nichrome ribbon that acts as the heating element. By making a small hole (~0.5 mm) through the ribbon the column can be threaded through it, which enables heating of a small spot (~1 mm) at any point on the capillary.

The excess material not fritted is removed by back flushing the column. The mechanical stability of the frit can be tested by applying pressure to it. The frit must be porous enough to allow for solvent flow, but mechanically stable enough to withstand the stress applied during packing. As an alternative to the sintered temporary frit, one can attach a union having a metallic frit at the end of the column to prevent the packing material from exiting the column during packing. After packing, the capillary is rinsed with water and the retaining frits are fabricated while rinsing. The use of organic solvents to flush the capillary column during frit formation is not recommended, because carbonaceous material from the solvent can be formed at the frit. The outlet frit is formed at the interface between the packed section and the open tube, then the inlet frit is created

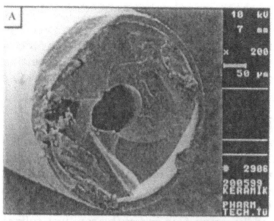

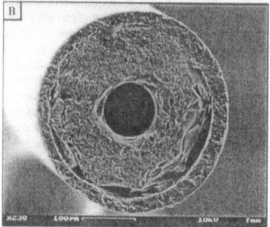

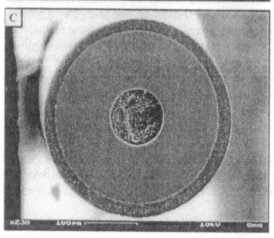

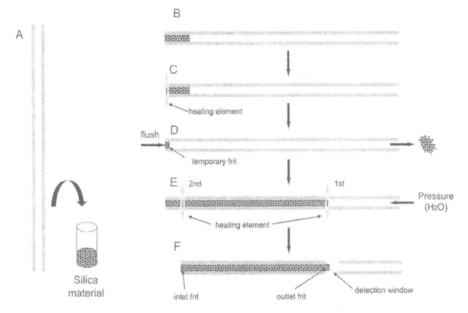

Fig. 23 CEC column fabrication process: (A) tapping empty capillary into a pile of wet silica; (B) silica material at capillary end to fabricate temporary frit; (C) formation of temporary frit by heating; (D) removing the excess of silica material after forming the temporary frit; (E) column is packed, pressurized with water, and the retaining frits are created with the heating element; (F) the fabricated column with retaining frits and detection window in place. (Reprinted from Ref. 197 with permission. Copyright Elsevier 2000.)

at a predetermined distance, creating the desired packed bed length (normally 20–30 cm). Once the frits are fabricated, the temporary frit is removed and the excess packing material inside the capillary is flushed out using a pump. During fabrication of the outlet frit at the interface between the open and packed segments of the column, a portion of polyimide coating on the capillary on the open segment of the column is also removed. The exposed section of the fused silica

Fig. 22 Scanning electron micrograph (SEM) of capillary ends (A) cut with a ceramic tile, (B) prepared by silicon carbide wet grinding with paper P 320, and (C) prepared by silicon carbide wet grinding with paper P 4000. (Reprinted from Ref. 195 with permission. Copyright Elsevier 2000.)

can serve as the optical window for spectroscopic detection methods. The column must be rinsed with well-degassed mobile phase prior to use for analysis. It is recommended to equilibrate the column with the mobile phase by applying voltage across the column in a stepwise manner. The equilibration step can aid in consolidating a packed bed that may contain voids [198]. One method of conditioning the CEC column is to apply 5 kV with the mobile phase in the column, followed by voltage ramping in 5 kV steps until 25–30 kV is reached. A stable electrical current at each step is indicative of an equilibrated column.

D. Column Packing

Even with several detailed methodologies reported for CEC column fabrication [30,33,36,43,46,199–202], column packing may still be a difficult task. The reliability and reproducibility of packed columns is ultimately dependent on the column packing. Columns that are poorly packed tend to exhibit low efficiency, asymmetric peak shapes, and poor resolution. Packing CEC columns is an elaborate process that requires skill and experience.

In one of the initial approaches to packed columns for CEC, thick-walled Pyrex tubing was packed with underivatized packing material and pulled at high temperatures to a desired diameter using a glass-drawing device, after which the packing material was derivatized with the stationary phase [27,78]. This column preparation procedure has since been abandoned because of the low success rate in fabricating the columns. Several other methods have been developed to pack capillary columns for CEC. The approach most commonly used is the slurry packing method adopted from HPLC column packing [30,33,36], although several other approaches have been developed, including packing by centripetal forces [201,203], making use of supercritical CO_2 [46,59], electrokinetic packing [199,200,204], and packing by gravity [202]. Details on these packing procedures have been discussed recently [197,205]. The preference for one method over another seems to depend on the familiarity with a given procedure. There seems to be indication that the traditional pressure packing approach can render columns with the lowest performance, unless one is very well experienced in the art [197]. Commercially available CEC columns are usually packed either by slurry pressure packing or by electrokinetic methods.

IV. MONOLITHIC COLUMNS

The concept of monolithic columns has been identified as an alternative to particle-packed columns for more than a decade [206,207], attracting considerable attention in CEC [15,48,50–58,60–62, 153,208–216]. In general terms, a monolithic column consists of a porous structure fabricated inside of a tube (or channel), forming a continuous bed of chromatographic material. The monolithic structure simplifies column technology by eliminating difficulties encountered with frit fabrication associated with packed columns, often responsible for bubble formation in CEC. Several approaches have been followed to fabricate monolithic capillary columns. Organic or inorganic monoliths have been synthesized by *in situ* polymerization. The rodlike structures thus created are porous in nature and can be either rigid structures [50–58,60–62,208,212,216–223] or soft gels [47,48,55,209,213,215,224–229]. In-depth reviews on in situ polymerization to form monolithic structures have been reported [222,230,231]. Another approach to monolithic columns is to consolidate a packed structure inside the column [56–60,63, 208,232].

A. Silica Monoliths

The most common approach to fabricating silica-based monolithic columns for CEC is the acid-catalyzed sol-gel process [231]. The sol-gel process is illustrated in Fig. 24. After an initial hydrolysis step, polymerization takes place through water- or alcohol-forming condensation reactions, which involves hydrolyzed and/or unhydrolyzed species. Commonly used precursors are tetraethoxysilane (TEOS) and tetramethoxysilanes (TMOS). Monolithic columns are typically prepared by mixing the monomeric precursors under acidic conditions at room temperature before introduction into the capillary column. A porogenic reagent is also added to the reaction mixture (e.g., polyethylene glycol, poly(ethylene) oxide); adjusting its quantity allows tailoring of the pore structure of the monolith [231]. Tanaka's group [53,54] has fabricated macroporous monolithic structures by hydrolysis and polymerization of TMOS using poly(ethylene) oxide as a porogen and reacting overnight inside the column (100 μm I.D.). The monolith was treated with aqueous ammonium hydroxide after it had been washed with water, followed by rinsing with ethanol.

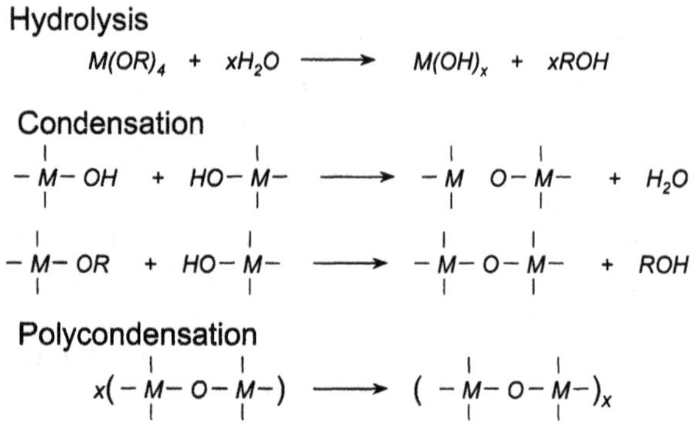

Hydrolysis

$$M(OR)_4 + xH_2O \longrightarrow M(OH)_x + xROH$$

Condensation

$$-\overset{\underset{|}{|}}{M}-OH + HO-\overset{\underset{|}{|}}{M}- \longrightarrow -\overset{\underset{|}{|}}{M}\ O-\overset{\underset{|}{|}}{M}- + H_2O$$

$$-\overset{\underset{|}{|}}{M}-OR + HO-\overset{\underset{|}{|}}{M}- \longrightarrow -\overset{\underset{|}{|}}{M}-O-\overset{\underset{|}{|}}{M}- + ROH$$

Polycondensation

$$x(-\overset{\underset{|}{|}}{M}-O-\overset{\underset{|}{|}}{M}-) \longrightarrow (-\overset{\underset{|}{|}}{M}-O-\overset{\underset{|}{|}}{M}-)_x$$

Fig. 24 Schematic representation of the silica reactions involved in the sol-gel process.

After drying the monolithic columns in an oven for 24 h at 330°C, the stationary phase was bonded at the surface of the silica material by introducing octadecyldimethyl-N, N-diethylaminosilane in toluene into the column and allowing it to react for two hours. Figure 25 shows SEM of the cross-section of a column fabricated by this means. One can observe small aggregates of silica connected to each other forming the monolithic structure. Pores as large as 8 μm were reported. Evaluation of the columns in the pressure driven and electrically driven modes suggested that the monolithic columns were more appropriate for CEC than for HPLC in terms of column efficiency (see Fig. 26). This was attributed to the improved A-term in the van Deemter equation for the electrically driven system. The monolithic columns prepared by these protocols have shown higher permeability than columns packed with particles of about 3 μm.

Other protocols have been reported to fabricate monolithic columns using sol-gel technology [55,62,95,227,233,234]; most of them are very similar. One approach worth mentioning is the one in which the stationary phase as well as its silica skeleton are fabricated in one single step, as originally proposed for open tubular formats [93]; this eliminates the secondary derivatization reactions to bond the stationary phase. Using such an approach, Malik's group has prepared a monolithic column with anodic EOF by mixing a selected mixture of precursors [95]. In their method, the sol-gel solution was prepared by mixing 100 μL of TMOS with 100 μL of N-octadecyldimethyl[3-(trimethoxysilyl)propyl] ammonium chloride (C$_{18}$-TMS),

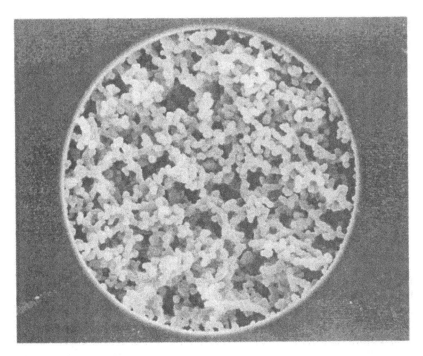

Fig. 25 Scanning electron micrograph (SEM) of a silica monolith inside a 100 μm i.d. capillary. (Reprinted from Ref. 53 with permission. Copyright American Chemical Society 2000.)

10 μL of Phenyldimethylsilane (PheDMS), and 100 μL of 99% trifluoroacetic acid (TFA). The mixture was vortexed for 5 minutes and centrifuged at 13,000 rpm for 5 min. The supernatant was used for the fabrication of the monolithic bed. The solution was introduced into the capillary tube that had been rinsed with water and then dried under helium flow while heating from 40°C to 250°C. The column was sealed at both ends and submitted to thermal conditioning from 35°C to 150°C at a rate of 0.2°C/min. The monolithic structure obtained is depicted in Fig. 27. Efficiencies in the range of 150,000 plates per meter were reported for benzene derivatives.

B. Soft Gels

Acrylamide-based gels have been used as monolithic columns for CEC [47,48,55,224–229]. These columns are not suitable for HPLC;

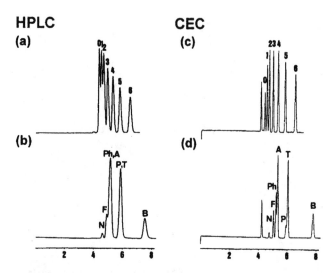

Fig. 26 Separation of alkylbenzenes (a,c) and polycyclic aromatic hydrocarbons (b,d) by HPLC (a,b; 90% acetonitrile/10% water) and CEC (c, d; 90% acetonitrile: 10% 50 mM Tris-HCl, pH = 8) in a monolithic silica column. (Reprinted from Ref. 53 with permission. Copyright American Chemical Society 2000.)

however, the nature of electrically driven flows allows their use in CEC. In most cases, the gels are prepared incorporating a functional monomer that provides for the generation of EOF (e.g., 2-acrylamido-2-propanesulfonic acid and vinylsulfonic acid). The incorporation of species such as butyl methacrylate and stearyl methacrylate increases the hydrophobicity of these aqueous gels [48,229]. Performing the polymerization in organic solvents facilitates the solubility of the hydrophobic monomers. Alternatively, hydrophobicity can be increased by derivatization with an organic ligand (e.g., C18) after polymerization [224]. The use of β-cyclodextrin-bonded positively charged polyacrylamide gels has been shown to be effective for enantiomeric separations [235]. It is important to realize that some of these monolithic structures are swollen gels with little solid material, while others are highly cross-linked structures. Using highly cross-linked acrylamide-base monolithic columns, separation efficiencies of up to 400,000 plates/m have been reported [215]. These columns were prepared by copolymerizing acrylamide mono-

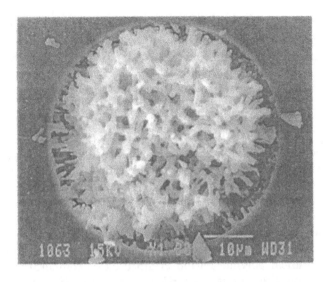

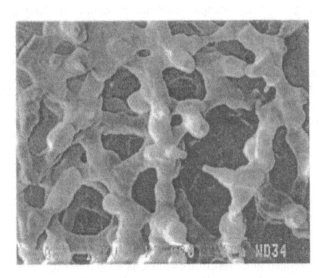

Fig. 27 Monolithic structure in which the stationary phase as well as its silica skeleton were fabricated in one single step: (A) 1800× magnification, (B) 7000× magnification. (Reprinted from Ref. 243 with permission. Copyright American Chemical Society 2000.)

mers, some of which contained an alkyl acrylate (butyl, hexyl, or lauryl) ligand, in a formamide aqueous solution using bisacrylamide as cross-linker. In addition, acrylic or vinylsulfonic acid and poly(ethylene glycol) were added for EOF generation and as a porogen, respectively. The macroporous polymeric structures were used to separate alkyl phenones, peptides, and carbohydrates.

C. Rigid Polymeric Structure

Monolithic columns for CEC have also been fabricated using polystyrene or polymethacrylate polymers [50–52,216–218,221,223]. Column selectivity is tailored by selecting the appropriate monomeric units to be incorporated in the polymeric structure. In general, a monolithic column is prepared by filling a desired portion of a fused silica capillary with a polymerizing mixture that leads to the porous polymer. In the case of polyacrylate-based monolithic polymers [50–52,216,221], radical polymerization is initiated by temperature or UV radiation and the porous properties of the monolith can be adjusted by using a ternary porogenic solvent system consisting of water, 1-propanol, and 1,4-butanediol [50]. A few examples of monomers and cross-linking agents used in methacrylate-based monolithic columns are shown in Fig. 28 [236]. Columns with reversed-phase [51,52,219] and chiral [216] characteristics have been reported to be stable through a wide pH range (i.e., 2–12). Figure 29, for example, shows separations performed in metacrylate-based monolithic columns at high and low pH. Sulfonic acid functionalities provide the required charges necessary for EOF generation. Monoliths with optimized porous characteristics have shown efficiencies of 210,000 plates/m [219].

Highly porous monoliths have also been prepared by copolymerizing chloromethylstyrene, divinylbenzene, and azobisisobutyronitrile in the presence of a suitable porogenic solvent inside fused-silica capillaries silanized with 3-(trimethoxysilyl)propyl methacrylate [217]. The resulting rod-like monolithic structure, which appears as tiny clumps fused at different locations, is shown in Fig. 30. By reacting the chloromethyl moieties with N, N-dimethyloctylamine, a positive charge was introduced at the surface, which led to an EOF from cathode to anode. Angiotensin-type peptides were separated using these columns. Monolithic columns with poly(styrene-co-divinylbenzene-co-methacrylic acid) have also been fabricated, showing

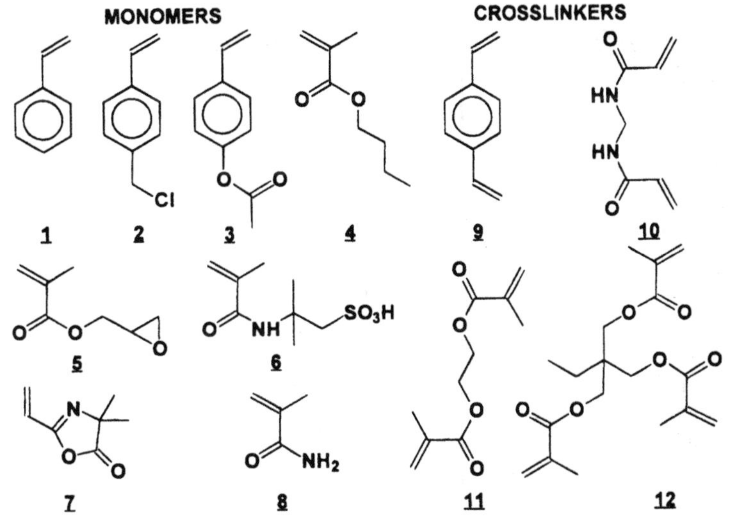

Fig. 28 Chemical structure of some monomers and cross-linking agents that can be used to form porous rigid organic monolithic structures. (Reprinted from Ref. 236 with permission. Copyright American Chemical Society 1999.)

similar retention mechanism as reversed-phase packing materials [8].

Molecular Imprint

The molecular recognition obtained with molecular imprinting has attracted interest in producing stationary phases for CEC [220]. The polymeric structure with the imprinted recognition site is synthesized in situ by assembling monomeric units (e.g., styrene and methacrylic acid) around a desired template molecule; the assembly takes place through molecular interactions (covalent or noncovalent) with the template molecule. Proper selection of the template molecule allows preparation of stationary phases with a predetermined selectivity. The template molecule is removed after polymerization, leaving a porous, rigid polymeric structure. The synthesis of a noncovalent molecular imprint polymer is schematically represented in Fig. 31. The use of molecular imprinted monolithic polymers has shown to be more attractive for CEC than HPLC because of the excessive

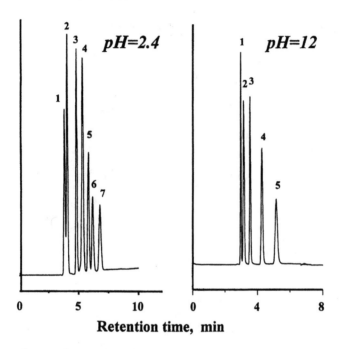

Fig. 29 Separation of aromatic acids (a) and anilines (b) on monolithic capillary columns. Conditions: butyl methacrylate–ethylene dimethacrylate stationary phase with 0.3 wt.% 2-acrylamido-2-methyl-1-propanesulfonic acid. Separation conditions: voltage, 25 kV; injection, 5 kV for 3 seconds; UV detection at 215 nm; pressure in vials, 0.2 MPa. (a) Capillary column, 30 cm (25 cm active length) × 100 μm I.D.; mobile phase, acetonitrile–5 mM phosphate buffer, pH 2.4 (60:40, v/v). Peaks: 3,5-dihydroxybenzoic acid (1), 4-hydroxybenzoic acid (2), benzoic acid (3), 2-toluic acid (4), 4-chlorobenzoic acid (5), 4-bromobenzoic acid (6), 4-iodobenzoic acid (7). (b) Capillary column, 28 cm (25 cm active length) × 100 μm I.D; mobile phase, acetonitrile–10 mM NaOH, pH 12 (80:20, v/ v). Peaks: 2-aminopyridine (1), 1,3,5-collidine (2), aniline (3), N-ethylaniline (4), N-butylaniline (5). (Reprinted from Ref. 222 with permission. Copyright Elsevier 2000.)

tailing observed in HPLC [212]. Most CEC applications using columns with molecular imprint stationary phases have focused on chiral separations [209,210,212,237]. The separation of propranolol enantiomers have been achieved in a monolithic column where (R)-propranolol was used as the template molecule for the imprint

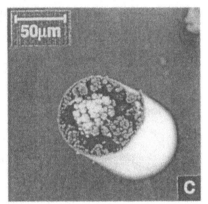

Fig. 30 Scanning electron micrograph (SEM) of a styrene-based monolithic structure in a 75 μm I.D. capillary column. (Reprinted from Ref. 217 with permission. Copyright Elsevier 1999.)

(see Fig. 32). The imprint recognized one of the enantiomers over the other, resulting in a stronger chromatographic retention. Although separation is achieved, relatively low efficiencies are obtained with molecular imprint monoliths. Higher efficiencies are achieved using OT-CEC [220,237]. In general, molecular imprints as stationary phases are a relatively new area that is under development; further research should lead to improved peak shape and efficiency.

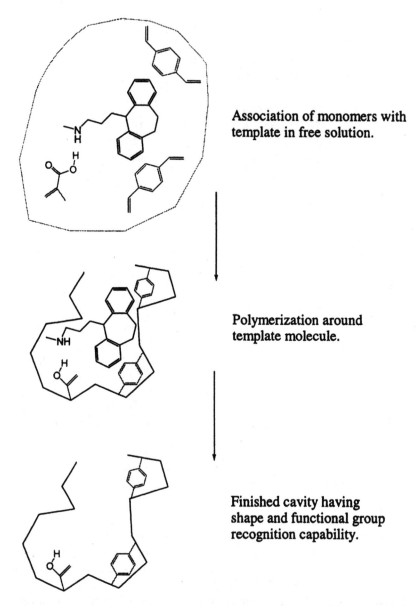

Association of monomers with template in free solution.

Polymerization around template molecule.

Finished cavity having shape and functional group recognition capability.

Fig. 31 Schematic representation for the synthesis of a noncovalent molecular imprint polymer. (Reprinted from Ref. 244 with permission. Copyright Elsevier 2000.)

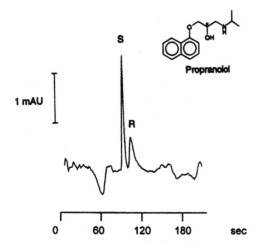

Fig. 32 CEC separation of propranolol enantiomer obtained on a monolithic column containing imprints of (R)-propranolol. (Reprinted from Ref. 209 with permission. Copyright American Chemical Society 1997.)

D. Particle Confinement

Monolithic columns can also be fabricated by immobilizing conventional packing materials inside the capillary column. In general, a column is packed with particles that are fixed in place by means of entrapment in a polymeric matrix or by a heat treatment, hence eliminating retaining frits. Since the packing material is fixed, the potential of gap formation within the packed bed is avoided. The packing material can be immobilized using silica-based (e.g., sol-gel processing) [57–61,208] or organic polymer matrices [232], and via sintering [56]. Two approaches have been used to immobilize packing material via sol-gel processing. In one approach, the particles are added to a sol-gel solution, which contains alkoxysilanes, ethanol, and hydrochloric acid forming a suspension [57,60,238]. This suspension is introduced into a capillary column; after drying, the particles are embedded in the sol-gel matrix. It must be emphasized that the concentration of particles in the suspension is critical for the homogeneous distribution of particles within the monolith. It was reported that the use of 300 mg of ODS particles per mL of TEOS rendered the optimum suspension. The efficiencies reported for these columns are lower than the ones obtained with traditionally packed columns. In addition, the reproducibility of column preparation seems to be problematic.

The other approach to immobilizing packing material via sol-gel processing is by introducing the entrapping solution into a packed column [58,59,61,208] that contains temporary retaining frits. The column is dried and the temporary retaining frits are removed. Remcho and coworkers filled packed columns with silicate sol solutions, which with a subsequent heating allowed for entrapment [58]. The columns were further cured with 0.1 M ammonium hydroxide and dried at 160°C. The monolithic columns showed efficiencies similar to those of packed columns but with reduced retentive properties. In a similar approach, columns have been packed using a CO_2 slurry packing method and then filled with a sol solution prepared with tetramethoxysilane and ethyltrimethylsilane [59,61,208]. Drying of the column was accomplished by supercritical CO_2 for 5 hours at 40°C, which was followed by another drying step at 120°C and at 250°C for the same amount of time. Figure 33 shows how the particles seem to be "glued" to each other when using this approach. Matrices entrapping 7 μm (4000 Å) and 3 μm (80 Å) ODS particles showed efficiencies of 220,000 and 175,000 plates/m, respectively. Reproducibility and column efficiency of monoliths fabricated by entrapping the packing material with the sol-gel solutions after the column has been packed are higher than for the columns in which the particles and the sol solution are loaded together into the columns.

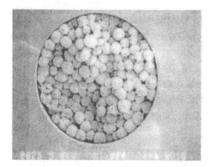

Fig. 33 Scanning electron micrograph (SEM) of sol-gel bonded particles (7 μm, 4000 Å pores) forming a monolithic structure. (Reprinted from Ref. 208 with permission. Copyright John Wiley & Sons 1999.)

The *in situ* polymerization of organic polymers to entrap particles inside capillary columns for CEC was reported by Remcho and
coworkers [232]. In their approach, columns were packed with 5 μm
ODS particles and dried. Then the dried packed column was filled
with a polymerization solution containing ethylene dimethacrylate,
the free radical initiator 2,2′-azobisisobutyronitrile, and a mixture
of solvents (water, 1-propanol, and 1,4 butanediol) to impart a porosity to the monolithic structure. Methyl-, ethyl, or butyl methacrylate
and 2-acrylamido-2-methyl-1-propanesulfonic acid could also be
added to increase hydrophobicity and to tune the EOF. The capillary
ends were sealed and the column was placed in an oven for 48 h at
60°C, allowing polymerization to proceed through the entire column.
After polymerization, the sealed ends were cut off and the column
was coupled to an open tube segment containing a detection window
for optical detection. This approach rendered columns with lower
retention than packed columns without entrapment. The access of
the analytes to the ODS functionality is presumably restricted because the polymerizing solution is likely to fill the pores of the silica
particle, reducing retention. Using an alkylmethacrylate in the polymerizing mixture aided in maintaining retention characteristics
similar to the packed bed column.

Another variation of the above method is to use the particles
as porosity templates for the monolithic structure [240]. In such an
approach, a capillary column having the walls silanized with 3-(trimethoxysilyl)propyl methacrylate is slurry-packed with a desired
particle size. The column is flushed with nitrogen and then the inter-
stitial space is filled with the solution to be polymerized. After the
in situ thermal polymerization of divinylbenzene with styrene or
ethylene dimethacrylate with butyl methacrylate, the column is
flushed with NaOH to dissolve the silica particles, rendering an organic polymer monolithic column with a porosity dictated by the size
of the silica particles. To pack the column, it is essential to use a
slurry solvent with high surface tension to fill the pores of the silica
packing material. This facilitates the flushing of the column with
NaOH to dissolve the silica material. Solutions of polyethylene glycol
or glycerol are suitable for this purpose. This approach eliminates
the search for a porogenic reagent in the polymerization reaction,
facilitating control of the monolith's porosity. The authors suggested
that by selecting particles with an appropriate surface chemistry, it
might be possible to influence the arrangement of the monomers be-

fore polymerization, dictating the orientation of the resulting polymer [240]. The entire fabrication process, however, is lengthy, particularly during the removal of the silica particles by flushing with NaOH at a velocity of 50 cm/h.

Monoliths have also been fabricated by sintering silica-based chromatographic particles inside a capillary column [56,63]. In one method, the packed capillary is submitted to rinses with 0.1 NaHCO$_3$ and acetone, drying, and a thermal treatment (two steps, 120°C and 360°C), resulting in sintering of the particles. Re-silanization of the column is performed to incorporate the stationary phase that may have been damaged during the process. The monoliths produced have been reported to be very stable and produced reproducible separations [56]. The multistep procedure used to fabricate the column is a drawback of this approach, because the process can be time-consuming, including the reattachment of the stationary phase. In another similar method, reversed-phase silica material was immobilized by a hydrothermal treatment [63]. The heating step was automated to reproducibly move a heating element along the capillary length with controlled speed and constant temperature (in the range of 300–400°C) for several cycles. Under optimized conditions, this method provided columns with the same column performance as packed columns having frits to retain the packing material [63].

V. MICROFABRICATED DEVICES

A recent approach to column technology in CE and CEC is the use of microfabricated structures or microchips [64–77,241]. This approach offers the opportunity of having very well controlled microfabricated channels to perform separations in a miniaturized format. Specific details on the fabrication of these devices are beyond the scope of this review; herein we give a brief overview as it is applied to CEC. For fabrication details, the reader can consult the literature cited within this section. Early work on CEC in a glass microchip was performed in the simplest approach to CEC, the open channel format [64]. The channels were fabricated using standard photolithographic and etching techniques with channel size of 5.6 µm deep by 66 µm wide. Chemical attachment of octadecylsilane to the surface of the channel provided the stationary phase. Sample loading was accomplished reproducibly via electrokinetic injection. Plate heights of 4.1 µm and 5.0 µm were reported for unretained and retained components, respectively. More recently, Ramsey and coworkers have utilized microchips for open-channel electrochromatography with gradient elution and in conjunction with capillary electrophoresis and voltage programming [65,66]. The channels were fabricated on crown glass or other glass substrate via photolithography and wet chemical etching. The stationary phase is attached by reacting the glass with octadecyltrimethoxysilane or octadecylisobutyl(dimethylamino)silane. Very fast (~20 s) and highly efficient separations of coumarin mixtures were demonstrated under gradient conditions [242]. A separation of fluorescently labeled tryptic digests of β-casein in a two-dimensional open channel CEC/CE format highlighted a second report [66], demonstrating a 50% increase in obtainable information going from a one- to a two-dimensional separation scheme. Figure 34 shows an image of a microchip used for two-dimensional open channel CEC/CE separations.

Packed and/or monolithic structures have also been fabricated in microchips [73–77]. Harrison and coworkers packed ODS packing materials (1.5–4 μm diameter) in glass microchips using electroosmotic pumping [76]. The particles were trapped in a cavity formed by two weirs, forming a "dam." The cavity was accessed by a separate channel in order to introduce the packing material; this channel also allowed the replacement of the packing material. The chromato-

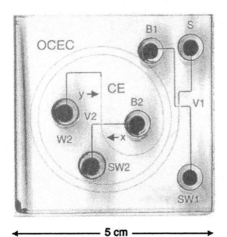

Fig. 34 Image of a microchip used for two-dimensional open channel CEC/CE separations. (Reprinted from Ref. 66 with permission. Copyright American Chemical Society 2001.)

graphic bed was utilized as a concentrator device providing concentration enhancement of up to 500-fold. Using a packed bed of 200 μm, the dyes BODIPY and fluorescein were separated isocratically in less than 20 s, with an aqueous/acetonitrile solution, achieving plate height of 2 μm.

An approach to filling microfabricated channels with continuous polymer beds has been developed by Hjertén and coworkers [73]. Quartz microchips were fabricated by a photolithographic procedure and polycrystalline silicon was deposited on the quartz. Upon mapping of the microchip geometry by dry and wet etching procedures, the polycrystalline silicon was then oxidized, followed by coating with silicon dioxide by low-pressure chemical vapor deposition. This was performed to create 40 μm channels with a thin oxide layer. The channels were washed with acetone, treated with 0.2 M NaOH and 0.2 M HCl for 30 minutes each, and rinsed with water before covalent attachment of [3-(methacryloyloxy)propyl]trimethoxysilane. This procedure allowed the monolithic bed to be anchored onto the walls of the channels. The activated channels were then filled with an acrylamide-based monomer solution containing mixtures of sulfonic and isopropyl groups, providing for EOF generation and

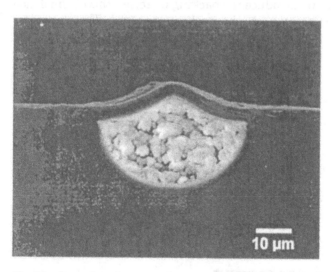

Fig. 35 Scanning electron micrograph (SEM) of an in situ polymerized continuous bed inside a microchip. (Reprinted from Ref. 73 with permission. Copyright American Chemical Society 2000.)

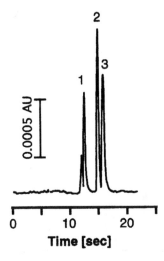

Fig. 36 CEC separations on an 18 mm length monolithic bed in a microchip; peaks: (1) uracil, (2) phenol, and (3) benzyl alcohol. (Reprinted from Ref. 73 with permission. Copyright American Chemical Society 2000.)

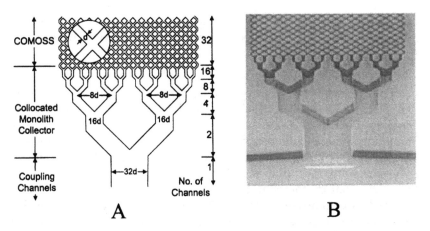

Fig. 37 (A) Design layout and (B) Scanning electron micrograph (SEM) of the entrance of a collocated monolith support structure. (Reprinted from Ref. 75 with permission. Copyright American Chemical Society 1998.)

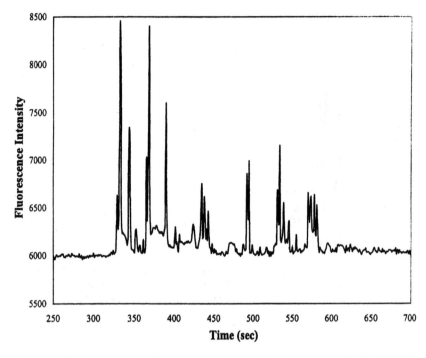

Fig. 38 CEC separation of a tryptic digest of ovalbumin on a C_{18}-COMOSS. (Reprinted from Ref. 77 with permission. Copyright Elsevier 1999.)

reversed-phase type of interactions, respectively. Figure 35 shows an SEM of an in situ continuous bed polymerized inside the microchip. Less than 20-s separations (see Fig. 36) were achieved in 18 mm length monolithic beds with separation efficiencies of ~350,000 plates/m. The electrochromatographic separation of the two antidepressants amitriptyline and nortriptyline and of a related quaternary ammonium compound (methyl amitriptyline) was also demonstrated. A comparison between the continuous bed microchips and a monolithic capillary column prepared with the same monomers showed similar separation efficiencies.

As an alternative to packing microchips, Regnier and coworkers designed a completely different microchip architecture that eliminates the need for particle packing or in situ polymerization, applicable to liquid and electrically driven separations on a chip [74,75,77].

Convective transport channels and collocated support structures were microfabricated on a microchip using ion etching, resembling a well-assembled monolithic structure. Figure 37 shows an example of a design layout and the microfabricated device, which has been named collocated monolith support structures (COMOSS). The architecture makes use of sequential binary splitting at the entrance of the column to distribute the mobile phase through the microchannels. When the microchips were operated in the reverse-phase mode, after incorporating the dimethyloctadecyl silane bonded phase at the surface, separation efficiencies of 777,000 plates/m were achieved for rhodamine 123 using laser-induced fluorescence as the detection system. The microfabricated devices were used to separate tryptic digests using acetonitrile/aqueous phosphate buffer as the mobile phase (see Fig. 38).

VI. FINAL REMARK

Column technology has been one of the most studied subjects in CEC, leading to the development of several column architectures. Materials specifically designed for CEC have started to emerge, as there is an intrinsic demand for a surface capable of generating EOF. The challenges imposed by the retaining frits during column fabrication have led to the development of monolithic structures. CEC on microfabricated devices is a viable alternative to incorporate chromatography into chemical analysis on a microchip, an area of continuous growth. There is no doubt that the CEC field has grown considerably in the past few years. Improvement in column technology has definitively played a key role in growth of the technique and has led the way for new avenues relevant to HPLC. CEC has not found widespread application yet, and its establishment as a complementary technique to HPLC and CE depends on it. Column technology, however, will continue to develop because advances in this field will benefit not just CEC but liquid chromatography in general.

ACKNOWLEDGMENT

We acknowledge financial support from The National Science Foundation (CHE-9614947 and CHE-0138114).

REFERENCES

1. U. Pyell, *Adv. Chromatogr.*, *41*: 1 (2001).
2. L. A. Colon, G. Burgos, T. D. Maloney, J. M. Cintron, and R. L. Rodriguez, *Electrophoresis*, *21*: 3965 (2000).
3. L. A. Colon, K. J. Reynolds, R. Alicea-Maldonado, and A. M. Fermier, *Electrophoresis*, *18*: 2162 (1997).
4. L. A. Colon, Y. Guo, and A. Fermier, *Anal. Chem.*, *69*: 461A (1997).
5. M. G, Cikalo, K. D. Bartle, M. M. Robson, P. Myers, and M. R. Euerby, *Analyst (Cambridge, U. K.)*, *123*: 87R (1998).
6. J. J. Pesek and M. T. Matyska, *Electrophoresis*, *18*: 2228 (1997).
7. Z. Liu, H. Zou, J. Y. Ni, and Y. Zhang, *Anal. Chim. Acta*, *378*: 73 (1999).
8. J. J. Pesek, M. T. Matyska, and S. Menezes, *J. Chromatogr. A*, *853*: 151 (1999).
9. J. J. Pesek, M. T. Matyska, and S. Cho, *J. Chromatogr. A*, *845*: 237 (1999).
10. M. T. Matyska, J. J. Pesek, J. E. Sandoval, U. Parkar, and X. Liu, *J. Liq. Chromatogr. Relat. Technol.*, *23*: 97 (2000).
11. J. J. Pesek and M. T. Matyska, *J. Capillary Electrophor.*, *4*: 213 (1997).
12. J. J. Pesek, M. T. Matyska, S. Swedberg, and S. Udivar, *Electrophoresis*, *20*: 2343 (1999).
13. M. T. Matyska, J. J. Pesek, and A. Katrekar, *Anal. Chem.*, *71*: 5508 (1999).
14. X. Huang, J. Zhang, and C. Horvath, *J. Chromatogr. A*, *858*: 91 (1999).
15. H. Sawada and K. Jinno, *Electrophoresis*, *20*: 24 (1999).
16. Z. J. Tan and V. T. Remcho, *J. Microcolumn Sep.*, *10*: 99 (1998).
17. W. Xu and F. E. Regnier, *J. Chromatogr. A*, *853*: 243 (1999).
18. J. W. Jorgenson and K. D. Luckas, *J. Chromatogr.*, *218*: 209 (1981).
19. B. Behnke and E. Bayer, *J. Chromatogr. A*, *680*: 93 (1994).
20. E. R. Verheij, U. R. Tjaden, W. M. A. Niessen, and J. van der Greef, *J. Chromatogr. A*, *554*: 339 (1991).
21. S. E. van den Bosch, S. Heemstra, J. C. Kraak, and H. Poppe, *J. Chromatogr. A*, *755*: 165 (1996).

22. F. Lelievre, C. Yan, R. N. Zare, and P. Gareil, *J. Chromatogr. A, 723*: 145 (1996).
23. M. M. Dittmann and G. P. Rozing, *J. Chromatogr. A, 744*: 63 (1996).
24. H. Yamamoto, J. Baumann, and F. Erni, *J. Chromatogr. A, 593*: 313 (1992).
25. C. Yan, D. Schaufelberger, and F. Erni, *J. Chromatogr. A, 670*: 15 (1994).
26. H. Rebscher and U. Pyell, *J. Chromatogr. A, 737*: 171 (1996).
27. J. H. Knox and I. H. Grant, *Chromatographia, 32*: 317 (1991).
28. H. Rebscher and U. Pyell, *Chromatographia, 38*: 737 (1994).
29. S. Kitagawa and T. Tsuda, *J. Microcol., Sep. 6*: 91 (1994).
30. R. J. Boughtflower, T. Underwood, and C. J. Paterson, *Chromatographia, 40*: 329 (1995).
31. S. Kitagawa and T. Tsuda, *J. Microcol. Sep., 7*: 59 (1995).
32. T. M. Zimina, R. M. Smith, and P. Myers, *J. Chromatogr. A, 758*: 191 (1997).
33. N. W. Smith and M. B. Evans, *Chromatographia, 38*: 649 (1994).
34. C. Yan, R. Dadoo, H. Zhao, R. N. Zare, and D. J. Rakestraw, *Anal. Chem., 67*: 2026 (1995).
35. M. R. Euerby, C. M. Johnson, K. D. Bartle, P. Myers, and S. C. P. Roulin, *Anal. Commun., 33*: 403 (1996).
36. R. J. Boughtflower, T. Underwood, and J. Maddin, *Chromatographia, 41*: 398 (1995).
37. D. B. Gordon, G. A. Lord, and D. S. Jones, *Rapid Commun. Mass Spectrom., 8*: 544 (1994).
38. S. J. Lane, R. Boughtflower, C. Paterson, and M. Morris, *Rapid Commun. Mass Spectrom., 10*: 733 (1996).
39. C. Yan, R. Dadoo, R. N. Zare, D. J. Rakestraw, and D. S. Anex, *Anal. Chem., 68*: 2726 (1996).
40. H. Rebscher and U. Pyell, *Chromatographia, 42*: 171 (1996).
41. J. Ding and P. Vouros, *Anal. Chem., 69*: 379 (1997).
42. D. B. Gordon, L. W. Tetler, and C. M. Carr, *J. Chromatogr. A, 700*: 27 (1995).
43. B. Behnke, E. Grom, and E. Bayer, *J. Chromatogr. A, 716*: 207 (1995).
44. M. T. Dulay, C. Yan, D. J. Rakestraw, and R. N. Zare, *J. Chromatogr. A, 725*: 361 (1996).

45. K. Schmeer, B. Behnke, and E. Bayer, *Anal. Chem.*, *67*: 3656 (1995).
46. M. M. Robson, S. Roulin, S. M. Shariff, M. W. Raynor, K. D. Bartle, A. A. Clifford, P. Myers, M. R. Euerby, and C. M. Johnson, *Chromatographia*, *43*: 313 (1996).
47. C. Fujimoto, *Anal. Chem.*, *67*: 2050 (1995).
48. J.-L. Liao, N. Chen, C. Ericson, and S. Hjerten, *Anal. Chem.*, *68*: 3468 (1996).
49. S. Fields, *Anal. Chem.*, *68*: 2709 (1996).
50. E. C. Peters, M. Petro, F. Svec, and J. M. J. Frechet, *Anal. Chem.*, *69*: 3646 (1997).
51. E. C. Peters, M. Petro, F. Svec, and J. M. J. Frechet, *Anal. Chem.*, *70*: 2288 (1998).
52. E. C. Peters, M. Petro, F. Svec, and J. M. J. Frechet, *Anal. Chem.*, *70*: 2296 (1998).
53. N. Ishizuka, H. Minakuchi, K. Nakanishi, N. Soga, H. Nagayama, K. Hosoya, and N. Tanaka, *Anal. Chem.*, *72*: 1275 (2000).
54. N. Tanaka, H. Nagayama, H. Kibayashi, T. Ikegami, K. Hosoya, N. Ishizuka, H. Minakuchi, K. Nakanishi, K. Cabrera, and D. Lubda, *J. High Resolut. Chromatogr.*, *23*: 111 (2000).
55. C. Fujimoto, *J. High Resolut. Chromatogr.*, *23*: 89 (2000).
56. R. Asiaie, X. Huang, D. Farnan, and C. Horvath, *J. Chromatogr. A*, *806*: 251 (1998).
57. M. T. Dulay, R. P. Kulkarni, and R. N. Zare, *Anal. Chem.*, *70*: 5103 (1998).
58. G. Chirica and V. T. Remcho, *Electrophoresis*, *20*: 50 (1999).
59. Q. Tang, B. Xin, and M. L. Lee, *J. Chromatogr. A*, *837*: 35 (1999).
60. C. K. Ratnayake, C. S. Oh, and M. P. Henry, *J. High Resolut. Chromatogr.*, *23*: 81 (2000).
61. Q. Tang and M. L. Lee, *J. High Resolut. Chromatogr.*, *23*: 73 (2000).
62. S. Suzuki, Y. Kuwahara, K. Makiura, and S. Honda, *J. Chromatogr. A*, *873*: 247 (2000).
63. T. Adam, K. K. Unger, M. M. Dittmann, and G. P. Rozing, *J. Chromatogr. A*, *887*: 327 (2000).
64. S. C. Jacobson, R. Hergenroder, L. B. Koutny, and J. M. Ramsey, *Anal. Chem.*, *66*: 2369 (1994).
65. J. P. Kutter, S. C. Jacobson, N. Matsubara, and J. M. Ramsey, *Anal. Chem.*, *70*: 3291 (1998).

66. N. Gottschlich, J. P. Jacobson, C. T. Culbertson, and J. M. Ramsey, *Anal. Chem.*, *73*: 2669 (2001).
67. D. J. Harrison, K. Flurik, K. Seiler, Z. H. Fan, C. S. Effenhauser, and A. Manz, *Science*, *261*: 895 (1993).
68. D. J. Harrison, A. Manz, Z. Fan, H. Ludi, and H. M. Widmer, *Anal. Chem.*, *64*: 1926 (1992).
69. K. Seiler, D. J. Harrison, and A. Manz, *Anal. Chem.*, *65*: 1481 (1993).
70. A. T. Woolley, G. F. Sensabaugh, and R. A. Mathies, *Anal. Chem.*, *69*: 2181 (1997).
71. Y. Liu, R. S. Foote, S. C. Jacobson, R. S. Ramsey, and J. M. Ramsey, *Anal. Chem.*, *72*: 4608 (2000).
72. S. C. Jacobson, R. Hergenroder, L. B. Koutny, R. J. Walmack, and J. M. Ramsey, *Anal. Chem.*, *66*: 1114 (1994).
73. C. Ericson, J. Holm, T. Ericson, and S. Hjerten, *Anal. Chem.*, *72*: 81 (2000).
74. F. E. Regnier, *J. High Resolut. Chromatogr.*, *23*: 19 (2000).
75. B. He, N. Tait, and F. Regnier, *Anal. Chem.*, *70*: 3790 (1998).
76. R. D. Oleschuk, L. L. Shultz-Lockyear, Y. Ning, and D. J. Harrison, *Anal. Chem.*, *72*: 585 (2000).
77. B. He, J. Ji, and F. E. Regnier, *J. Chromatogr. A*, *853*: 257 (1999).
78. T. Tsuda, K. Nomura, and G. Nakagawa, *J. Chromatogr.*, *248*: 241 (1982).
79. G. J. M. Bruin, P. P. H. Tock, J. C. Kraak, and H. H. Poppe, *J. Chromatogr.*, *517*: 557 (1990).
80. O. Van Berkel, J. C. Kraak, and H. H. Poppe, *Chromatographia*, *24*: 739 (1987).
81. S. Folestad, B. Josefson, and M. Larson, *J. Chromatogr.*, *391*: 347 (1987).
82. K. Göhlin and M. Larson, *J. Chromatogr.*, *645*: 41 (1993).
83. Y. Ruan, J. C. Kraak, and H. H. Poppe, *Chromatographia*, *35*: 597 (1993).
84. O. Van Berkel, J. C. Kraak, and H. H. Poppe, *J. Chromatogr.*, *449*: 345 (1990).
85. K. Göhlin and M. Larson, *J. Microcolumn Sep.*, *3*: 547 (1991).
86. Z. J. Tan and V. T. Remcho, *Anal. Chem.*, *69*: 581 (1997).
87. J. J. Pesek and M. T. Matyska, *J. Chromatogr, A*, *736*: 255 (1996).

88. J. J. Pesek, M. T. Matyska, J. E. Sandoval, and E. J. Williamsen, *J. Liq. Chromatogr. Relat. Technol.*, *19*: 2843 (1996).
89. J. J. Pesek, M. T. Matyska, and L. Mauskar, *J. Chromatogr. A*, *763*: 307 (1997).
90. J. J. Pesek and M. T. Matyska, *J. Chromatogr. A*, *736*: 313 (1996).
91. J. J. Pesek and M. T. Matyska, *Spec. Publ. R. Soc. Chem.*, *235*: 97 (1999).
92. J. J. Pesek and M. T. Matyska, *J. Chromatogr. A*, *887*: 31 (2000).
93. Y. Guo and L. A. Colón, *Anal. Chem.*, *67*: 2511 (1995).
94. Y. Guo and L. A. Colón, *J. Microcol. Sep.*, 7: 485 (1995).
95. J. D. Hayes and A. Malik, *Anal. Chem.*, *73*: 987 (2001).
96. J. D. Hayes and A. Malik, *J. Chromatogr. B*, *695*: 3 (1997).
97. S. Constantin and R. Freitag, *J. Chromatogr. A*, *887*: 253 (2000).
98. S. A. Rodriguez and L. A. Colón, *Anal. Chim. Acta*, *397*: 207 (1999).
99. P. Narang and L. A. Colon, *J. Chromatogr. A*, *773*: 65 (1997).
100. S. Fanali, *J. Chromatogr. A*, *735*: 77 (1996).
101. V. Schurig and D. Wistuba, *Electrophoresis*, *20*: 2313 (1999).
102. D. W. Armstrong, Y. Tang, T. Ward, and M. Nichols, *Anal. Chem.*, *65*: 1114 (1993).
103. J. Szemán and K. Ganzler, *J. Chromatogr. A*, *668*: 509 (1994).
104. V. Schurig, M. Jung, S. Mayer, M. Fluck, S. Negura, and H. Jakubetz, *J. Chromatogr. A*, *694*: 119 (1995).
105. S. Mayer and V. Schurig, *Electrophoresis*, *15*: 835 (1994).
106. R. B. Kotia, L. Li, and L. B. McGown, *Anal. Chem.*, *72*: 827 (2000).
107. M. A. Rehder and L. B. Mc Gown, *Electrophoresis*, *22*: 3759 (2001).
108. A. S. Rathore and C. Horvath, *Anal. Chem.*, *70*: 3069 (1998).
109. M. G. Cikalo, K. D. Bartle, and P. Myers, *J. Chromatogr. A*, *836*: 25 (1999).
110. M. Schmid, F. Bauml, A. P. Kohne, and T. Welsch, *J. High Resolut. Chromatogr.*, *22*: 438 (1999).
111. R. A. Carney, M. M. Robson, K. D. Bartle, and P. Myers, *J. High Resolut. Chromatogr.*, *22*: 29 (1999).

112. I. S. Lurie, T. S. Conver, and V. L. Ford, *Anal. Chem.*, *70*: 4563 (1998).
113. P. D. A. Angus, E. Victorino, K. M. Payne, C. W. Demarest, T. Catalano, and J. F. Stobaugh, *Electrophoresis*, *19*: 2073 (1998).
114. M. R. Euerby, C. M. Johnson, S. F. Smyth, N. Gillott, D. A. Barrett, and P. N. Shaw, *J. Microcolumn Sep.*, *11*: 305 (1999).
115. L. C. Sander, M. Pursch, B. Maerker, and S. A. Wise, *Anal. Chem.*, *71*: 3477 (1999).
116. L. Roed, E. Lundanes, and T. Greibrokk, *Electrophoresis*, *20*: 2373 (1999).
117. M. M. Dittmann and G. P. Rozing, *J. Microcolumn Sep.*, *9*: 399 (1997).
118. T. Tanigawa, T. Nakagawa, K. Kimata, H. Nagayama, K. Hosoya, and N. Tanaka, *J. Chromatogr. A*, *887*: 299 (2000).
119. T. Adam, S. Ludtke, and K. K. Unger, *Chromatographia*, *49*: S49 (1999).
120. R. M. Seifar, W. T. Kok, J. C. Kraak, and H. Poppe, *Chromatographia*, *46*: 131 (1997).
121. R. M. Seifar, S. Heemstra, W. T. Kok, J. C. Kraak, and H. Poppe, *Biomed. Chromatogr.*, *12*: 140 (1998).
122. C. G. Bailey and C. Yan, *Anal. Chem.*, *70*: 3275 (1998).
123. L. Zhang, W. Shi, H. Zou, J. Ni, and Y. Zhang, *J. Liq. Chromatogr. Relat. Technol.*, *22*: 2715 (1999).
124. L. Zhang, Y. Zhang, J. Zhu, and H. Zou, *Anal. Lett.*, *32*: 2679 (1999).
125. S. Li and D. K. Lloyd, *Anal. Chem.*, *65*: 3684 (1993).
126. S. Li and D. K. Lloyd, *J. Chromatogr. A*, *666*: 321 (1994).
127. D. K. Lloyd, S. Li, and P. Ryan, *J. Chromatogr. A*, *694*: 285 (1995).
128. M. Lammerhofer and W. Lindner, *J. Chromatogr. A*, *829*: 115 (1998).
129. C. Wolf, P. L. Spence, W. H. Pirkle, D. M. Cavender, and E. M. Derrico, *Electrophoresis*, *21*: 917 (2000).
130. D. Wistuba and V. Schurig, *Electrophoresis*, *20*: 2779 (1999).
131. C. Wolf, P. L. Spence, W. H. Pirkle, E. M. Derrico, D. M. Cavender, and G. P. Rozing, *J. Chromatogr. A*, *782*: 175 (1997).
132. D. K. Lloyd, A.-F. Aubry, and E. De Lorenzi, *J. Chromatogr. A*, *792*: 349 (1997).

133. D. Wistuba, H. Czesla, M. Roeder, and V. Schurig, *J. Chromatogr. A, 815*: 183 (1998).
134. K. Otsuka, C. Mikami, and S. Terabe, *J. Chromatogr. A, 887*: 457 (2000).
135. M. Girod, B. Chankvetadze, and G. Blaschke, *J. Chromatogr. A, 887*: 439 (2000).
136. A. S. Carter-Finch and N. W. Smith, *J. Chromatogr. A, 848*: 375 (1999).
137. C. Karlsson, H. Wikström, D. W. Armstrong, and P. K. Owens, *J. Chromatogr. A, 887*: 349 (2000).
138. H. Wikstrom, L. A. Svensson, A. Torstensson, and P. K. Owens, *J. Chromatogr. A, 869*: 395 (2000).
139. A. Dermaux, F. Lynen, and P. Sandra, *J. High Resolut. Chromatogr., 21*: 575 (1998).
140. M. Lämmerhofer, E. Tobler, and W. Linder, *J. Chromatogr. A, 887*: 421 (2000).
141. A. Dermaux and P. Sandra, *Electrophoresis, 20*: 3027 (1999).
142. M. M. Robson, M. G. Cikalo, P. Myers, M. R. Euerby, and K. D. Bartle, *J. Microcolumn Sep., 9*: 357 (1997).
143. Q.-L. Luo and J. D. Andrade, *J. Microcolumn Sep., 11*: 682 (1999).
144. Q.-H. Wan, *J. Phys. Chem. B, 101*: 8449 (1997).
145. Q.-H. Wan, *J. Phys. Chem. B, 101*: 4860 (1997).
146. Q.-H. Wan, *Anal. Chem., 69*: 361 (1997).
147. G. Choudhary and C. Horvath, *J. Chromatogr. A, 781*: 161 (1997).
148. A. S. Rathore, E. Wen, and C. Horvath, *Anal. Chem., 71*: 2633 (1999).
149. J. C. Giddings, in *Dynamics of Chromatography*, Marcel Dekker, New York, 1965.
150. J. H. Knox and I. H. Grant, *Chromatographia, 24*: 135 (1987).
151. C. L. Rice and R. Whitehead, *J. Phys. Chem., 69*: 4017 (1965).
152. K. J. Reynolds and L. A. Colon, *J. Liq. Chromatogr. Relat. Technol., 23*: 161 (2000).
153. S. Luedtke, T. Adam, and K. K. Unger, *J. Chromatogr. A, 786*: 229 (1997).
154. S. Luedtke, Th. Adam, N. von Doehren, and K. K. Unger, *J. Chromatogr. A, 887*: 339 (2000).
155. R. Stol, W. T. Kok, and H. Poppe, *J. Chromatogr. A, 853*: 45 (1999).

156. D. Li and V. T. Remcho, *J. Microcolumn Sep.*, *9*: 389 (1997).
157. C. Yang and Z. E. Rassi, *Electrophoresis*, *20*: 18 (1999).
158. R. Stol, H. Poppe, and W. Th. Kok, *J. Chromatogr. A*, *887*: 199 (2000).
159. Patrick T. Vallano and Vincent T. Remcho, *J. Phys. Chem. B*, 3223 (2001).
160. R. Stol, H. Poppe, and W. Th. Kok, *J. Chromatogr. A*, *887*: 199 (2000).
161. E. Wen, R. Asiaie, and C. Horvath, *J. Chromatogr. A*, *855*: 349 (1999).
162. S. Kitagawa, A. Tsuji, H. Watanabe, M. Nakashima, and T. Tsuda, *J. Microcolumn Sep.*, *9*: 347 (1997).
163. D. Li, H. H. Knobel, and V. T. Remcho, *J. Chromatogr. B Biomed. Sci. Appl.*, *695*: 169 (1997).
164. M. G. Cikalo, K. D. Bartle, and P. Myers, *Anal. Chem.*, *71*: 1820 (1999).
165. E. F. Hilder, M. Macka, and P. R. Haddad, *Anal. Commun.*, *36*: 299 (1999).
166. M. C. Breadmore, M. Macka, and P. R. Haddad, *Electrophoresis*, *20*: 1987 (1999).
167. P. Huang, X. Jin, Y. Chen, J. R. Srinivasan, and D. M. Lubman, *Anal. Chem.*, *71*: 1786 (1999).
168. M. Ye, H. Zou, Z. Liu, and J. Ni, *J. Chromatogr. A*, *869*: 385 (2000).
169. M. Ye, H. Zou, Z. Liu, and J. Ni, *J. Chromatogr. A*, *887*: 223 (2000).
170. M. Zhang, C. Yang, and Z. El Rassi, *Anal. Chem.*, *71*: 3277 (1999).
171. M. Zhang and Z. El Rassi, *Electrophoresis*, *20*: 31 (1999).
172. M. Zhang and Z. El Rassi, *Electrophoresis*, *19*: 2068 (1998).
173. K. Walhagen, K. K. Unger, A. M. Olsson, and M. T. W. Hearn, *J. Chromatogr. A*, *853*: 263 (1999).
174. N. Smith and M. B. Evans, *J. Chromatogr. A*, *832*: 41 (1999).
175. T. Adam and M. Kramer, *Chromatographia*, *49*: S35 (1999).
176. V. Spikmans, S. J. Lane, and N. W. Smith, *Chromatographia*, *51*: 18 (2000).
177. W. Wei, G. Luo, and C. Yan, *Am. Lab.*, *30(1)*: 20C (1998).
178. K. Walhagen, K. K. Unger, and M. T. W. Hearn, *J. Chromatogr. A*, *887*: 165 (2000).

179. M. Zhang, G. K. Ostrander, and Z. El Rassi, *J. Chromatogr. A*, *887*: 287 (2000).
180. W. Wei, G. A. Luo, G. Y. Hua, and C. Yan, *J. Chromatogr. A*, *817*: 65 (1998).
181. W. Wei, Y. M. Wang, G. A. Luo, R. J. Wang, Y. H. Guan, and C. Yan, *J. Liq. Chromatogr. Relat. Technol.*, *21*: 1433 (1998).
182. A. M. Enlund, R. Isaksson, and D. Westerlund, *J. Chromatogr. A*, *918*: 211 (2001).
183. A. Karcher and Z. El Rassi, *Electrophoresis*, *20*: 3280 (1999).
184. R. M. Seifar, J. C. Kraak, W. T. Kok, and H. Poppe, *J. Chromatogr. A*, *808*: 71 (1998).
185. H. Poppe, *J. Chromatogr. A*, *778*: 3 (1997).
186. M. M. Dittmann, G. P. Rozing, G. Ross, T. Adam, and K. K. Unger, *J. Capillary Electrophor.*, *4*: 201 (1997).
187. E. F. Hilder, C. W. Klampfl, M. Macka, P. R. Haddad, and P. Myers, *Analyst (Cambridge, U. K.)*, *125*: 1 (2000).
188. Y. Chen, G. Gerhardt, and R. Cassidy, *Anal. Chem.*, *72*: 610 (2000).
189. M. Mayer, E. Rapp, C. Marck, and G. J. M. Bruin, *Electrophoresis*, *20*: 43 (1999).
190. B. Behnke, J. Johansson, S. Zhang, E. Bayer, and S. Nilsson, *J. Chromatogr. A*, *818*: 257 (1998).
191. J.-R. Chen, M. T. Dulay, R. N. Zare, F. Svec, and E. Peters, *Anal. Chem.*, *72*: 1224 (2000).
192. Chen Jing-Ran, R. N. Zare, E. C. Peters, F. Svec, and J. J. Frechet, *Anal. Chem.*, *73*: 1987 (2001).
193. H. J. Cortes, C. D. Pfeiffer, T. S. Stevens, and B. E. Richter, *J. High Resolut. Chromatogr.*, *10*: 446 (1987).
194. G. Choudhary, C. Horvath, and J. F. Banks, *J. Chromatogr. A*, *828*: 469 (1998).
195. E. Rapp and E. Bayer, *J. Chromatogr. A*, *887*: 367 (2000).
196. G. A. Lord, D. B. Gordon, P. Myers, and B. W. King, *J. Chromatogr. A*, *768*: 9 (1997).
197. L. A. Colón, T. D. Maloney, and A. Fermier, *J. Chromatogr. A*, *887*: 43 (2000).
198. P. D. A. Angus, C. W. Demarest, T. Catalano, and J. F. Stobaugh, *J. Chromatogr. A*, *887*: 347 (2000).
199. C. Yan, Sept. 26 (1995).
200. M. Inagaki, S. Kitagawa, and T. Tsuda, *Chromatography*, *14*: 55R (1993).

201. A. M. Fermier and L. A. Colon, *J. Microcolumn Sep.*, *10*: 439 (1998).
202. K. J. Reynolds and L. A. Colon, *Analyst (Cambridge, U. K.)*, *123*: 1493 (1998).
203. T. D. Maloney and L. A. Colon, *Electrophoresis*, *20*: 2360 (1999).
204. R. Stol, M. Mazereeuw, U. R. Tjaden, and J. van der Greef, *J. Chromatogr. A*, *873*: 293 (2000).
205. L. A. Colón, T. D. Maloney, and A. M. Fermier, in *Capillary Electrochromatography* (Deyl, Zdenek and Svec, Frantisek, Eds.), Elsevier, Amsterdam, 2001.
206. S. Hjerten, J. Liao, and R. Zhang, *J. Chromatogr.*, *473*: 273 (1989).
207. F. Svec and J. M. J. Frechet, *Anal. Chem.*, *64*: 820 (1992).
208. Q. Tang, N. Wu, and M. L. Lee, *J. Microcolumn Sep.*, *11*: 550 (1999).
209. L. Schweitz, L. I. Andersson, and S. Nilsson, *Anal. Chem.*, *69*: 1179 (1997).
210. L. Schweitz, L. I. Andersson, and S. Nilsson, *Chromatographia*, *49*: S93–S94 (1999).
211. L. Schweitz, L. I. Andersson, and S. Nilsson, *J. Chromatogr. A*, *792*: 401 (1997).
212. L. Schweitz, L. I. Andersson, and S. Nilsson, *J. Chromatogr. A*, *817*: 5 (1998).
213. M. R. Schure, R. E. Murphy, W. L. Klotz, and W. Lau, *Anal. Chem.*, *70*: 4985 (1998).
214. N. A. Mischuk and P. V. Takhistov, *Colloids Surf. A: Physicochem. Eng. Asp.*, *95*: 119 (1995).
215. A. Palm and M. V. Novotny, *Anal. Chem.*, *69*: 4499 (1997).
216. E. C. Peters, K. Lewandowski, M. Petro, F. Svec, and J. M. J. Frechet, *Anal. Commun. 35*, 83 (1998).
217. I. Gusev, X. Huang, and C. Horvath, *J. Chromatogr. A*, *855*: 273 (1999).
218. B. Xiong, L. Zhang, Y. Zhang, H. Zou, and J. Wang, *J. High Resolut. Chromatogr.*, *23*: 67 (2000).
219. C. Yu, F. Svec, and J. M. J. Frechet, *Electrophoresis*, *21*: 120 (2000).
220. V. T. Remcho and Z. J. Tan, *Anal. Chem.*, *71*: 248A (1999).
221. S. M. Ngola, Y. Fintschenko, W.-Y. Choi, and T. J. Shepodd, *Anal. Chem.*, *73*: 849 (2001).

222. F. Svec, E. C. Peters, D. Sykora, and J. M. J. Frechet, *J. Chromatogr. A, 887*: 3 (2000).
223. S. Zhang, X. Huang, J. Zhang, and C. Horvath, *J. Chromatogr. A, 887*: 465 (2000).
224. C. Ericson, J.-L. Liao, K. Nakazato, and S. Hjerten, *J. Chromatogr. A, 767*: 33 (1997).
225. C. Fujimoto, M. Sakurai, and Y. Muranaka, *J. Microcolumn Sep., 11*: 693 (1999).
226. C. Fujimoto, Y. Fujise, and E. Matsuzawa, *Anal. Chem., 68*: 2753 (1996).
227. C. Fujimoto, *Analusis, 26*: M49–M52 (1998).
228. S. Hjerten, A. Vegvari, T. Srichaiyo, H.-X. Zhang, C. Ericson, and D. Eaker, *J. Capillary Electrophor., 5*: 13 (1998).
229. C. Ericson and S. Hjerten, *Anal. Chem., 71*: 1621 (1999).
230. F. Svec, E. C. Peters, D. Sykora, G. Yu, and J. M. Frechet, *J. High Resolut. Chromatogr., 23*: 3 (2000).
231. N. Tanaka, H. Kobayashi, K. Nakanishi, H. Minakuchi, and N. Ishizuka, *Anal. Chem., 73*: 421A (2001).
232. G. S. Chirica and V. T. Remcho, *Anal. Chem., 72*: 3605 (2000).
233. Z. Chen and T. Hobo, *Anal Chem., 73*: 3348 (2001).
234. M. T. Dulay, J. P. Quirino, B. D. Bennett, M. Kato, and R. N. Zare, *Anal. Chem., 73*: 3921 (2001).
235. T. Koide and K. Ueno, *J. High Resol. Chromatogr., 23*: 59 (2000).
236. F. Svec and J. M. J. Frechet, *Ind. Eng. Chem. Res., 38*: 34 (1999).
237. Z. J. Tan and V. T. Remcho, *Electrophoresis, 19*: 2055 (1998).
238. C. K. Ratnayake, C. S. Oh, and M. P. Henry, *J. Chromatogr. A, 887*: 277 (2000).
239. K. D. Bartle, A. A. Clifford, P. Myers, M. M. Robson, K. Seale, D. Tong, D. N. Batchelder, and S. Cooper, in *Unified Chromatography* (Parcher, J. F. and Chester, T. L., Eds.), American Chemical Society, Washington, DC, 2000.
240. G. S. Chirica and V. T. Remcho, *J. Chromatogr. A, 924*: 223 (2001).
241. A. Manz, D. J. Harrison, E. Verpoorte, and H. M. Widmer, *Adv. Chromatogr., 33*: 1 (1993).
242. C. D. Bevan and I. M. Mutton, *Methodol. Surv. Bioanal. Drugs, 24*: 299 (1996).
243. J. D. Hayes and A. Malik, *Anal. Chem., 72*: 4090 (2000).
244. P. T. Vallano and V. T. Remcho, *J. Chromatogr. A, 887*: 125 (2000).

3

Gas Chromatography with Inductively Coupled Plasma Mass Spectrometric Detection (GC-ICP MS)

Brice Bouyssiere, Joanna Szpunar, Gaëtane Lespes, and Ryszard Lobinski *CNRS-Université de Pau et des Pays de l'Adour, Pau, France*

I. INTRODUCTION

Inductively coupled plasma mass spectrometry (ICP MS) is rapidly conquering the field of element-selective detection in gas chromatography (GC) despite the presence of a number of cheaper, sometimes even more efficacious, but less versatile competitive techniques, such as flame photometry, atomic fluorescence, atomic absorption, or atomic emission spectrometry [1]. The pioneering works on the GC-ICP MS coupling go back to the mid-1980s with the landmark papers of Van Loon et al. [2] and Chong and Houk [3]. Since then, the ICP quadrupole MS has been undergoing a constant improvement leading to a wider availability of more sensitive, less interference-prone, smaller and cheaper instruments, which favors its use as a chromatographic detector. The combination of capillary GC with ICP MS has become an ideal methodology for speciation analysis [4] for organometallic compounds in complex environmental and industrial samples because of the high resolving power of GC and of the sensitivity and specificity of ICP MS [5–10]. Indeed, the features of ICP MS—such as low detection limits reaching the one femtogram (1 fg) level, high matrix tolerance allowing the direct analysis of complex samples (such as gas condensates), or the isotope ratios measurement capacity enabling accurate quantification by isotope dilution—position ICP MS at the lead of all the GC element specific detectors.

The introduction of ICP time-of-flight MS increased the speed of data acquisition, allowing multi-isotope measurement of milli-second-wide chromatographic peaks and improving precision of iso-tope ratios determination [11–15]. An even better precision was an-nounced by the preliminary reports on the use of magnetic sector multicollector instruments [16,17]. These instrumental develop-ments go in parallel with the miniaturization of GC hardware, allowing the time-resolved introduction of gaseous analytes into an ICP (e.g., based on microcolumn multicapillary GC) and sample preparation methods including microwave-assisted, solid-phase microextraction (SPME) or purge and capillary trap automated sam-ple introduction systems [18]. In 2001 the first GC-ICP MS inter-face became commercially available, proving the recognition by the analytical instrument industry of the maturity of this coupled tech-nique.

The objective of this chapter is to present the state of the art of GC-ICP MS in terms of instrumental developments and applica-tions.

II. FEASIBILITY OF AN ANALYSIS BY GC-ICP MS

The prerequisites for a successful application of GC-ICP MS include the volatility and thermal stability of analytes, and the effective ion-ization of a target element in an inductively coupled plasma. The first condition limits the area of application to organometallic com-pounds that are thermally stable in the native form or can be con-verted to a gas chromatographable form by means of derivatization. The second condition is valid for most of elements in the periodic system; it should be noted, however, that effective ionization of S, F, and Cl in an ICP is extremely difficult.

The species of interest have been summarized in Fig. 1. There are mostly organometallic anthropogenic contaminants and prod-ucts of their degradation or environmental transformation. The targeted gaseous metal(loid) compounds have included dimethyl-mercury (Me_2Hg), dimethyl selenide (Me_2Se), methyl-, butyl-, and phenyl tin ($Bu(Ph)_nSn^{(4-n)+}$, $n \leq 4$), trimethylstibine (Me_3Sb), tri-methylbismuthine (Me_3Bi), methylated arsines (Me_xAsH_y, $x + y =$

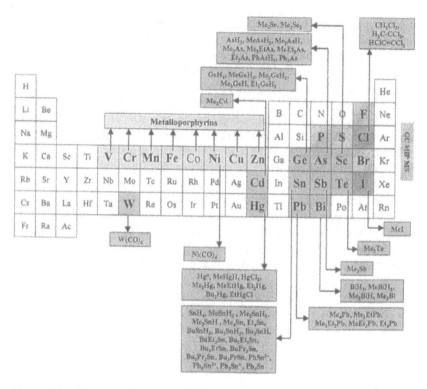

Fig. 1 Species amenable to analysis by gas chromatography with inductively coupled plasma mass spectrometry (GC-ICP MS).

3), dimethyltelluride (Me$_2$Te), alkylated lead (Et$_x$Me$_y$Pb, x + y = 4), and the molybdenum and tungsten hexacarbonyls (Mo(CO)$_6$, W(CO)$_6$).

III. SAMPLE PREPARATION FOR GC-ICP MS

Sample preparation in speciation analysis has been a subject in two of our recent reviews [19,20], and here we will focus on recent developments relevant to GC-ICP MS. The most important include the introduction of NaBPr$_4$ for the derivatization of organometallic species [21] and the use of headspace solid-phase microextraction

(SPME) [22–28], stir bar sorptive extraction [29], and purge and capillary trapping for analyte recovery and preconcentration [30,31].

A. Derivatization Techniques

The position of tetraalkylborates allowing the derivatization in the aqueous phase, such as $NaBEt_4$ for organomercury and for organotin speciation analysis, and the newly introduced $NaBPr_4$ [21] for organolead analysis is well established. The possibility of the simultaneous determination of Sn, Hg, and Pb following propylation has been demonstrated [21]. Artifact formation with hydride generation of antimony has been discussed [32].

In a comparison study, anhydrous butylation using a Grignard reagent, aqueous butylation by means of $NaBEt_4$ and aqueous propylation with $NaBPr_4$ were discussed for mercury speciation [33]. The absence of transmethylation during the sample preparation was checked using a 97% enriched [202]Hg inorganic standard [33]. In another comparison study, esterification of the carboxylic selenomethionine group using 2-propanol followed by the acylation of the amino group with trifluoroacid anhydride and simultaneous esterification and acylation with ethyl chloroformate-ethanol were investigated for the determination of selenomethionine by GC-ICP MS [34].

B. Purge-and-Trap Using Capillary Cryofocusing

A semiautomated compact interface for time-resolved introduction of gaseous analytes from aqueous solutions into an ICP MS without the need for a full-size GC-oven has been described [31,35]. The working principle (Fig. 2) is based on purging the gaseous analytes with an inert gas, drying the gas stream using a 30-cm tubular Nafion membrane, and trapping the compounds in a thick film-coated capillary tube followed by their isothermal separation on a multicapillary column. Recoveries were reported to be quantitative up to a volume of 50 mL [30,31,35]. Fig. 3 shows an example chromatogram obtained using the above-described device coupled to ICP MS for a mercury speciation analysis in a crab tissue extract.

C. Solid-Phase Micro-Extraction (SPME)

SPME is a solvent-free technique that offers numerous advantages such as simplicity, the use of a small amount of liquid phase, low

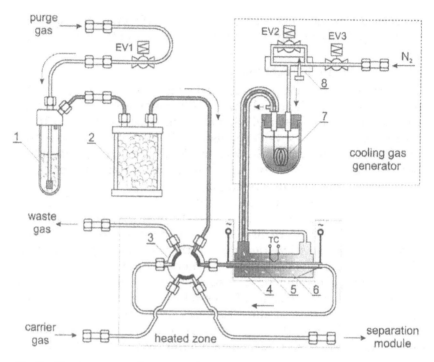

Fig. 2 Semi-automated purge-and-trap–multicapillary GC interface for the time-resolved sample introduction into ICP MS (1) purge vessel with sample inside, (2) permeation type drier, (3) six-port chromatographic valve equipped with microelectric actuator, (4) capillary trap, (5) trap heater, (6) cooling chamber, (7) copper coil immersed in liquid nitrogen, (8) split valve (EV = electromagnetic valves, TC = thermocouple).

cost, and compatibility with on-line analytical procedures. SPME is based on an equilibrium between the analyte concentrations in the headspace (or in the aqueous phase) and in the solid-phase fiber coating. Low extraction efficiencies are hence sufficient for quantification but the amount of the analyte available may be very small. Thus, the interest in the high sensitivity of GC-ICP MS combined with SPME. This makes the combination SPME-GC-ICP MS an emerging analytical tool for elemental speciation in environmental and biological samples [24].

SPME-GC-ICP MS was first used by Moens et al. [23] for the simultaneous speciation of organomercury, -lead, and -tin com-

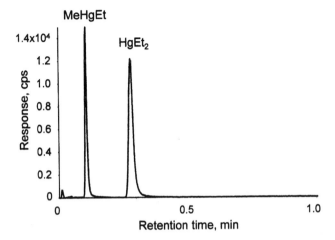

Fig. 3 A chromatogram of a crab sample extract analyzed by purge-and-trap multicapillary GC-ICP MS.

pounds ethylated in situ with $NaBEt_4$. Headspace SPME at no-equilibrium conditions was optimized as an extraction/preconcentration method for triphenyltin residues in tetramethylammonium hydroxide (TMAH) and KOH-EtOH extracts of potato and mussel samples [22]. Derivatization was carried out with $NaBEt_4$ for 10–20 min [22]. Direct SPME (from the aqueous phase) was studied, but the sensitivity was 10 times lower. A detection limit of 2 pg L^{-1} was reported for an aqueous standard but a value of 125 pg L^{-1} was given for the sample extract corresponding to a DL in the low ng/g range (dry weight) [22]. Slightly lower detection limits (0.6 −20 pg 1^{-1}) were reported in another work in which the SPME was performed in equilibrium conditions from the aqueous phase. The precision for standards was in the range of 8% to 25% [27]. An example chromatogram is shown in Fig. 4.

Similar in principle to SPME is an extraction technique using stir bars coated with a relatively thick (0.3–1 mm) layer of poly(dimethylsiloxane): stir bar sorptive extraction (SBSE) [36]. The stir bar (1–4 cm in length) is added to an aqueous sample for stirring and extraction. After a certain stirring time, the bar is removed and thermally desorbed into a GC. Owing to the much larger volume of the stationary phase, the extraction efficiency in SBSE is by far

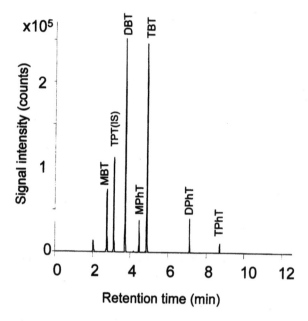

Fig. 4 A chromatogram obtained for a mixture of organotin standards in water (50 ng L^{-1}) by SPME-capillary GC-ICP MS.

superior to that of SPME. The instrumental detection limits reported for organotin compounds were exceptionally good (10 fg L^{-1}); in practice, however, values of 0.1 pg L^{-1} could be achieved [29].

D. Gas Chromatography of Organometallic Species

Packed column GC was used in the early studies on GC-ICP MS coupling [2,3]. Packed columns can, by design, handle high flow rates and large sample sizes, but the separation efficiency and detection limits are compromised because of the high dispersion of the analytes on the column. Packed column GC-ICP MS is a favorable technique to follow hydride generation purge-and-trap because of easier handling of highly volatile species at temperatures below −100°C [37,38].

Capillary GC-ICP MS, first described by Kim et al. [39,40], offers improved resolving power over packed column GC-ICP MS, which is of importance for the separation of complex mixtures of organometallic compounds found in many environmental samples. The reduced sample size and the high dilution factor with the detector's makeup gas necessary to match the spectrometer's optimum flow rate result in a loss of sensitivity. Recently, a number of articles appeared on rapid (flash) GC employing columns that consist of a bundle of 900–2000 capillaries of small (20–40 μm) internal diameter, referred to as multicapillary columns (for a review, see Ref. 41). Such a bundle allows one to eliminate the deficiencies associated with the use of capillary and packed columns while the advantages of both are preserved. Multicapillary GC features high flow rates that minimize the dilution factor and facilitate the transport of the analytes to the plasma. The coupling between MC GC and ICP MS offered 0.08 pg detection limits for Hg speciation [42].

An interesting feature is the use of multicapillary microcolumns for sample introduction into an ICP. Rodriguez et al. [43] showed isothermal separations of organometallic species using a 50-mm column, which opens the way to the miniaturization of GC sample introduction units, possibly making the classic GC oven redundant. A chromatogram (Fig. 5) demonstrates the potential of this technique for rapid multielement speciation analysis.

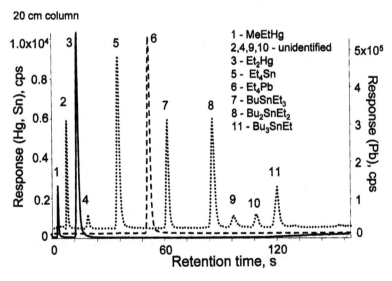

Fig. 5 A multi-element multispecies multicapillary GC-ICP MS chromatogram of a fish extract. (From Ref. 43.)

Gas chromatography using chiral stationary phases, such as L-valine-*tert*-butylamide [44], was recently shown to allow the resolution of *N*-trifluoroacetate-*O*-isopropyl derivatives of DL-selenomethionine enantiomers in selenium nutritional supplements [44].

IV. INTERFACE BETWEEN GC AND ICP MS

The basic requirement for an interface is that analytes be maintained in gaseous form during transportation from the GC column to the ICP, in a way that condensation in the interface is prevented. This can be achieved either by efficiently heating the transfer line up to the end, avoiding the cold spots, or by using an aerosol carrier. This results in two basic design types of the GC-ICP MS interface. In the first type, the spray chamber is removed and the transfer line is inserted part of the way up the central channel of the torch. The required efficiency of heating depends on the species to be analyzed. In the second type, the GC effluent is mixed in the spray chamber with the aqueous aerosol prior to introduction into the plasma. Regardless of the type of interface, an addition of oxygen to the plasma gas is essential to prevent carbon deposition (and sometimes metal entrapment) and to reduce the solvent peak.

A. Interface Designs Based on Direct Connection of Transfer Line to Torch

The effluent from the GC (a few milliliters per minute) requires a makeup carrier gas to achieve sufficient flow to get the analytes into the central channel of the plasma. Plasma operating conditions are optimized with the Xe gas added at the constant mass flow rate to the argon nebulizer gas using a mass flow controller via a T-piece. The xenon (the intensity of which is measured during the chromatogram) also serves as an internal standard. The advantage is the absence of aerosol in the plasma, which limits energy losses for desolvation and vaporization, reduces polyatomic interference, and results in a high sensitivity. The basic features differentiating the interface designs within this type is whether and how the heating of the transfer line is accomplished and how easily the coupling and decoupling can be realized.

In the simplest case—an analysis for volatile species cryotrapped and released by thermal desorption—a simple glass piece, connected to the end of the U-shaped glass chromatographic column, assured the transport of methyl and ethyl species of Hg, Pb and Sn, As and Ge [38,45] A similar design was used for the transfer of butyltin hydrides thermally desorbed from a packed column [46]. A high velocity Ar carrier makeup gas flow was added post-column to prevent the condensation of the tin species in the transfer line. Peak profiles for mono- and dibutyl hydrides were satisfactory, but pronounced tailing was observed for the tributyltin hydride [46]. A heating block at the exit from the oven (at the mixing point with the makeup argon) was necessary. In the other design, a multicapillary column was interfaced to ICP MS for the analysis of ethylmercury derivatives upon ethylation, liquid-gas extraction, and capillary cryotrapping [31,42]. This design produced negligible peak broadening for the ethylated mercury species; less volatile compounds than Et_2Hg produced broad peaks unappropriate for quantification.

In order to improve the transport of less volatile analytes to the plasma, interfaces heated on a section of their length were designed. The idea behind this is that the GC effluent efficiently heated during part of its way through the interface will not condense on the non-heated parts. Kim et al. [39,40,47,48] proposed the use of a 60-cm long Al rod, wound in a heating tape, and housing the capillary in a longitudinal slot. Smaele et al. [49] used resistively heated stainless tubing [49]. Argon gas was made to flow around the transfer capillary in the heated transfer line. The interface showed apparently no peak tailing even for fairly volatile species, such as Bu_3SePe, and was readily mountable and demountable. A similar design with a transfer line made of quartz and resistively heated to 250°C produced no peak tailing for species with boiling points as low as that of Bu_3SnEt [50]. In another design, a sole heated block was placed at the exit of the gas chromatograph [51,52]. The transport of the analytes to the plasma was achieved with a flexible 1.5 mm I.D. PTFE 80-cm tube at room temperature. The PTFE tubing was inserted into the central channel of the ICP MS torch [51]. In order to prevent condensation in the tube, a high-velocity flow of Ar was introduced externally to the column.

In the above-discussed interfaces, the part of the transfer line inserted in the ICP injector could not be efficiently heated. This resulted in a possible occurrence of cold spots, especially for the analysis for high-boiling-point compounds. Therefore, interface designs in which the capillary was extended close to the plasma and heated up to its very end inside the torch were developed. The earliest design [53] developed for the analysis of metalloproteins with retention indices above 6000 employed a stainless steel tube inserted concentrically in a 3-mm I.D. ICP injector up to 24 mm from the plasma. The tube was resistively heated up to 400°C and housed the GC capillary. The makeup argon was preheated using a heated (to red) Ni-Cr wire inside a silica tube. Because of the low heat capacity of Ar gas, aggressive heating was needed and the tubing between the Ar heater and ICP MS injector should be reduced to minimum. In another design a 1-m long copper tube was heated with heating tape and insulated. A resistively heated teflon-coated Ni-Cr wire was positioned within the transfer line for the additional temperature control 200°C [13]. In the rf field the NiCr wire was kept perpendicular to the field to avoid conduction of rf energy [54,55].

The commercialized design (Fig. 6) was based on the 1-m flexible resistively heated (and thermally insulated) transfer line housing the capillary. This was completed with a 10-cm rigid transfer line inserted up to the end of the central channel of the ICP MS torch

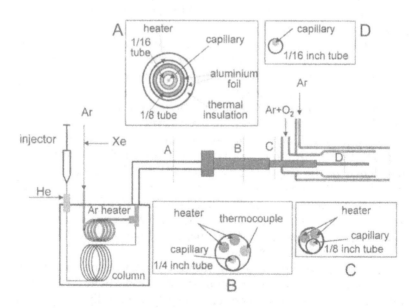

Fig. 6 Scheme of the prototype of a commercialized GC-ICP MS interface. (From Ref. 43, copyright (2002) American Chemical Society.)

[43,56–59]. The rigid part of the transfer line encapsulated an additional heater and a thermocouple to measure the temperature. The heaters extended to 5 cm from the end of the capillary. The end of the rigid part was placed inside the ICP MS torch in place of the conventional injector. The makeup argon gas was preheated by passing through a 1-m 1/16-inch coil placed inside the chromatographic oven. It was made to flow between the internal wall of the heated transfer line and the external wall of the capillary tube to merge with the GC carrier gas just before the plasma [43].

The stiffness of the metallic designs spurred some work on flexible, easily removed interfaces. Teflon tubing coiled with NiCr wire housing a capillary and the NiCr wire proved to have sufficient heat tolerance to temperatures below 200°C [60].

B. Interface Designs via a Spray Chamber

The major inconvenience of the transfer line-to-torch interfaces is the need to dis- and reassemble the system when the instrument also serves for routine elemental analyses of solutions. The conditions in plasma are readily affected by changes in the composition of the matrix exiting the chromatographic column. An internal standard (e.g., Xe gas added to the makeup argon) is required to correct for the varying ionization conditions [43,57,58] and improving the repeatability by a factor of 2.5 [27]. However, correction in this way for mass fractionation in isotope dilution GC-ICP MS analysis is difficult. In order to alleviate the dis- and reassembling problems, an interface was designed that allowed the selection of either the GC effluent or the sample aerosol using a "zero-dead volume valve" and requires no system reconfiguration when switching between the two modes [61,62]. The makeup gas was added using a sheathing device in a way that it surrounded the sample (rather then completely mixing with it), maintaining the analyte in the central channel of the plasma [61,62].

Feldmann et al. proposed a design in which the GC effluent was mixed with the aqueous aerosol, ensuring stable plasma conditions [63]. The transfer capillary was inserted into the torch injector, and an aqueous solution was continuously aspirated into the plasma via a T-piece connection using a conventional Meinhard concentric nebulizer via a Scott double-pass spray chamber. In this way, the aerosol was transfered into the plasma together with the analyte gas. In a modified design, a low-volume water-cooled cyclonic spray chamber was used [17]. An internal standard (e.g., T1 solution for mass frac-tionation correction) could thus be added to the aqueous solution. Prohaska et al. [64] proposed an interface in which the end of the GC-capillary was introduced into the sample inlet tube of a conven-tional Meinhard nebulizer, which was connected to the end of the torch by a PTFE tube. ^{129}Xe was measured as internal standard. Even if condensation of the separated species might occur, the aero-sol formed was transferred to the ICP without being adsorbed on the surface of the tube. Therefore, no additional heating of the tube was required.

The disadvantage of the systems including a spray chamber is the relatively low sensitivity because of the loss of energy for the desolvation of wet aerosol. Also, the applicability of this type of inter-face to less volatile species (Kovats index above 5000) needs to be verified. The design fails with complex matrices such as gas conden-sates, the matrix condensing in the spray chamber [65].

V. LOW-POWER ICP AS A TUNABLE IONIZATION SOURCE FOR GC EFFLUENTS

The conventional (>1 kW) atmospheric pressure argon ICP offers high sensitivity and element specificity but provides no molecular information because atomic ions only are obtained. The analyte iden-tification is solely based on a comparison of retention time with stan-dards. Another disadvantage is the limited ability to ionize non-metals (e.g., sulfur and halides), which are important elements in organic geochemical and environmental samples. For these reasons, interest in low-pressure ICP has been growing.

A low-pressure helium ICP at powers between 4 and 40 W and at a pressure of 1 mbar was used for the production of electron-impact-like mass spectra for a series of organometallic and haloge-nated species in GC effluents [66,67]. By tuning the power and pres-sure, various degrees of fragmentation could be obtained. The initial detection limits at the 1 ng level were nothing spectacular, but an improved design offered detection limits below 10 pg for halogenated compounds while the helium consumption remained below 0.5 l min^{-1} at a rf power of approximately 100 W [68]. A GC–low pressure ICP MS instrument sustaining a helium (6 ml min^{-1}) plasma at 6 W was designed [69,70]. By altering the composition of the plasma gas alone, it was possible to utilize the low-pressure ICP as a soft ionization source or as a harsh ionization source, the latter providing

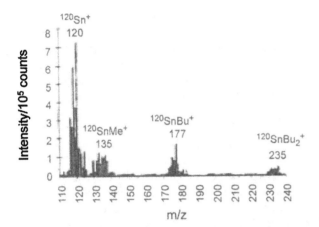

Fig. 7 Mass spectrum obtained at the peak apex in GC-low pressure ICP MS of a TBT standard (1.6 ng). (From Ref. 72. Reproduced by permission of the Royal Society of Chemistry.)

elemental information only [70]. The source could be operated in a tunable mode, between hard and soft ionization regimes. Detection limits were 2–76 pg in the atomic mode and 11–230 pg in the molecular mode for the halogeno (Cl, Br, I) benzenes. The problems associated with the poor linear calibration range and high detection limits for the molecular ions were alleviated by the appropriate choice of reagent gases. The technique was applied to the detection of Et_4Pb in standard reference fuel with a detection limit of 7 pg [71].

Another generation of plasma sources based on a low-power, reduced-pressure, helium ICP, operated at an estimated incident rf power of 12–15 W and pressures of approximately 1 mbar was shown to provide sub-picogram detection limits in the atomic mode. Species-characteristic mass spectra were obtained, but for analyte concentrations 1000 times higher [72]. The system was further applied successfully for the determination of organotin compounds in CRMs [73]. An example of GC–low pressure ICP MS spectra obtained in the soft ionization conditions is shown in Fig. 7.

VI. MASS SPECTROMETRIC DETECTION IN GAS CHROMATOGRAPHY

Quadrupole mass analyzers have predominantly been used; their sensitivity has improved by a factor of 10 during the past decade.

Tao et al. [58] reported an excellent instrumental detection limit of 0.7 fg by operating the shield torch at normal plasma conditions using the HP 4500 instrument. Among other types of analyzers, TOF MS has been extensively studied as a GC detector during the past few years [13–15,74]. Applications of sector-field analyzers [16,17, 64,75], also with multicollectors [16,17], are slowly appearing.

A. GC-ICP Time-of-Flight MS

The measurement of a time-dependent, transient signal by sequential scanning using a quadrupole or a sector-field (single collector) mass spectrometer results in two major types of difficulties. The first one is the limited number of isotope intensity measurements that can be carried out within the time span of a chromatographic (especially from capillary or multicapillary GC) peak. The other one is the quantification error known as spectral skew [76], which arises during the measurement of adjacent mass-spectral peaks at different times along a transient signal. Alleviating these difficulties requires increasing the number of measurement points per time unit and the simultaneous measurement of the isotopes of which the ratio is investigated.

The ability to produce complete mass spectra at a high frequency (typically $>20,000$ s^{-1}) makes TOF MS nearly ideal for the detection of transient signals produced by high-speed chromatographic techniques. The simultaneous extraction of all m/z ions for mass analysis in TOF MS eliminates the quantification errors of spectral skew, reduces multiplicative noise, and makes TOF MS a valuable tool for determining multiple transient isotopic ratios [13–15,74]. An example of data acquisition during GC-ICP TOF MS is shown in Fig. 8. ICP TOF MS suffers, however, from the lower sensitivity in the monoelemental mode in comparison with the latest generation of ICP quadrupole mass spectrometers. A minimum of approximately 500 pg of each species is necessary for the measurement of isotope ratios with a precision better than 0.5% [15]. The limitations of the pulse counting system are clearly seen, with peak heights of more than 2000 counts reaching saturation (for an integration time of 100 ms) [15]. On the other hand, Heisterkamp et al. [14] reported a DL of 10–15 fg for alkyllead compounds, a value that is comparable with ICP Q MS. It should be emphasized that the loss of sensitivity in the monoelemental mode is compensated by the fact that the number of isotopes determined during one chromatographic run is no longer

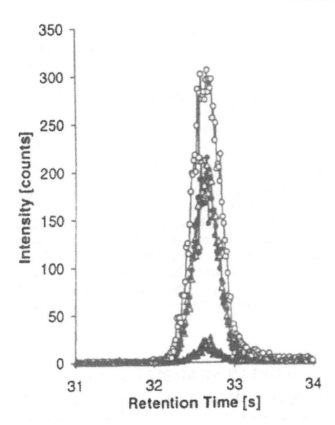

Fig. 8 GC-ICP TOF MS of a chromatographic peak with the 12.5 ms acquisition frequency for three tin isotopes (▲ ^{204}Pb △ ^{206}Pb ⬤ ^{207}Pb ⊖ ^{208}Pb). (From Ref. 14 by permission of Springer-Verlag.)

limited by peak definition (as in ICP MS employing a quadrupole analyzer), because the number of data points per chromatographic peak is independent of the number of measured isotopes.

B. GC-ICP MS Using Sector Field Mass Analyzers

Published reports in this field have been scarce [16,17,64,75]. Prohaska et al. used an ICP sector-field double-focusing mass spectrometer for the analysis of arsenic hydrides and organoarsenic compound in gaseous emissions from a microcosm, after their separation on a capillary column [64]. Neither instrumental detection limits were reported nor chromatograms shown, however. The Sanz-Medel group proposed the same technique for the detection of [32]S in GC analysis of volatile sulfur compounds in bad breath and reported absolute DLs in the low nanogram range under cold plasma conditions, when a guard electrode was used [75].

The other two reports concerned the precise measurement of isotopic ratios by GC-ICP multicollector MS [16,17]. Four lead, [203]Tl, [205]Tl (for the correction of the mass bias), and [202]Hg (for the correction of the [204]Pb isobaric overlap) were monitored simultaneously. A double-focusing instrument applied for this purpose allowed a detection limit of 1 pg for [207]Pb (introduced as Et_4Pb), which is poorer than values reported with other ICP mass spectrometers [17]. The minimum time resolution was limited by the Axiom software to 50 ms; 60 points could define a full peak width of 3 seconds [17]. Much lower detection limits were obtained by GC coupled to a single magnetic sector instrument equipped with a hexapole collision cell. A value of 2.9 fg was reported for the most abundant [208]Pb isotope [16].

VII. GC-ICP MS STUDIES USING STABLE ISOTOPES

The isotopic specificity of ICP MS opens the way to the use of stable isotopes or stable isotope–enriched species for studies of transformations and of artifact formation during sample preparation. The wider implementation of the use of isotope dilution for quantification in GC-ICP MS had, until recently, been limited by the nonavailability of organometallic species with the isotopically enriched element. However, standards for isotopically enriched Me[201]Hg [77], MBT, DBT, and TBT [78] have been recently synthesized and applications are being developed. The prerequisite of the use of stable dilution techniques is the precise and accurate measurement of the isotopic ratios. The to-date applications to real-world samples have been exclusively carried out with ICP Q MS, but precision and accuracy values for the measurement of isotope ratios in standard compounds by ICP TOF MS [13] and by sector-field multicollector [16,17] instruments have been reported. Tracer experiments can be performed with ICP Q MS, but for studies of natural fractionation its precision may not be sufficient.

A. Precision and Accuracy of an Isotope Ratio Measurement in GC-ICP MS

A precision of 1% was reported for the Hg isotope ratios determined for MeEtHg eluted from a packed column by GC-ICP Q MS [79]. In capillary GC Q ICP MS, a precision of 0.5% was reported for Se (derivatized as piazselenol) [80]. A tin isotope-ratio measurement accuracy of 0.28% and a precision of 2.88% were calculated for a 1-s wide GC peak of Me_4Sn [13]. Haas et al. reported that a minimum of 0.5 ng of an organometallic species was necessary for the measurement of isotope ratios with a precision better than 0.5%; the best value (0.34%) was attained for Me_2SnH_2 [15]. ICP TOF MS gave 10 times better precision than ICP Q MS in comparable conditions when natural gas samples were analyzed [15]. The precision of the determination of isotopic ratios is improved by the use of simultaneous multi-elemental ion extraction [14].

The precision values reported for the measurement of major Pb isotopes ratio with a double-focusing ICP multicollector MS were below 0.07% (for a 3-s transient signal) and corresponded to an accuracy of 0.35% [17]. When a magnetic sector instrument (with a hexapole collision cell and multicollector detection) was used, the precision was in the range of 0.02% to 0.07% for ratios of high-abundance isotopes and injections of 5–50 pg [16]. After mass bias correction the accuracy was within 0.02–0.15% [17].

For accurate determinations by the isotope dilution technique, the mass discrimination effect (ca. 0.5% per mass unit) must be taken into account. The ways to measure and to correct for the mass bias included the sequential measurement of the isotope ratios in the sample and the standard [81] or the addition of an internal standard, such as Cd [14,15] or Tl [16,17], and the simultaneous measurement of the $^{111}Cd/^{113}Cd$ or $^{203}Tl/^{205}Tl$ isotopic ratios, respectively. From the practical viewpoint, the latter system requires the simultaneous delivery of the analyte and of the internal standard to the ICP MS, which can be done only via a spray-chamber interface in view of the involatility of Cd and Tl species. An alternate measurement of the isotope ratio of the analyte element in a standard was possible owing to a diffusion cell containing pure chemical of the element to be determined [81]. It consisted of a glass vial covered by a membrane that allowed diffusion of the volatile calibrant species into the flow cell. If the isotope ratios of the element in the calibration compound are known, the measured isotope ratio of the separated species in the sample could be corrected [81].

B. Monitoring of Artifacts During Sample Preparation Procedure

Enriched isotopes and isotope-enriched species provide an important diagnostic tool for the validation development of new analytical methods. The formation of MeHg$^+$ during water vapor distillation procedure, detected owing to isotopically enriched Hg^{2+} [79,82], shed a new light on the accuracy of methylmercury determinations in sediment samples [83]. Stable isotope-labeled mercury species allowed a GC-ICP MS study of the simultaneous Hg^{2+} biomethylation and CH$_3$Hg$^+$ demethylation at ambient trace levels with the sensitivity superior to that of radiotracer techniques [84].

Using a similar approach and a Me^{201}Hg$^+$ spike, MeHg$^+$ was transformed into elemental mercury (Hg^0) in the presence of chloride and bromide during the derivatization by NaBEt$_4$ but not by NaBPr$_4$ [77]. Isotope enriched Hg^{2+} allowed the observation of an artifactual formation of methylmercury when the water vapor distillation was applied for aqueous rain samples containing visible particles [85]. No artifact formation of methylmercury during sample preparation was observed following the addition of a ^{201}Hg^{2+} isotope standard [60].

C. Speciated Isotope Dilution Analysis

Fundamentals of ID GC-ICP MS for species-specific analysis were extensively discussed by Gallus and Heuman [81]. The speciated ID analysis is only possible for element species well defined in their structure and composition. The species must not undergo interconversion and isotope exchange prior to separation. The equilibration of the spike and analyte, attainable in classical ID MS by multiple sequential dissolution and evaporation-to-dryness cycles, cannot be guaranteed to be achieved for speciated ID analysis in solid samples. Consequently, the prerequisite of the ID method—that the spike is added in the identical form as the analyte—is extremely difficult, not to say impossible, to attain. Nevertheless, some advantages, such as the inherent corrections for the loss of analyte during sample preparation, for the incomplete derivatization yield, and for the intensity suppression/enhancement in the plasma are evident. In particular, the ID quantification seems to be attractive in speciation analysis of complex matrices (e.g., gas condensates) when the different organic constituents of the sample modify continuously the conditions in the plasma and thus the sensitivity [86].

Isotopically enriched species represent the ultimate means for specific accurate and precise instrumental calibration. Not only are they are useful for routine determination by speeding analysis, but they also assist in the testing and diagnostics of new analytical methods and techniques. To date, the application examples of speciated ID GC-ICP MS have been relatively scarce. The determination of dibutyltin in sediment was carried out by ID analysis using a ^{118}Sn-enriched spike. No recovery corrections for aqueous ethylation or extraction into hexane were necessary and no rearrangement reactions were evident from the isotope ratios [87]. A mixed spike containing ^{119}Sn enriched mono-, di-, and tributyltin was prepared by direct butylation of ^{119}Sn metal and characterized by reversed isotope dilution analysis by means of natural mono-, di-, and tributyltin standards. The spike characterized in this way was used for the simultaneous determination of the three butyltin compounds in sediment CRMs [78]. Isotopically labeled Me_2Hg, $MeHgCl$ and $HgCl_2$ species were prepared and used for the determination of the relevant species in gas condensates with detection limits in the low pg range [86].

VIII. SELECTED APPLICATIONS

The application areas in speciation that have most benefited from the GC-ICP MS coupling include the analysis of air, landfill gas, and energy-related samples because of the need for the high sensitivity in the multielemental mode while the complexity of the matrix is responsible for interference in other techniques such as GC-AAS, GC-AFS, and GC-MIP AED.

A. Analysis of Air and Atmospheric Samples

Some volatile metal(loid) compounds are very toxic, so their concentrations in ambient air or workplace air can be of special concern and may require continuous monitoring. Early research by Hirner et al. on volatile metal and metalloid species in gases from municipal waste deposits [88,89] have been creatively developed [37,38,63,90–97].

Sampling remains the most critical point in the analysis of air for organometallic species. Solvent or absorbing liquids cannot be used for sampling because the interactions needed for the analyte to be caught in the absorbent risk changing its chemical form. The use of adsorbing stationary phases suffers from irreversible adsorption, degradation during the desorption process, and artifact formation [98]. Therefore, cryotrapping using a chromatographic packing (type SP-2100 on a 60/80 mesh stationary phase) followed by thermal desorption into an ICP MS, extensively developed by Feldmann et al. [63,90,91,93,94,96,97], is gradually gaining success among other groups [37,38,95]. It is despite the disadvantages of using large amounts of liquid nitrogen on site and the need for a pump and a power supply. Usually a second trap is needed for cryofocusing in order to allow a narrow analyte band (for enhanced resolution) to enter the GC column. The increasing sensitivity of ICP MS instruments allows one to decrease the volume of samples. In this context, sampling of volatile metal(loid) compounds such as hydrides, methylated, and permethylated species of arsenic, antimony, and tin using Tedlar bags described prior to cryotrapping GC-ICP MS is worth emphasizing [98].

The analyses are usually run in the multielemental mode. A method capable of identifying 29 species (b.p. in the range of $-90°C$ to $250°C$) of 12 elements in gaseous, liquid, and solid samples was developed [99]. The detection limits were at the 0.1 to 2.5 pg levels but air quantities up to 100 L could be sampled, which resulted in concentration detection limits down to 1 pg m^3 [38]. A large volume hydride generation system (samples of $0.5-1.0$ L processed at a rate of 3 h^{-1}) was developed; the hydrides were cryotrapped in a field-packed column at $-196°C$. The desorption was carried out in the laboratory prior to analysis by GC-ICP MS [45]; the detection limits were in the 1 to 20 pg L^{-1} range [45]. With the precision achieved, the combination of cryotrapping GC and ICP TOF MS is a powerful tool for monitoring volatile multi-element species in multi-tracer experiments and isotope dilution methodology [15].

In terms of quantification, the main drawback is the nonavailability of standard gas mixtures, which could be used for calibration. Only a few standard compounds are commercially available, the preparation of volatile metal(loid) compounds in gas samples being difficult. The thermodynamic stability of gaseous standards is very low in comparison with the same species in the environment while handling of gas mixtures at very low concentrations (pg l^{-1}) is almost impossible because of adsorption/desorption effects. A method of quantification by mixing the gas sample with Rh-containing aerosol was proposed [94].

B. Water Samples

Rain, snow, seawater, and geothermal waters were analyzed for volatile metal(loid) species by purge and trap GC-ICP MS [37,45, 77,100,101]. Ionic species were volatilized prior to liquid-gas extraction by hydride generation (multielemental analysis) [45] or ethylation (Hg) [77]. The detection limits depend on the sample volume and can be lower than 1 pg/L for a 1-L sample volume.

ICP MS detection in GC enabled considerable progress in ultratrace speciation analysis of tin. Whereas the classical approach based on the extraction of butyltins by DDTC in pentane followed by Grignard propylation offered detection limit of 20 pg L^{-1} for a 0.5-L sample of harbor water [102], the application of *in situ* derivatization with NaBEt$_4$ followed by stirbar SPE allowed to reduce DLs down to 0.1 pg/l for a 30 mL using similar equipment [29]. The use of NaBEt$_4$ derivatization followed by extraction into hexane and in-liner preconcentration technique allowed a detection limit of 0.01 pg L^{-1} in a 1-L sample of ocean water [58].

For ultratrace organolead analysis in atmospheric deposits and tap water extraction with DDTC followed by Grignard [103] or NaBPr$_4$ [14] derivatization was reported with the detection limits down to 50 pg L^{-1} (quadrupole) and 500 pg L^{-1} (TOF), respectively. Alternatively, DDTC-complexes of Me$_3$Pb$^+$, Me$_2$Pb^{2+}, Et$_3$Pb$^+$ and Et$_2$Pb^{2+} in rain water were sorbed on C$_{60}$ (fullerene) beads and eluted with NaBPr$_4$ hexane solution prior to GC-TOF ICP MS to obtain a detection limit of 3 pg L^{-1} [74].

C. Soil, Sediments, and Particulate Matter

Mono-, di-, and tributyltin compounds were leached from sediments with water [50,104], glacial acetic acid [27], acetic acid–water [43] or acetic acid/methanol [21,23,52,78,87] mixtures. Chelation with tropolone followed by solvent extraction was reported [59]. The organotin compounds were usually derivatized with NaBEt$_4$ [23, 27,43,50,59,78,87,104], but *in situ* derivatization with NaBPr$_4$ [21] or by hydride generation [46] were also reported. Solvent extraction followed by capillary GC was the most common approach [21,43,50,59,78,87,104] but SPME [23,27] was also used. Organotin hydrides were purged, cryotrapped, and released from a packed column by thermal desorption [20]. Detection limits in the 0.1–1 ng/g (dry weight) range are routinely quoted [46].

The synthesis of NaBPr$_4$ offered a convenient derivatization method able to discriminate between methyl and ethyllead derivatives [14], but extraction of ionic organolead complexes with DDTC followed by butylation using a Grignard reagent (BuMgCl) remains a valid approach [52]. Organolead compounds were determined by GC-ICP MS in airborne particulate matter [52] and in road dust CRM [14,52]. The excess of inorganic lead makes its complexation with EDTA prior to extraction necessary.

Methylmercury in sediment was determined by distillation followed by NaBEt$_4$ ethylation, purge and trap onto Tenax and thermodesorption from a packed column [79]. Hydride generation followed by cryotrapping and thermal desorption into an ICP MS was proposed for the determination of many species of As, Hg, Se, Pb and other elements in soil, sediment and waste deposits [99,101,105,106].

D. Biota

The basic applications of GC-ICP MS for biological tissues concern the determination of methylmercury [33,42,48,60]. Methylmercury was recovered by leaching with KOH [48] or HCl [33,60] or by solubilization with TMAH [42]. Derivatization with NaBEt$_4$ followed by purge and trap capillary GC [42] or by solvent extraction into toluene [33] or nonane [60] was recommended. Detection limits are 0.2 ng/g for the purge-and-trap GC-ICP MS [42] and 2.6 ng/g for the solvent extraction approach [60].

For the purpose of organotin speciation analysis, biological tissues were solubilized in TMAH [22,29,43] or leached with methanol-HCl mixture by sonication [27]. Derivatization with NaBEt$_4$ was solely used [22,27,29,43]. Extraction into isooctane is the classic approach [43], but novel techniques such as headspace SPME [22,27] or stir bar SPE [29] have been increasingly used.

E. Energy-Related Samples

An ICP is more robust than an MIP and tolerates a higher load of organic compounds [107]. In terms of petroleum matrices ICP MS

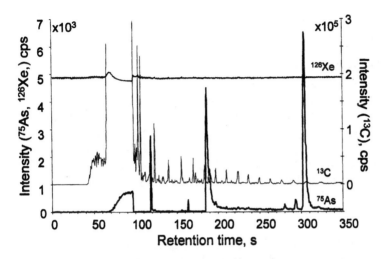

Fig. 9 Speciation analysis for organo-arsenic in a gas condensate sample by GC-ICP MS. (From Ref. 108. Reproduced by permission of the Royal Society of Chemistry.)

has been successfully used as GC detector of organolead in fuel [39], metalloporphyrin in shale [47,53] and organomercury [57,86] and organoarsenic [108] in gas condensates. An example chromatogram showing speciation of organoarsenic species in a gas condensate sample is shown in Fig. 9. Oxygen, introduced to the plasma auxiliary gas, is mandatory. It reduces matrix interferences and was reported to improve twice the sensitivity [86] To date, the GC-ICP MS coupling has been the only one to produce species-specific information on the whole range of dialkylmercury species in gas condensate samples at the µg l⁻¹ and sub-µg l⁻¹ level [57].

F. Conclusions and Perspectives

Gas chromatography with ICP MS detection shows comparable figures of merit with those of GC-MIP AED for standard applications including speciation of organomercury, organolead, and organotin in the environment. However, it offers a number of advantages in cases where extremely low sensitivity, multi-elemental screening, precise isotope ratio measurements, or the analysis of complex matrices are required. The detection limits down to subfemtogram and low fg mL⁻¹ levels make GC-ICP MS ideal for the determination of background and contamination levels in pristine environments. The multi-isotope capacity and the reasonably high precision of isotope ratio measurements during narrow transient signals favors ICP TOF MS as the detection technique of choice for multi-element screening of atmospheric samples and for stable isotope tracer studies by GC. Sector field ICP multicollector MS, owing to its high isotope ratio measurement precision, is likely to become a technique of choice for monitoring natural isotope fractionation processes related to biogeochemical cycles of trace elements. GC-ICP MS is the only technique to date enabling a direct speciation analysis of As and Hg in natural gases and gas condensates at the μg L^{-1} level.

REFERENCES

1. R. Lobinski, *Appl. Spectrosc.*, *51*: 260A (1997).
2. J. C. Van Loon, L. R. Alcock, W. H. Pinchin, and J. B. French, *Spectrosc. Lett.*, *19*: 1125 (1986).
3. N. S. Chong and R. S. Houk, *Appl. Spectrosc.*, *41*: 66 (1987).
4. D. M. Templeton, F. Ariese, R. Cornelis, L. G. Danielsson, H. Muntau, H. P. Van Leeuven, and R. Lobinski, *Pure Appl. Chem.*, *72*: 1453 (2000).
5. S. J. Hill, M. J. Bloxham, and P. J. Worsfold, *J. Anal. At. Spectrom.*, *8*: 499 (1993).
6. N. P. Vela, L. K. Olson, and J. A. Caruso, *Anal. Chem.*, *65*: 585 (1993).
7. F. A. Byrdy and J. A. Caruso, *Environ. Sci. Technol.*, *28*: 528A (1994).
8. P. C. Uden, *J. Chromatogr. A.*, *703*: 393 (1995).
9. K. Sutton, R. M. C. Sutton, and J. A. Caruso, *J. Chromatogr. A*, *789*: 85 (1997).
10. G. K. Zoorob, J. W. McKiernan, and J. A. Caruso, *J. Anal. At. Spectrom.*, *128*: 145 (1998).
11. B. W. Pack, J. A. C. Broekaert, J. P. Guzowski, J. Poehlman, and G. M. Hieftje, *Anal. Chem.*, *70*: 3957 (1998).
12. F. Vanhaecke and L. Moens, *Fresenius' J. Anal. Chem.*, *364*: 440 (1999).
13. A. M. Leach, M. Heisterkamp, F. C. Adams, and G. M. Hieftje, *J. Anal. At. Spectrom.*, *15*: 151 (2000).
14. M. Heisterkamp and F. C. Adams, *Fresenius J. Anal. Chem.*, *370*: 597 (2001).

15. K. Haas, J. Feldmann, R. Wennrich, and H. J. Stärk, *Fresenius J. Anal. Chem.*, *370*: 587 (2001).
16. E. M. Krupp, C. Pécheyran, S. Meffan-Main, and O. F. X. Donard, *Fresenius J. Anal. Chem.*, *370*: 573 (2001).
17. E. M. Krupp, C. Pécheyran, H. Pinaly, M. Motelica-Heino, D. Koller, S. M. M. Young, I. B. Brenner, and O. F. X. Donard, *Spectrochim. Acta*, *56B*: 1233 (2001).
18. R. Lobinski, I. R. Pereiro, H. Chassaigne, A. Wasik, and J. Szpunar, *J. Anal. At. Spectrom.*, *13*: 859 (1998).
19. J. Szpunar, B. Bouyssiere, and R. Lobinski, in *Elemental Speciation—New Approaches for trace element analysis* (J. A. Caruso, K. L. Sutton, and K. L. Ackley, Eds.), Elsevier, Amsterdam, 2000, p. 7.
20. B. Bouyssiere, J. Szpunar, M. Potin Gautier, and R. Lobinski, in *Handbook of Elemental Speciation* (R. Cornelis, Ed.), 2002.
21. T. De Smaele, L. Moens, R. Dams, P. Sandra, J. Van der Eycken, and J. Vandyck, *J. Chromatogr. A.*, *793*: 99 (1998).
22. J. Vercauteren, A. De Meester, T. De Smaele, F. Vanhaecke, L. Moens, R. Dams, and P. Sandra, *J. Anal. At. Spectrom.*, *15*: 651 (2000).
23. L. Moens, T. De Smaele, R. Dams, P. Van Den Broek, and P. Sandra, *Anal. Chem.*, *69*: 1604 (1997).
24. Z. Mester, R. Sturgeon, and J. Pawliszyn, *Spectrochim. Acta*, *B 56*: 233 (2001).
25. T. De Smaele, L. Moens, R. Dams, and P. Sandra, in *Applied Solid Phase Microextraction* (J. Pawliszyn, Ed.), Royal Society of Chemistry, Cambridge, 1999, p. 296.
26. T. De Smaele, L. Moens, P. Sandra, and R. Dams, *Mikrochim. Acta*, *130*: 241 (1999).
27. S. Aguerre, G. Lespes, V. Desauziers, and M. Potin-Gautier, *J. Anal. At. Spectrom.*, *16*: 263 (2001).
28. Z. Mester, J. Lam, R. Sturgeon, and J. Pawliszyn, *J. Anal. At. Spectrom.*, *15*: 837 (2000).
29. J. Vercauteren, C. Pérès, C. Devos, P. Sandra, F. Vanhaecke, and L. Moens, *Anal. Chem.*, *73*: 1509 (2001).
30. A. Wasik, I. Rodriguez Pereiro, and R. Lobinski, *Spectrochim. Acta B*, *53*: 867 (1998).
31. A. Wasik, I. Rodriguez Pereiro, C. Dietz, J. Szpunar, and R. Lobinski, *Anal. Commun.*, *35*: 331 (1998).
32. I. Koch, J. Feldmann, J. Lintschinger, S. V. Serves, W. R.

Cullen, and K. J. Reimer, *Appl. Organomet. Chem.*, *12*: 129 (1998).

33. R. G. García Fernandez, M. M. Bayon, J. I. G. Alonso, and A. Sanz-Medel, *J. Mass Spectrom.*, *35*: 639 (2000).
34. M. Vazquez Pelaez, M. Montes Bayon, J. I. Garcia Alonso, and A. Sanz-Medel, *J. Anal. At. Spectrom.*, *15*: 639 (2000).
35. A. Wasik, R. Lobinski, and J. Namiesnik, *Inst. Sci. Techn.*, *29*: 393 (2001).
36. E. Baltussen, P. Sandra, F. David, and C. Cramers, *J. Microcolumn Sep.*, *11*: 737 (1999).
37. D. Amouroux, E. Tessier, C. Pecheyran, and O. F. X. Donard, *Anal. Chim. Acta*, *377*: 241 (1998).
38. C. Pecheyran, C. R. Quetel, F. M. Martin, and O. F. X. Donard, *Anal. Chem.*, *70*: 2639 (1998).
39. A. W. Kim, M. E. Foulkes, L. Ebdon, S. J. Hill, R. L. Patience, A. G. Barwise, and S. J. Rowland, *J. Anal. At. Spectrom.*, *7*: 1147 (1992).
40. A. Kim, S. Hill, L. Ebdon, and S. Rowland, *J. High. Res. Chromatogr.*, *15*: 665 (1992).
41. R. Lobinski, V. Sidelnikov, Y. Patrushev, I. Rodriguez, and A. Wasik, *Trends Anal. Chem.*, *18*: 449 (1999).
42. S. Slaets, F. Adams, I. R. Pereiro, and R. Lobinski, *J. Anal. At. Spectrom.*, *14*: 851 (1999).
43. I. Rodriguez, S. Mounicou, R. Lobinski, V. Sidelnikov, Y. Patrushev, and M. Yamanaka, *Anal. Chem.*, *71*: 4534 (1999).
44. S. Perez Mendez, M. Montes Bayon, E. Blanco Gonzalez, and A. Sanz-Medel, *J. Anal. At. Spectrom.*, *14*: 1333 (1999).
45. C. M. Tseng, D. Amouroux, I. D. Brindle, and O. F. X. Donard, *J. Environ. Mon.*, *2*: 603 (2000).
46. E. Segovia Garcia, J. I. Garcia Alonso, and A. Sanz Medel, *J. Mass Spectrom.*, *32*: 542 (1997).
47. L. Ebdon, E. H. Evans, W. G. Pretorius, and S. J. Rowland, *J. Anal. At. Spectrom.*, *9*: 939 (1994).
48. H. E. L. Armstrong, W. T. Corns, P. B. Stockwell, G. O'Connor, L. Ebdon, and E. H. Evans, *Anal. Chim. Acta*, *390*: 245 (1999).
49. T. De Smaele, P. Verrept, L. Moens, and R. Dams, *Spectrochim. Acta*, *50B*: 1409 (1995).
50. A. Prange and E. Jantzen, *J. Anal. At. Spectrom.*, *10*: 105 (1995).
51. M. Montes Bayon, M. Gutierrez Camblor, J. I. Garcia Alonso, and A. Sanz-Medel, *J. Anal. At. Spectrom.*, *14*: 1317 (1999).

52. I. A. Leal-Granadillo, J. I. G. Alonso, and A. Sanz-Medel, *Anal. Chim. Acta*, *423*: 21 (2000).
53. W. G. Pretorius, L. Ebdon, and S. J. Rowland, *J. Chromatogr.*, *646*: 369 (1993).
54. J. Poehlman, B. W. Pack, and G. M. Hieftje, *Am. Lab.*, *30*: 50C (1998).
55. B. W. P. Poehlman and G. M. Hieftje, *Int. Lab.*, *29*: 26 (1999).
56. C. R. Quétel, H. Tao, M. Tominaga, and A. Miyazaki, *ICP Inf. Newsl.*, *21*: 77 (1996).
57. H. Tao, T. Murakami, M. Tominaga, and A. Miyazaki, *J. Anal. At. Spectrom.*, *13*: 1085 (1998).
58. H. Tao, Ramaswamy Babu Rajendran, C. R. Quetel, T. Nakazato, M. Tominaga, and A. Milyazaki, *Anal. Chem.*, *71*: 4208 (1999).
59. R. B. Rajendran, H. Tao, Nakazato, T. Nakazato, and A. Miyazaki, *Analyst*, *125*: 1757 (2000).
60. Q. Tu, J. Qian, and W. Frech, *J. Anal. At. Spectrom.*, *15*: 1583 (2000).
61. G. R. Peters and D. Beauchemin, *J. Anal. At. Spectrom.*, *7*: 965 (1992).
62. G. R. Peters and D. Beauchemin, *Anal. Chem.*, *65*: 97 (1993).
63. J. Feldmann, R. Gruemping, and A. V. Hirner, *Fresenius J. Anal. Chem.*, *350*: 228 (1994).
64. T. Prohaska, M. Pfeffer, M. Tulipan, G. Stingeder, A. Mentler, and W. W. Wenzel, *Fresenius J. Anal. Chem.*, *364*: 467 (1999).
65. B. Bouyssiere, unpublished.
66. E. H. Evans and J. A. Caruso, *J. Anal. At. Spectrom.*, *8*: 427 (1993).
67. E. H. Evans, W. G. Pretorius, L. Ebdon, and S. J. Rowland, *Anal. Chem.*, *66*: 3400 (1994).
68. T. M. Castillano, J. J. Giglio, E. H. Evans, and J. A. Caruso, *J. Anal. At. Spectrom.*, *9*: 1335 (1994).
69. G. O'Connor, L. Ebdon, E. H. Evans, H. Ding, L. K. Olson, and J. A. Caruso, *J. Anal. At. Spectrom.*, *11*: 1151 (1996).
70. G. O'Connor, L. Ebdon, and E. H. Evans, *J. Anal. At. Spectrom.*, *12*: 1263 (1997).
71. G. O'Connor, L. Ebdon, and E. H. Evans, *J. Anal. At. Spectrom.*, *14*: 1303 (1999).
72. J. W. Waggoner, M. Belkin, K. L. Sutton, J. A. Caruso, and H. B. Fannin, *J. Anal. At. Spectrom.*, *13*: 879 (1998).
73. J. W. Waggoner, L. S. Milstein, M. Belkin, K. L. Sutton,

J. A. Caruso, and H. B. Fannin, *J. Anal. At. Spectrom.*, *15*: 13 (2000).

74. J. R. Baena, M. Gallego, M. Valcarcel, J. Leenaers, and F. C. Adams, *Anal. Chem.*, *73*: 3927 (2001).
75. J. Rodriguez-Fernandez, M. Montes-Bayon, R. Pereiro, and A. Sanz-Medel, *J. Anal. At. Spectrom.*, *16*: 1051 (2001).
76. J. F. Holland, C. G. Enke, J. Allison, J. T. Stilts, J. R. Pinkston, B. Newcome, and J. T. Watson, *Anal. Chem.*, *55*: 997 (1983).
77. N. Demuth and K. G. Heumann, *Anal. Chem.*, *73*: 4020 (2001).
78. J. R. Encinar, M. I. Monterde Villar, V. G. Santamaria, J. I. Garcia Alonso, and A. Sanz-Medel, *Anal. Chem.*, *73*: 3174 (2001).
79. H. Hintelmann, R. D. Evans, and J. Y. Villeneuve, *J. Anal. At. Spectrom.*, *10*: 619 (1995).
80. K. G. Heumann, S. M. Gallus, G. Rädlinger, and J. Vogl, *Spectromchim. Acta B*, *53*: 273 (1998).
81. S. M. Gallus and K. G. Heumann, *J. Anal. At. Spectrom.*, *11*: 887 (1996).
82. H. Hintelmann, R. Falter, G. Ilgen, and R. D. Evans, *Fresenius J. Anal. Chem.*, *358*: 363 (1997).
83. P. Quevauviller, F. Adams, J. Caruso, M. Coquery, R. Cornelis, O. F. X. Donard, L. Ebdon, M. Horvat, R. Lobinski, R. Marabito, H. Muntau, and M. Valcarcel, *Anal. Chem.* http://pubs.acs.org/journals/anchem/announcements/letter991202.htm (1999).
84. H. Hintelmann and R. D. Evans, *Fresenius J. Anal. Chem.*, *358*: 378 (1997).
85. J. Holz, J. Kreutzmann, R.-D. Wilken, and R. Falter, *Appl. Organomet. Chem.*, *13*: 789 (1999).
86. J. P. Snell, I. I. Stewart, R. E. Sturgeon, and W. Frech, *J. Anal. At. Spectrom.*, *15*: 1540 (2000).
87. J. R. Encinar, J. I. Garcia Alonso, and A. Sanz-Medel, *J. Anal. At. Spectrom.*, *15*: 1233 (2000).
88. A. V. Hirner, J. Feldmann, R. Goguel, S. Rapsomanikis, R. Fisher, and M. O. Andreae, *Appl. Organomet. Chem.*, *8*: 65 (1994).
89. A. V. Hirner, *GIT Fachz Lab.*, *39*: 524 (1995).
90. J. Feldmann and A. V. Hirner, *Int. J. Environ. Anal. Chem.*, *60*: 339 (1995).

91. J. Feldmann, T. Riechmann, and A. V. Hirner, *Fresenius J. Anal. Chem.*, *354*: 620 (1996).

92. E. B. Wickenheiser, K. Michalke, C. Drescher, A. V. Hirner, and R. Hensel, *Fresenius J. Anal. Chem.*, *362*: 498 (1998).

93. J. Feldmann and W. R. Cullen, *Environ. Sci. Technol.*, *31*: 2125 (1997).

94. J. Feldmann, *J. Anal. At. Spectrom.*, *12*: 1069 (1997).

95. P. Andrewes, W. R. Cullen, and E. Polishchuk, *Appl. Organomet. Chem.*, *13*: 659 (1999).

96. J. Feldmann, *J. Environ. Monit.*, *1*: 33 (1999).

97. J. Feldmann, E. M. Krupp, D. Glindemann, A. V. Hirner, and W. R. Cullen, *Appl. Organomet. Chem.*, *13*: 739 (1999).

98. K. Haas and J. Feldmann, *Anal. Chem.*, *72*: 4205 (2000).

99. U. M. Grüter, J. Kresimon, and A. V. Hirner, *Fresenieus J. Anal. Chem.*, *368*: 67 (2000).

100. A. V. Hirner, J. Feldmann, E. Krupp, R. Grümping, R. Goguel, and W. R. Cullen, *Org. Geochem.*, *29*: 1765 (1998).

101. A. V. Hirner, E. Krupp, F. Schulza, M. Koziolb, and W. Hofmeisterb, *J. Geochem. Explor.*, *64*: 133 (1998).

102. T. De Smaele, L. Moens, R. Dams, and P. Sandra, *Fresenius J. Anal. Chem.*, *355*: 778 (1996).

103. M. Heisterkamp, T. De Smaele, J.-P. Candelone, L. Moens, R. Dams, and F. C. Adams, *J. Anal. At. Spectrom.*, *12*: 1077 (1997).

104. E. Jantzen and A. Prange, *Fresenius J. Anal. Chem.*, *353*: 28 (1995).

105. A. V. Hirner, U. M. Grüter, and J. Kresimon, *Fresenius J. Anal. Chem.*, *368*: 263 (2000).

106. E. M. Krupp, R. Grümping, U. R. R. Furchtbar, and A. V. Hirner, *Fresenius J. Anal. Chem.*, *354*: 546 (1996).

107. B. Bouyssiere, F. Baco, F. Savary, and R. Lobinski, *Oil Gas Sci. Technol.*, *55*: 639 (2000).

108. B. Bouyssiere, F. Baco, F. Savary, H. Garraud, D. Gallup, and R. Lobinski, *J. Anal. At. Spectrom.*, *16*: 1329 (2001).

4

GC-MS Analysis of Halocarbons in the Environment

Filippo Mangani, Michela Maione, and Pierangela Palma *Institute of Chemical Sciences, University of Urbino, Urbino, Italy*

I. INTRODUCTION

A. Halocarbons

The broad family of halocarbons includes chemical compounds that, in addition to carbon and hydrogen, contain fluorine, bromine, chlorine, or iodine. This group of compounds includes substances of both biogenic and anthropogenic origin that are ubiquitous trace constituents of the oceans and of the atmosphere. They are involved in a number of atmospheric reactions, among which the destruction of stratospheric ozone is the most remarkable.

Chlorofluorocarbons and Hydrogenated Substitutes

Chlorofluorocarbons (CFCs) started to draw the attention of the scientific, industrial, and political world soon after the 1974 appearance of Rowland and Molina's theory on stratospheric ozone depletion [1,2]. CFCs can be considered as derivatives of the simpler hydrocarbons (methane, CH_4, and ethane, C_2H_6) obtained by substitution of the hydrogen atoms with chlorine and fluorine atoms. These compounds are also known by the trade name of Freons, and they are named with the initials CFC or F followed by a three-digit number or a two-digit number, if the first one is zero. The first digit indicates the number of carbon atoms minus one, the second digit indicates the number of hydrogen atoms plus one, and the third one refers to the number of fluorine atoms. Therefore, CFC-11 and CFC-113 stand for $CFCl_3$ (trichlorofluoromethane) and $C_2F_3Cl_3$ (trichlorotrifluoromethane), respectively.

CFCs were first developed in the 1930s to replace ammonia (flammable and toxic) and sulfur dioxide (toxic) refrigerants. In fact, because of the high chemical stability of CFCs, they are noncorrosive, nonexplosive, nonflammable, and very low in toxicity and thus promote worker and consumer safety. These characteristics, coupled with several peculiar physical properties, such as low vapor phase thermal conductivity, convenient vapor pressure and temperature characteristics, desirable solubility properties, compatibility with many construction materials, and high stability, have led to the widespread use of CFCs in consumer product.

Main uses of CFCs are in the refrigeration of perishable foods and medical supplies, as aerosol propellants, in air conditioning (house, office, cars), in electronic and mechanical components cleaning, and in the construction of plastic insulating foams.

Estimated world production rose from approximately 181 Gg year^{-1} in 1960 to 905 Gg year^{-1} in 1974 while the world consumption by applications reached in the same year 916 Gg year^{-1}. Due to the ability of CFCs to deplete the ozone layer, they are currently regulated under the Montreal Protocol [3], and their global production has dramatically declined since 1990, resulting in substantial reductions in their annual rate of increase [4–9].

Furthermore, CFCs are characterized by high global warming potential (GWP) values, being able to absorb IR radiation emitted by the Earth's surface within the atmospheric windows [10,11]. However, the direct warming effect of CFCs may be offset by indirect cooling effects caused by their destruction of stratospheric ozone, which is itself a greenhouse gas [12]. Furthermore, stratospheric ozone depletion may also exert an indirect cooling effect via its effect on the oxidation state of the atmosphere [13]. However, net GWPs of CFCs tend to be positive.

More recently, CFCs containing hydrogen atoms have been grouped in hydrochlorofluorocarbons (HCFCs) and hydrofluorocarbons (HFCs). These compounds are now in production or under development to replace CFCs.

Hydrochlorofluorocarbons (HCFCs) are essentially CFCs that include one or more hydrogen atoms. The presence of hydrogen makes the resulting compounds susceptible to photodecomposition and to attack by OH radicals, ubiquitous in the troposphere. As a consequence, HCFCs have much shorter atmospheric lifetimes than CFCs and consequently are less likely to migrate to the stratosphere where they would destroy ozone. The Copenhagen Amendments to the Montreal Protocol [14] placed HCFCs under control, with elimination of HCFC-22 planned by 2020 and all others by 2030. HCFCs still have high GWPs, and these are aggravated by the fact that they have weaker indirect cooling effects than CFCs.

HFCs contain only carbon, hydrogen, and fluorine, and therefore they do not destroy ozone. Consequently, they are a desirable CFC replacement. Ironically, the characteristic that makes HFCs a desirable replacement from an ozone perspective also makes them potent greenhouse gases with no indirect cooling effects.

Bromofluorocarbons

Bromofluorocarbons are similar to CFCs except that they contain at least one bromine atom. They are usually classified as halons and the initial H is followed by four digits indicating the number of carbon, fluorine, chlorine, and bromine atoms, respectively. Thus H-1301 and H-2402 stand for $CBrF_3$ and $CBrF_2CBrF_2$, respectively. The halons are used as fire-extinguishing agents, both in built-in systems and in portable fire extinguishers. Halons are particularly destructive to stratospheric ozone, and consequently they are regulated under the Montreal Protocol. Their indirect cooling effect, related to the ability to destroy ozone, reduces their net GWPs, which tend to be negative. Emissions of halons has always been low, although the exact figure is uncertain. Atmospheric data show that growth rates of the halons have been declining since the early 1990s [15].

Methyl Halides

Methyl halides are those compounds consisting of one carbon atom, three hydrogen atoms, and either one chlorine, bromine, or iodine atom. They include methyl chloride (CH_3Cl), methyl bromide (CH_3Br), and methyl iodide (CH_3I) and are characterized by atmospheric lifetimes on the order of weeks. Unlike CFCs, which are only anthropogenic, methyl halides are formed through both man-made and natural sources. Man-made sources of methyl halides are the burning of biomass and coal and the use of fumigants in agriculture. Natural sources include salt marshes on the edge of the ocean and certain wood fungi on trees. However, only half of the methyl chloride and two-thirds of the methyl bromide sources are known. A recent study [16] gives direct evidence that some of the unknown sources of these halides could be plants, particularly rice paddies.

Perfluorinated Halocarbons

Perfluorinated halocarbons (perfluorocarbons, PFCs) are man-made compounds composed of carbon and fluorine. Among these, perfluoromethane (CF_4), perfluoroethane (C_2F_6), perfluoropropane (C_3F_8), and perfluorobutane (C_4F_{10}) are characterized by high global warming potentials (6,000 to 10,000, relative to that of carbon dioxide) and long atmospheric lifetimes (up to 50,000 years). Since PFCs do not contain chlorine and bromine atoms, they do not deplete the ozone layer. Therefore, they do not possess the indirect cooling effects of CFCs. As a result, they are unambiguously greenhouse gases. Principal source of PFCs is as a by-product of aluminum smelting. They are also used as cleaning agents in semiconductor manufacturing and as replacement for CFCs and halons. Small amounts are produced during uranium enrichment processes. Emission of these compounds are not regulated and are difficult to assess. However, PFCs are likely to increase because of the great expansion of the semiconductor industry. Recently the U.S. Environmental Protection Agency (EPA) launched the *PFC Emissions Reduction Partnership for the Semiconductor Industry*, a voluntary, nonregulatory program that supports the industry's efforts to reduce PFC emissions [17].

Halogenated Solvents

This group of man-made compounds includes methylene chloride (CH_2Cl_2), chloroform ($CHCl_3$), methylchloroform (CH_3CCl_3), carbon tetrachloride (CCl_4), trichloroethene ($CHCl{=}CCl_2$), and perchloroethylene ($CCl_2{=}CCl_2$).

Methylene chloride is used in foam blowing, paint and varnish remover, plastics processing, aerosols, metal degreasing and cleaning, and as a process solvent. It is a weak greenhouse gas, and its relatively short atmospheric lifetime, of less than one year, probably prevents it from reaching the stratosphere where it would be damaging to ozone. Being a potential carcinogen, methylene chloride emissions are regulated and the compound is listed in the EPA's Toxics Release Inventory (TRI).

Chloroform is used primarily as a feedstock for HCFC-22, with secondary use as a solvent. Its environmental impact lies in that it is a weak greenhouse gas. As a carcinogen, chloroform is reported to the EPA's TRI.

Methylchloroform is a volatile liquid insoluble in water, formerly used as a solvent or degreaser in industry. It is a rather stable compound, with an atmospheric lifetime of about 5 years. It is regulated under the Montreal Protocol as an ozone-depleting chemical. It has indirect cooling effects and a rather low GWP. Consequently, its net effect is likely to promote global cooling [18]. Emissions have declined rapidly in past years to negligible levels.

Carbon tetrachloride was widely used as a raw material in many industrial uses, including the production of CFCs, and as a solvent. Solvent use ended when it was discovered to be carcinogenic, being regulated by the Clean Air Act Amendments. Carbon tetrachloride is also regulated under the Montreal Protocol as an ozone-depleting chemical. Carbon tetrachloride has a high direct GWP (1,400), which is offset by its indirect cooling effect of destroying ozone. Its net effect favors global cooling [19]. Its production ceased in January 1996; meanwhile emissions started to decline at the beginning of the 1990s.

Trichloroethene's greatest use is to remove grease from fabricated metal parts and from textiles. Major releases to atmosphere are due to emissions from metal degreasing plants. Since being introduced to the dry-cleaning industry in the late 1930s, trichloroethylene has replaced most other solvents because of its relatively low toxicity and nonflammability. Its other major uses are as a metal cleaning and degreasing solvent, as a solvent in automotive aerosols, and as a chemical intermediate in the production of several fluorinated compounds.

Multihalogen Compounds

Multihalogen compounds, such as dibromomethane, tribromomethane (bromoform), bromodichloromethane, bromochloromethane, chloroiodomethane, and so on, are primarily released into the atmosphere by biogenic sources as a result of photosynthetic processes [20–22]. They have atmospheric lifetimes on the order of several weeks, and their tropospheric concentrations are at parts per trillion (ppt) levels [23].

In water, bromodichloromethane and tribromomethane are mostly formed as a by-product when chlorine is added to drinking water to kill disease-causing organisms. Small amounts are also made in chemical plants for use in laboratories or in making other chemicals. A very small amount (less than 1% of the amount coming from human activities) is formed by algae in the ocean.

B. The Environmental Issue

Stratospheric Ozone Depletion

The chemical stability of some halocarbons is at the base of rising environmental concerns, among which is stratospheric ozone depletion.

Mainly responsible for the stratospheric ozone destruction are the CFCs, which are of exclusively anthropogenic origin and are destroyed only in the stratosphere. However, there are other brominated and chlorinated compounds able to deliver chlorine and bromine in the stratosphere, where bromine is about 50 times more efficient in depleting ozone than chlorine. It was assessed that the synergistic effect of chlorine and bromine species accounts for approximately 20% of the polar stratospheric ozone depletion. Disregarding the chlorofluorocarbons, methyl chloride and methyl bromide are believed to be the largest single contributors of halogens to the atmosphere. It is possible however, that other rather short-living chlorinated and brominated compounds could reach the stratosphere as well.

As a matter of fact, for the man-made CFCs and Halons, there are no known mechanism for the destruction of CFCs in the troposphere, where pollutant removal is generally provided. Thus, these compounds would remain in the troposphere until, as postulated by Rowland and Molina [1,2], they would enter into stratosphere and would be photolyzed by solar ultraviolet (UV) radiation (at wavelengths shorter than 220 nm) at 2530 km altitude. Photolysis would lead to the release of chlorine atoms (1) which, in turn would lead, through a chain reaction, to the destruction of ozone (O_3). The fundamental set of reactions is the following:

$$CCl_3F \xrightarrow{\text{UV light}} Cl + CCl_2F \tag{1}$$

$$CCl_2F_2 \xrightarrow{\text{UV light}} Cl + CClF_2 \tag{2}$$

$$Cl + O_3 \longrightarrow ClO + O_2 \tag{3}$$

$$\underline{ClO + O \longrightarrow Cl + O_2} \tag{4}$$

$$O_3 + O \longrightarrow O_2 + O_2 \tag{5}$$

The chain reaction (3) and (4) transform one ozone molecule and one oxygen atom into two O_2 molecules while the Cl atom remains unchanged and can start the process again.

The ClO_x cycle is interconnected with the concentration of other species present in the stratosphere such as the nitrogen oxides (NO_x cycle), the hydrogen oxides (HO_x cycle), and methane, as shown in Fig. 1.

The free radical chain reaction can be terminated by reactions leading to temporary reservoirs such as HCl, HOCl, and $ClONO_2$ from which free radicals can be regenerated through other routes. Since O_3 absorbs most of the "dangerous" solar UV radiation in the 280–310 nm region (UVb), a net decrease in the total ozone column causes an increase of the UVb radiation reaching the Earth's surface. Although many uncertainties remain regarding potential effects, increases in UVb could lead to adverse effects on plants, ani-

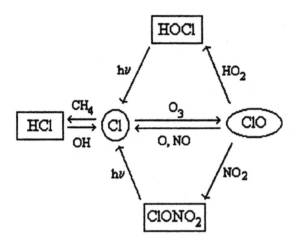

Fig. 1 Schematic representation of the key reactions and species in the stratospheric chemistry of chlorine. (From Ref. 24.)

mals, and humans (increase of erythema, skin cancer cases, etc.). There were insufficient data to test the theory at the time it was proposed; however, under the lead of government agencies, scientists from academia, government, and industries undertook the enormous task of developing the knowledge of atmospheric chemistry to predict future (long-term) depletion of the stratospheric ozone layer.

The United Nations Environmental Programme (UNEP) launched in 1977 the "Coordinating Committee on the Ozone Layer" with the goal of preparing periodic scientific assessments. The Vienna Convention for the Protection of the Ozone Layer, designed so that protocols could be added requiring specific control measures, was adopted in March 1985 [25].

A breakthrough in the CFCs-ozone controversy was represented by the discovery in 1985 [26] of massive springtime losses of ozone over the Halley Bay station, in Antarctica, since the mid- to late 1970s. This phenomenon, known as "ozone hole" was confirmed by Total Ozone Mapping Spectrometer (TOMS) measurements on the Nimbus-7 satellite and by balloon sonda measurements.

It was soon established, by means of three polar expeditions in 1986–87 and of a high-flying aircraft mission in 1987 that the ozone hole was caused by the presence of CFCs. In fact, the peculiar conditions of the lower Antarctic stratosphere, characterized by winter temperature drops below $-80°C$ and consequent formation of polar stratospheric clouds, not normally found in the stratosphere of the Temperate Zone, allow heterogeneous reactions [Eqs. (6) and (7)] involving HCl and H_2O, which may transform $ClONO_2$, product of one of the main removal processes of the free radical ClO (see Fig. 1), into HNO_3.

$$HCl + ClONO_2 \longrightarrow Cl_2 + HNO_3 \tag{6}$$

$$H_2O + ClONO_2 \longrightarrow HOCl + HNO_3 \tag{7}$$

Since the HNO_3 from the reactions above remains in the clouds, a denitrification process of the air mass, leaving a consequent low residual concentration of NO and NO_2, takes place.

Thus, ClO concentration rises to ppbv range and a significant reaction with other ClO radicals takes place [Eq. (8)]. At springtime (end of August-October in the Southern hemisphere), when solar radiation is again available, the dimer (ClOOCl) releases two chlorine atoms in reactions (9) and (10).

$$ClO + ClO + M \longrightarrow ClOOCl + M \tag{8}$$

$$ClOOCl \xrightarrow{\text{light}} Cl + ClOO \tag{9}$$

$$ClOO + M \longrightarrow Cl + O_2 + M \tag{10}$$

$$O_3 + O_3 \longrightarrow O_2 + O_2 + O_2 \tag{11}$$

The net result is the transformation of two O_3 molecules into three O_2 molecules [Eq. (11)] without the need for the O atom step of Eq. (4) required in the stratosphere at tropical to northern latitudes. It should be noted that bromine atoms released from halons and methyl bromide would lead to the same results and actually, bromine is 40–100 times more efficient than chlorine, on a atom-per-atom basis, in destroying stratospheric ozone.

The simultaneous revealing of the Antarctic ozone hole and of appreciable ozone decreases in the Northern Hemisphere drew again the attention of the public, and the Montreal Protocol, ratified by 27 nations on September 1987 and effective on January 1, 1989, required developed countries to reduce their consumption of CFCs to 50% of 1986 rates over a 10-year period [3]. The Montreal Protocol called also for periodic assessments and amendments to the controls

Table 1 List of Halocarbons Detected in the Atmosphere, 1970–2000

Species	Chemical formula	B. P. (°C)	$\overline{X}$ (ppt) (2000)	Lifetime (years)	ODP	Main sources
CFC-11	CCl_3F	23.8	265	50	1.00	Refrigerant, aerosol propellant, foam-blowing agent
CFC-12	CCl_2F_2	−29.8	540	102	0.86	Refrigerant, aerosol propellant, foam-blowing agent
CFC-113	CCl_2FCClF_2	47.7	75	85	0.90	Solvent
CFC-114	$CClF_2CClF_2$	3.7	18	300	0.85	Solvent
CFC-115	CF_3CClF_2	−38.7	~4**	1700	0.40	Propellant, refrigerant
CFC-116	CF_3CF_3	−78.2	~4**	1×10^4		Aluminum production
Methyl chloride Chloromethane	CH_3Cl	−24.2	600	1.5	0.02	Natural, biomass burning
Carbon tetrachloride	CCl_4	76.3	100	42	1.20	Solvent and CFC feedstock
Methyl chloroform 1,1,1-Trichloroethane	CH_3CCl_3	74.0	50	4.9	0.12	Solvent
Trichloromethane Chloroform	$CHCl_3$	61.2	~10*	0.55		HCFC feedstock
Methylene chloride Dichloromethane	CH_2Cl_2	40.0	~35*	0.41		Solvent

Compound	Formula					End use
Trichloroethene	$CHCCl_3$	87.0	~10*			Solvent
Trichloroethylene						
Tetrachloroethene	C_2Cl_4	121.0	~160	0.35		Solvent
Perchloroethylene						
Methyl iodide	CH_3I	42.4	~2*			Natural
Iodomethane						
H-1211	CF_2ClBr	−4.0	4	20	12.5	Fire retardant
H-1301	CF_3Br	−57.8	2	65	0.60	Fire retardant
Methyl bromide	CH_3Br	3.6	8–15	1.7–2.0	0.3–0.8	Natural 60%, fumigant 40%
Bromomethane						
Dibromomethane	CH_2Br_2	1.8	1–8*	0.40	0.17	Natural
Tribromomethane	$CHBr_3$		2–3*			Natural
Dibromoethane	$C_2H_4Br_2$		0.2			Natural
Dibromochloromethane	$CHClBr_2$		1*			Natural
Bromodichloromethane	$CHCl_2Br$		1–2*			Natural
Bromochloromethane	CH_2ClBr		140			
HCFC-22	CHF_2Cl	−40.8		12.1	0.045	Refrigerant, foam-blowing agent
HCFC-141b	CH_3CFCl_2	32.3	12	9.4	0.1	Foam-blowing agent
HCFC-142b	CH_3CF_2Cl	−9.8	12	18.4	0.056	Foam-blowing agent
HCFC-123	$CHCl_2CF_3$	28.0	1.2	1.4	0.017	Refrigerant
HCFC-124	$CHFClCF_3$	−11.8	1.5		0.030	Sterilant
HFC-23	CHF_3	−82.2	n.d.	264	$<4.0 \times 10^{-4}$	HCFC by-product
HFC-125	CHF_2CF_3	−48.4	3	32.6	$<3.0 \times 10^{-5}$	Specialized end use
HFC-134a	CH_2FCF_3	−25.9	18	14.6	$<5.0 \times 10^{-4}$	Refrigerant

and the 1990 London Amendment moved further to a total phase-out of CFCs by the year 2000. The 1992 Copenhagen Amendment brought the deadline for phase-out to January 1, 1996, and included compounds other than CFCs, such as CH_3CCl_3, CCl_4, and the halons $CBrF_3$, $CBrClF_2$, and $CBrF_2CBrF_2$ [14].

Table 1 reports an expanded list of halocarbons that have been detected in the atmosphere in the period 1970–2000. Average atmospheric (tropospheric) concentrations defined also as tropospheric mixing ratios in part per trillion (ppt = 10^{-12} v/v) are reported together with the main emission sources. Lifetime, and the Ozone Depletion Potential (ODP), defined as the steady-state depletion of ozone by unit mass emission of a given species relative to the same steady-state ozone reaction by unit mass emission of CFC-11, are reported as well.

Global Warming

Global warming can be defined as a phenomenon by which the average temperature on Earth is gradually increasing over its level in recent history. This is attributed to the increased concentration in the atmosphere of the so-called greenhouse gases (GHGs) that are able to absorb long-wave radiation emitted by the Earth's surface and reradiate it toward Earth. A measure of this ability to absorb the long-wave radiation is given by the global warming potential (GWP)—i.e., the ratio of global warming, or radiative forcing, from one unit mass of a greenhouse gas, relative to that of one unit mass of carbon dioxide, over a period of time. Global warming should not be mistaken with the natural "greenhouse effect," a process that is essential for life on Earth, and which is due to naturally occurring greenhouse gases (i.e., carbon dioxide, water vapor, methane, ozone, nitrous oxide, and other trace gases). The cause for concern is not the natural greenhouse effect, but the enhancement of this effect as a result of human activities, such as the burning of fossil fuels, large-scale fertilizer use, cattle production, and deforestation, which led to an increase of the amount of the naturally occurring GHGs in the atmosphere above normal levels. In addition, human activities led to the atmospheric emission of totally anthropogenic gases, characterized by extremely high GWP values such as CFCs, HCFCs, HFCs, halons, and PFCs [26,27].

High GWP values of the above-mentioned halocarbons are due not only to radiative properties of the different substances but also

Table 2 Global Warming Potentials of Selected Halocarbons Relative to CO_2, at 100 Years Time Horizon

Compound	GWP (relative to CO_2)	Compound	GWP (relative to CO_2)
CFC-11	1,320	HFC-23	11,700
CFC-12	6,650	HFC-125	2,800
CFC-113	9,300	HFC-134a	1,300
CFC-114	9,300	HFC-152a	140
CFC-115	9,300	HFC-227ea	2,900
HCFC-22	1,350	CF_4	6,900
HCFC-123	93	C_2F_6	9,200
HCFC-124	480	CCl_4	− 1,550
HCFC-141b	270	CH_3CCl_3	−700
HCFC-142b	1,650	CH_3Cl	9
Halon-1301	−31,400	$CHCl_3$	5

to the time-scale characterizing the removal of the substance from the atmosphere. Radiative properties control the absorption of radiation per kilogram of gas present at any instant, but the lifetime controls how long an emitted amount is retained in the atmosphere and hence is able to influence the thermal budget. HCs are able to absorb radiation strongly in the 8 to 13 μm atmospheric window region, i.e., the region of the spectrum where the natural substances in the Earth's atmosphere do not absorb. Furthermore, they are characterized by atmospheric lifetimes ranging from a few decades up to hundreds of years. However, the direct warming effect of some HCs, like CFCs, halons, CH_3CCl_3, and CCl_4, may be offset by the indirect cooling effects caused by their destruction of stratospheric ozone and by their indirect cooling effect (see Section I.A). GWP values (relative to CO_2) of selected halocarbons are reported in Table 2 [26].

Water Contamination

The occurrence of HCs in drinking water is due to (1) chlorine disinfection of drinking water supply, and (2) the widespread use of HCs as industrial solvents, which resulted in the contamination of rivers, lakes, and underwater water supplies. Major concerns are associated with their potential toxicity towards humans and animals.

   1. Trihalomethanes (THMs)—i.e., chloroform, bromodichloromethane, dibromochloromethane, and bromoform—are the predom-

Table 3 Chlorinated Solvents Regulated under the Safe Dinking Water Act

Compound	Maximum contaminant level (mg L^{-1})	Potential health effects from ingestion of water	Sources of contaminant in drinking water
Carbon tetrachloride	0.005	Liver problems; increased risk of cancer	Discharge from chemical plants and other industrial activities
1,2-Dichloroethane	0.005	Increased risk of cancer	Discharge from industrial chemical factories
cis-1,2-Dichloroethylene	0.07	Liver problems	Discharge from industrial chemical factories
trans-1,2-Dichloroethylene	0.1	Liver problems	Discharge from industrial chemical factories
Dichloromethane	0.005	Liver problems; increased risk of cancer	Discharge from pharmaceutical and chemical factories
Ethylene dibromide	0.00005	Stomach problems; reproductive difficulties; increased risk of cancer	Discharge from petroleum refineries
Tetrachloroethylene	0.005	Liver problems; increased risk of cancer	Discharge from factories and dry cleaners
1,1,2-Trichloroethane	0.005	Liver, kidney, or immune system problems	Discharge from industrial chemical factories
Trichloroethylene	0.005	Liver problems; increased risk of cancer	Discharge from petroleum refineries

inant chlorine disinfection by-products, accounting for about 10% of the total organic halogen compounds formed by water chlorination (the other being chlorinated acetic acids, halogenated acetonitriles, chloral hydrate, chlorophenols) [28]. Precursors of these halogenated compounds include natural humic and fulvic compounds and algal material. Some epidemiological studies suggest that there may be an association between exposure to trihalomethanes in drinking water and increased frequencies of cancers of the stomach, colon, rectum, or pancreas. However, these studies do not provide information whether the observed effects are due to THMs or to one or more of the other byproducts also present in chlorinated water [29].

2. Chlorinated solvents were developed, as stated above, as cleaning solutions for removing grease and carbon buildup from metal parts. For over 40 years they were widely used by industry for a variety of cleaning tasks. These solvents have created significant groundwater contamination, even although the most volatile species when released to water will primarily evaporate and have little potential for accumulating in water bodies. Table 3 reports a list of selected HCs regulated by the U.S. EPA under the Safe Drinking Water Act (SDWA), the main federal law regulating the quality of drinking water in the United States.

II. THE ANALYTICAL PROBLEM

Gas chromatography coupled to mass spectrometry (GC-MS) is the chosen analytical technique for the determination of HCs in environmental samples.

Because of the stratospheric ozone issue, HCs have been determined mostly in air samples, though measurements in water samples have been carried out as well, especially for biogenic compounds. The methods for water analysis differ only for the sample collection and the treatment step—the analytical step may be identical.

It should be pointed out that the necessity for an accurate determination of "background" HC tropospheric concentrations stems from the need to monitor the annual and long-term rates of accumulation of HCs in the atmosphere and to determine their atmospheric lifetimes. Atmospheric lifetimes are critical parameters for estimating how long HCs remain in the atmosphere after the emission step and for calculating the ODPs and GWPs. Annual and cumulative releases derived from the measurements can be used to control the production and release estimates of these compounds, and to assess country conformities to the Montreal Protocol and relative amendments. On the other hand, measurements in the upper troposphere and in the lower stratosphere (vertical profiles) provide a powerful tool for testing photochemical and diffusion models.

Determination of HCs in the oceans is of special concern because the oceanic flux is still an unknown factor in the atmospheric budgets of halogens. The production and the rate of loss of halocarbons in the oceans governs the concentration of the volatile halogenated compounds in seawater and thereby the flux of these compounds between the ocean and the atmosphere. The magnitude of the ocean flux could be sensitive to global climate change because both temperature and marine productivity are sensitive to the climate changes. Therefore, it is important to have specific knowledge of the formation and removal of compounds of biological origin in order to be able to make global halogen budget calculations.

Furthermore, the determination of anthropogenic halocarbons has become an important issue after the discovery that water springs and wells might be contaminated by industrial and civil activities in industrialized countries. This, in many cases, may be due to water chlorination disinfection.

Like any other method devoted to the determination of trace compounds, HC trace analysis methods require the careful evaluation of the following steps:

Sample treatment

GC-MS analysis of samples

Preparation and analysis of standards (calibration)

In HC determination, as in most analytical problems connected to pollution evaluation, the concentration of the contaminants is much too low to allow direct injection of the sample. Thus, preconcentration techniques are necessary in order to chromatographically analyze samples in which the concentration of the single components is in the range of parts per trillion.

Other major difficulties encountered in estimating global average concentrations of HCs arise from the variability in the monitoring data. At any given site, such variability may be due to the presence or proximity of local sources and relative seasonal variations, presence of interferences, and instrumental calibration problems.

A. Atmospheric Sample Treatment

Sampling and enrichment steps are perhaps the most critical points of the analytical chain and a correct use of the techniques involved implies a thorough knowledge and an intelligent evaluation of the various factors involved, in order to inject a representative sample into the chromatographic column.

For sampling and enrichment of HCs from atmospheric samples, several systems have been used, which fall in two broad categories: grab sampling in suitable containers, followed by sample enrichment prior to the chromatographic analysis; and sampling in adsorption tubes filled with suitable adsorbents. This second procedure allows one to simultaneously perform sampling and preconcentration of the atmospheric sample.

Grab Sampling

This is the easiest way to collect air and allows one to carrying out all other steps of the measurement procedure in the laboratory. It can be performed by simply opening and then closing the valve of a previously evacuated stainless steel container [30–41]. The container may also be filled with an inert gas (i.e., helium) after evacuation. It has been demonstrated that HCs are not stable if the inner surface of the stainless steel canister has not been passivated with some kind of process. One of the most popular is the stainless steel canister electropolished with the SUMMA® treatment, used by several authors and in the Atmospheric Lifetime Experiment (ALE) [42], an experiment designed in the 1980s, aimed at determining accurately the atmospheric concentrations of CFC-12, CFC-11, CCl_4, and CH_3CCl_3 [43].

Such canisters are equipped with metal bellows valves (all stainless steel Nupro® valves) since it is essential to avoid contact of the air sampling with polymeric seals, gaskets, and so forth during storage. Canisters are available in different volumes (0.85, 3.2, 15, and 32 L), according to the different sampling requirements. Fig. 2 reports a schematic diagram of a 850-mL SUMMA canister.

These canisters are commercially available, but several researchers prepare their own container by using a pretreatment and cleaning procedure that is, in any case, based on an electropolishing process followed by chemical or ultrasonic washing and vacuum baking (for detail, see Refs. 33, 36, 38, 44). Aculife®-treated (silanized) aluminum cylinders later became available and had the advantage

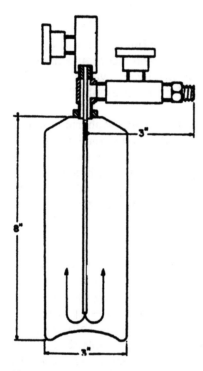

Fig. 2 Schematic diagram of a 850-mL passivated SUMMA® canister.

of being much lighter. Aculife® chemical vapor deposition treatment provides an internally inert surface that prevents HCs from reacting with the aluminum. In any case, both types of cylinder are suitable for sampling and storage of HCs.

The main disadvantage of this procedure is the limited sample volume, because these containers should be small and light to be easily transported in rather remote locations or on board ships or aircraft. Such a limitation, which does not allow the measurement of less detectable compounds, can be partially overcome by pressurizing the sample above ambient pressure. Care should be taken in the choice of the pump because it has been reported that common metal bellows pump may cause a slight contamination with some halocarbons in the parts per trillion range, probably because of the Teflon seal of the pump valves [35]. All-metal bellows pumps or diaphragm pumps [37] should then be used and air sample can be easily

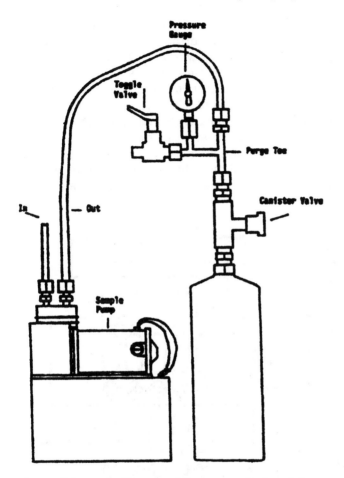

Fig. 3 Schematic diagram of the pressurized canister sampling setup.

pressurized up to 3–4 10^3 hPa. Fig. 3 reports a schematic diagram of the pump-canister setup.

A grab sampling system made of 14 stainless-steel 1.6-L canisters was utilized aboard an ER-2 aircraft for collecting whole air samples at altitudes spanning from 9 to 21 Km [45] in the Arctic atmosphere. The ER-2 aircraft carries no crew, so the system must be operated automatically, with minimal control by the pilot. A four-stage metal bellows pump was used to pressurize the canisters to 3 10^5 Pa. The stainless steel inlet port, located above the upper surface of the nose of the aircraft, was designed to remove most of the water or ice from the airflow to the pump. Furthermore, the pump was not started by the pilot until the aircraft was above tropospheric clouds (ca. 9 km). A schematic diagram of the system is depicted in Fig. 4.

Samples were then analyzed in two stages for detecting CH_4, CO, N_2O, CFC-12, CFC-11, CFC-113, CH_3CCl_3, and CCl_4, in order to collect valuable information aimed at evaluating motions and history of air masses in the Antarctic atmosphere.

Alternatively, "cryogenic sampling" can be carried out [42]. This procedure allows the condensation of large volumes of air in relatively small containers, because the partial pressure of air at the

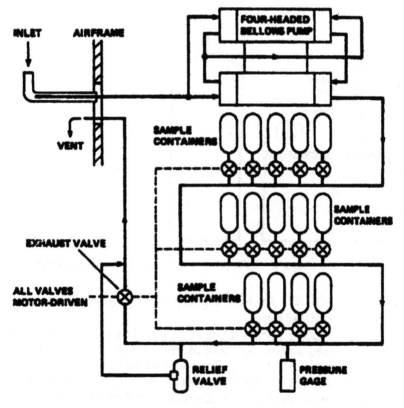

Fig. 4 Schematic diagram of ER-2 whole-air sampling system. The all stainless-steel system contains fourteen 1.6-L canisters. (From Ref. 45.)

temperature of liquid nitrogen is about one half of the atmospheric pressure (at ground level). The mechanical stability of the container is the only limitation to the maximum pressure. However, consistent amounts of liquid nitrogen are required, and this causes severe logistics problems for remote areas and on-board sampling.

Adsorption Tubes

The other way of collecting air samples is to accumulate halocarbons on adsorbent-filled cartridges, through which air is passed by means of a pump or a large volume syringe. Cartridges are small, lightweight, and inexpensive if compared with stainless steel cylinders, and can be reactivated and used several times. The main theoretical advantage is that the compounds of interest can be enriched "ad hoc," so that very low trace compounds can then be determined. To exploit the full potential of this sampling mode, several experimental factors must be evaluated, such as cartridge capacity, type of adsorbent, relative retention capability of the adsorbent (which is generally expressed by the parameter called break-through volume, or BTV), desorption temperature and volume, blank levels, limit of detection and detector range of linearity (determined by the compound and actual atmosphere concentration, and by the detector used). Two major inconveniences are the possible formation of artifacts during the desorption step, usually carried out by flash-heating at temperatures above room temperature, and the contamination from storage and from laboratory air during transfer to the thermal-desorbing/analyzing apparatus.

Several type of adsorbents have been evaluated, with two different purposes, which can be summarized as follows:

"Off line" sampling (i.e., remote areas) with traps returning to the laboratory for GC analysis. These methods can also be used for enriching grab samples just prior to GC analysis.

"On line" enrichment (using micro-traps), also suitable in automated sampling/analyzing apparati for *in situ* measurements.

The adsorbents used for sampling and pre-concentration of organic air pollutants, like HCs, are divided into two broad categories, according to their retention properties: weak or strong adsorbents. A weak adsorbent does not retain the most volatile HCs to a significant extent at room temperature, while under the same conditions it retains heavier compounds. A strong adsorbent retains even the most volatile compounds.

As stated above, one of the most critical parameters to be considered is the BTV,—the maximum air volume that can be sampled using a given quantity of the adsorbent, before the compound exists the trap. BTVs depend on many factors, among which temperature in crucial. Another very important factor that influences the BTV is the simultaneous presence of more than one compound and their concentration in the trapped air.

The adsorbent of choice in this HCs determination must have the following features: it must have a high BTV, possibly even at room temperature, it must allow a complete recovering of the adsorbed compounds during the thermal desorption step, and it must not give any decomposition of the compounds of interest. As a matter of fact, when weak adsorbents are used, the two last conditions are accomplished. However, during the enrichment step, subambient temperatures must be used in order to obtain acceptable BTV values for the most volatile compounds. On the other hand, strong adsorbents often need excessively high temperatures to completely release the adsorbates.

For this reason, multilayer adsorption tubes are frequently employed in the simultaneous determination of compounds of different volatility and different concentration levels. Adsorbent materials are chosen in order to set up an adsorption tube that possibly allows one to perform the pre-concentration step at ambient temperature, in order to avoid problems related to the utilization of cryogenic techniques that are critical, especially in field use.

Table 4 lists the most popular adsorbents, together with their characteristics.

Next, the main kind of adsorbents and combination of adsorbents used for the pre-concentration of HCs from air samples will be described.

Bruner and coworkers [46,47] evaluated the properties of graphitized carbon black conventional cartridges for use in conjunction with packed columns equipped with electron capture detector (ECD) and MS detection. U-shaped glass-lined stainless-steel traps (30 cm × 0.3 cm I.D.) packed with 800 mg of Carbopack C or 400 mg of Carbopack B showed break-through volumes at −90°C larger than 8 L for the least volatile compound, CH_3Cl. This is more than adequate to allow the sampling of smaller volumes so that the detection

Table 4 Adsorbent Material Main Characteristics

Commercial name	MTD* (°C)	SA** (m² g⁻¹)	Porosity[a]	Type[b]	H***[c]
Carboxen 1000	400	1200	MIP	CSM	L
Carbosieve SIII	400	840	MIP	CSM	L
Chromosorb 106	250	750	P	SP	H
Porapak Q	250	550	P	EVD-DVB-CP	H
Carbograph 5	>400	260	MAP	GCB	H
Carboxen 569	400	390	MIP	CMS	M
Chromosorb 102	250	350	P	S-DVB-CP	H
Porapak S	300	350	P	PVP	H
Carbotrap or Carbopack B	>400	100	MAP	GCB	H
Tenax TA	350	35	P	PDPPO	H
Tenax GR	350	35	P	PDPPO +GCB	H
Carbotrap or Carbopack C	>400	12	MAP	GCB	H
Carbopack F	>400	5	MAP	GCB	H
Glass Beads	400	<1	NONP	Silicate	H

[a] MIP = microporous (mean pore diameter 1–10nm); P = porous (mean pore diameter 20–100nm); MAP = mesoporous or macroporous (mean pore diameter 100–300nm); NONP = non-porous (mean pore diameter >300nm).
[b] CMS = Carbon Molecular Sieve; EVB-DVB-CP = Ethylvynilbenzene-divynilbenzene copolymer; GCB = Graphitized carbon black; PDPPO = Poly(2,6-diphenyl-p-phenylene oxide); PVP = polyvinylpiridine; S-DVB-CP = Styrene-divinylbenzene copolymer; SP = Polystyrene polymer.
[c] H = high; M = medium; L = Low
* MDT = Maximum desorption temperature;
** SA = Specific surface area;
*** H = Hydrophobicity

limit and limit of linearity of the detectors used are met. In practice, a 1-L sample was sufficient to measure the main CFCs and halocarbons present in remote sites air. Another important feature of these adsorbents is the low affinity for water, which is important when sampling at subambient temperature is carried out. However, a high desorption temperature (250°C) is then required for injection, so that decomposition of the strongly retained and/or less stable compounds may occur.

Evaluation of four different traps to be used in conjunction with capillary columns and ECD detection has been carried out by Frank and Frank [48] for CCl_3F, $C_2Cl_3F_3$, $CHCl_3$, CH_3CCl_3, CCl_4, C_2HCl_3, C_2Cl_4 (see Tables 5 and 6). BTVs for Tenax were too small for the more volatile compounds and the BTV ratios between the least and the most strongly retained compounds were very large. Charcoal showed higher capacity for the more volatile but showed large BTVs for chloroethanes and strong affinity for water; coating of the adsorbent attenuates but does not solve these problems. Better performance is shown by the mixed adsorbents (90% Tenax + 10% Charcoal) trap and by Graphtrap. However, considering the BTV values (sufficiently high for CCl_3F), the lowest BTV ratios (i.e., 140) for the most and the least strongly retained compounds, and the lower temperature required for complete desorption (see Table 6), Graphtrap seems to be the best adsorbent.

Other authors [49] used sorption tubes (15 cm × 0.32 cm O.D.) containing two adsorbent beds (Tenax TA + Carbosphere S): species of low volatility are adsorbed on Tenax TA, whereas the unretained,

Table 5 Break-Through Volumes (L) of Some Halocarbons on Four Adsorbents at 36°C (dimension of the adsorbent bed 90 × 3 mm, for Charcoal 41 × 3 mm; N number of theoretical plates)

Adsorbent combination	CCl_3F	$C_2Cl_3F_3$	$CHCl_3$	CH_3CCl_3	CCl_4	C_2HCl_3	C_2Cl_4
Tenax TA, 89 mg, 60–80 mesh, N = 5–50	0.007	0.005	0.44	0.04	0.04	1.14	4.47
Charcoal SK-4, coated with 1% SP-2100, 168 mg, 60–80 mesh, N = 10–30	90	391	196	62	546	58000	—
Mixture of 90% Tenax TA + 10% SP-2100 coated charcoal SK-4, 110 mg, N = 5–30	2	16	11	51	59	127	450
Graphtrap, 550 mg, N = 20–70	0.15	0.6	0.2	0.5	0.5	1.2	24

Source: From Ref 48.

Table 6 Temperatures (°C) for
Complete Desorption of the
Halocarbons with 20 mL H_2 from
Cartridges Filled with 110 mg
Tenax TA/Charcoal (A) or 550 mg
Graphtrap (B)

Compound	A	B
CCl_3F	180	110
$C_2Cl_3F_3$	210	150
$CHCl_3$	220	170
CH_3CCl_3	240	200
CCl_4	250	210
C_2HCl_3	250	230
C_2Cl_4	310	280

Source: From Ref. 48

more volatile species are subsequently retained on Carbosphere S. Thermal desorption is carried out with a reverse gas stream and efficient recoveries (T = −50°C) are obtained for CCl_4, C_2HCl_3, and C_2Cl_4.

Also, Porapack Q was used [50] as trapping material, and a temperature of −50°C is required for quantitative sampling of the above-mentioned compounds, whereas a temperature of −100°C allows the collection of the very volatile analytes. Alumina was disregarded because it is suspected to react with incompletely halogenated compounds at elevated temperatures.

A multilayer adsorption tube (30 cm long × 0.64 cm O.D., internal volume 6.7 mL) for the analysis of CFCs and their replacement compounds was developed by Sturges and Elkins [51]. The sequence of the adsorbents is the following: HayeSep D_B (a porous polymer), Carboxen 100, and Carbosieve S-II. This trap allows enrichment of up to 5 L air samples at the temperature of 25°C. Desorption was carried out at 200°C, using 500 mL of nitrogen at the flow rate of 50 mL min^{-1}. More recently, the following dual-layer adsorption tube was proposed [52]: a glass tube (10 cm long × 0.4 cm I.D.) filled with 100 mg of a lower-surface area (90 m g^{-1}) adsorbent (Carbograph 1) and 300 mg of Carbograph 5 (560 m^2 g^{-1}). Before use, tubes were conditioned under UHP nitrogen flow at 250°C for at least

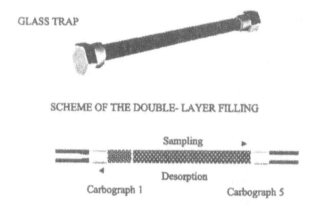

Fig. 5 Scheme of the dual-layer adsorption tube.

2 hours. A diagram of the multilayer adsorption tube is shown in Fig. 5.

The adsorption tube was connected, via a needle valve, to a SUMMA® canister filled at the pressure of 3 10^5 Pa with ambient air. The enriched volume was measured by means of a large volume glass syringe connected to the end of the adsorption tube. For field use, sample enrichment was directly performed by drawing a given air volume through the adsorbent bed by means of a portable pump or a large volume syringe. The adsorbed compounds were introduced into the gas chromatographic column by thermal desorption at 220°C for 5 mins, and the desorbed analytes were transferred into the gas chromatographic system at the flow rate of 6 mL min^{-1}, reversing the carrier gas flow direction with respect to the sampling flow direction.

In order to avoid the post desorption cryofocusing step (see Sample Injection section), necessary when capillary gas chromatographic columns are used, a dual-adsorbent microtrap containing Carboxen 1003 (6 mg) and Carboxen 1000 (5 mg) was used [53,54]. The design of the microtrap is the following: stainless-steel 306 tubing 24 cm long × 0.058 cm I.D., active length 14 cm. This microtrap allows quantitative trapping of a number of HCFCs and HFCs at the temperature of −50°C, sampling up to 3 L of actual air. The adsorbed compounds are rapidly desorbed by the direct ballistic heating of the microtrap to a temperature of 260°C in about 4 s.

For the investigation of HC content in volcanic gaseous discharges, the following approach was used. In a study aimed at deriving an upper limit for the global budget of natural emissions of CFCs [55], the following sample treatment procedure was described: gas samples were collected in 100-mL Giggenbach glass bottles previously evacuated, and 20-mL sample aliquots were pumped through a trap filled with an alkaline adsorbent (Ascarite) and then passed through a 0.7-mL stainless-steel pre-concentration loop filled with glass beads, kept at $-196°C$ by means of liquid nitrogen. Then the sample loop was heated at 80°C and its contents transferred to a stainless-steel capillary at $-196°C$. The analysis was performed in capillary GC-ion trap MS. More than 300 organic substances were detected in fumarolic and lava gases, among which were 100 chlorinated, 25 brominated, 5 fluorinated, and 4 iodinated compounds. The greatest variety of halocarbons was found in the lava gas from Mt. Etna (Italy), where the following HCs were detected at sub-ppb to ppb (part per billion) levels: CH_3Cl (0.61 to 84), CH_2Cl_2 (0.01 to 12), $CHCl_3$ (0.03 to 5.5), CCl_4 (0.04 to 9.2), CH_3Br (0.01 to 3.7), CH_3I (0.02 to 1.9), CHCCl (<0.01 to 50), CH_2CHCl (0.01 to 15), $CHClCCl_2$ (0.02 to 2.2), CCl_2CCl_2 (0.01 to 1.5), CFC-12 (0.11 to 0.58), CFC-11 (0.03 to 0.45). On the base of these data, the authors estimated that volcanic sources could account for less than 1 ppqv (part per quadrillion by volume) in the atmospheric budget of CFCs. Therefore, they stated that the contribution of natural CFC emissions from volcanoes is negligible as compared to the anthropogenic CFC burden.

Sample Injection

Direct injection of large volumes of air (20–50 mL) into packed column has been carried out in earlier stages. Aliquots of air were transferred from the canister into an evacuated loop using a syringe [34] and/or connecting both the canister and the loop to a vacuum line [33,56]. The content of the loop is then swept from the carrier gas onto the head of the column kept at subambient temperature. However, such a procedure allows only the quantitation of those compounds with concentration in the order of hundreds of ppts and with a limit of detection (LOD) well below actual atmospheric mixing ratios. Actually, this is possible with the development of ECDs: it has been found that 2–5-mL loops were sufficient to measure halogenated compounds, and this technique has been largely employed for *in situ* measurements, by flushing the injection loop with the air sampled through a pipeline. A suitable system of switching valves (manually or electronically actuated) allows the correct timing of the sampling injection steps. This procedure has been followed in ALE-GAGE monitoring stations and onboard ship measurements [6].

The use of capillary columns and/or the need of monitoring other trace halocarbons compulsorily require some pre-concentration procedure before injection. In fact, the sample size should be sufficiently reduced to allow injection and separation in the gas chromatographic system, and pre-separation of N_2, O_2, CO, and H_2O from the halocarbons may be required. Since water vapor may be the main interference and may inhibit the use of capillary columns, water has to be removed especially from samples collected under high humidity conditions. This can be performed by using drying agents such as $Mg(ClO_4)_2$ or K_2CO_3 [37,57]. It has been observed that such species may cause changes in sample composition, and several researchers prefer to use Nafion® drier and molecular sieves to partially remove water. On the other side, decomposition after long-term storage may occur for compounds such as CH_3CCl_3, CCl_4, C_2HCl_3, and such process is inhibited by water. For this reason, water is removed after sample collection and just before sample injection.

Direct coupling of cartridge or enrichment loops with chromatographic columns is possible only if both have similar diameters and fast desorption ensures narrow band width. It should also be pointed out that uneven temperature profiles along the cross section of the cartridge may cause distortion of the eluting band. Also, the bands of adsorbed compounds tend, if cartridges are stored, to broaden depending upon volatility and temperature. Thus, band compression by cryofocusing is generally necessary, especially if capillary columns are used. Cryogenic trapping can be performed (1) on the head of the chromatographic column itself, (2) on a pre-concentration trap (loop), (3) on empty open capillaries, and (4) by using microtraps (packed capillaries).

Small stainless-steel pre-columns packed with porous glass beads for pre-concentration of the air sample have been used by several authors [38,58]. At the temperature of liquid nitrogen, all the halocarbons are adsorbed and separated from N_2, O_2 (Ar and CO). Glass beads show two main advantages. The first one is that desorption temperatures are quite moderate (60–100°C), so that thermal decomposition is unlikely to occur. The other one is that glass beads can be easily cleaned and cause no detectable blank values or measurable memory effects. These pre-concentration columns have been used in conjunction either with a single capillary GC-ECD system for halocarbons measurement [58] or a triple capillary GC system equipped with two flame ionization detectors (FIDs) and one ECD for simultaneous non-methane hydrocarbons (NMHCs) and halocarbons measurements [38,59,60] with excellent precision and accuracy.

Uncoated fused silica capillaries (0.32–0.53 mm I.D.) kept at approximately −145°C have been also used for pre-concentration of air samples in the capillary GC-ECD [36] or GC-MS [52,61] analysis of halocarbons.

A stainless-steel microtrap (0.32 cm × 4 cm long) filled with 0.04 g of Tenax GC and kept at 0°C was used for the cryofocalization of air samples in the GC-MS analysis of organobromine compounds (CH_2BrCl, CH_2Br_2, $CHBrCl_2$, $CHBr_2Cl$, $CHBr_3$) [62]. The actual air samples (4 L) were previously enriched onto 0.7 g of Tenax GC.

Problems often encountered may be aerosol formation, broadening of the early eluting peaks and tailing of the later eluting ones, and blocking of the trap by water condensation. Fast heating of the cryogenic trap, absence of cold spots at the interface capillary trap/capillary column, and use of drying agents are compulsory for the solution of these problems.

On Line Apparati and *in Situ* Measurements

Drawbacks of off-line methods are that it is often difficult to obtain acceptable blank levels, while contamination or possible reactions may occur during the time necessary for storage and transportation. Further, traces of halocarbons may be taken up by the highly polluted laboratory air during the transfer of the cartridge to the thermal desorber, even if performed very rapidly. For these reasons, the most popular procedure now adopted is based on remote sampling using canisters and enrichment with sorption tubes "on line" with the GC apparatus.

Furthermore, on-line apparati, when automated, can be effectively used for continuous *in situ* measurements also in remote sites. Obviously, these kinds of apparati require the introduction and the integration of several changes in the hardware and software systems of the basic instrumentation, so that all the steps of the analysis (sampling, sample injection, and calibration) can be performed automatically. Lack of personnel control and maintenance even for long periods of time must be foreseen, and this may require the introduction of different but rugged technologies. An automated apparatus for repetitive sampling with sorbent-packed traps, followed by on-line thermal desorption, stationary phase focusing, capillary GC, and EC detection was developed by H. Frank and coworkers for analyzing airborne C_1-C_2 halocarbons [63]. The instrument consists of an electronic timing device, a heated ten-port valve, two packed microtraps, a gas chromatograph equipped with a thick-film capillary column, and ECD. Two glass capillary sorption traps (1.2 mm I.D. × 100 mm length) packed with spherical divinylbenzene polymer (14 mg) and Carbosphere (8.3 mg) are used for enriching 40-mL ambient air at room temperature. Thermal desorption is carried out at 200°C. Obviously, the small air volume sampled allowed only the quantitation of the most abundant HCs and/or of those HCs whose ECD response is high.

An automated analytical system for the determination of chlorinated VOC at ppb levels was developed for industrial hygiene applications [64]. The system is equipped with a variable temperature adsorption trap (VTAT) packed with a longer bed of Tenax TA and a shorter bed of activated charcoal. Aliquots of 100 mL of indoor air collected in dry-cleaner establishments and in shower rooms were flushed through the trap at ambient temperature and subsequently desorbed (at 250°C) onto a capillary column, using an EC detector. In dry-cleaner establishments, CH_2Cl_2, CCl_4, $CHCl_3$, and C_2Cl_4 were detected at ppb levels, whereas in shower room air also the brominated compounds (i.e., $CHCl_2Br$ and $CHBr_2Cl$), by-products of water chlorine disinfection, were determined.

The determination of biogenic HCs (i.e., $CHBr_3$, CH_2Br_2, $CHCCl_3$) has been carried out with an automated GC-MS system [65]. The on-line enrichment trap was a stainless-steel tube (5 cm × 4 mm O.D.) containing 0.08 g of Tenax TA, cooled with a Peltier device, that requires no cryogen and is therefore more suitable for field studies. Trap temperature was kept at 15°C, which was low enough for trapping the compounds of interest. Furthermore, no water filter was required, because of the low trapping efficiency of water at this temperature; 200-mL air samples were injected and no cryogenic cooling of the capillary column was required. The method allowed the simultaneous determination of the HCs, together with other biogenic compounds such as isoprene and dimethyl sulfide, with detection limits in 200-mL samples of approximately 2 ppt. In more recent years, the same research group performed the analysis of air samples collected during two cruises covering the western Pacific, Indian Ocean, southeast Asian Sea, and Southern Ocean [66]. Air samples were collected in two different kind of canisters (fused-silica lined and electro-chemical buffing) and subsequently analyzed using a pre-concentration/capillary GC-MS within a few months. The pre-concentration unit consisted of two traps containing glass beads and Tenax TA, respectively, a capillary trap (top of the capillary column), and three switching valves; 500-mL air samples were drawn through the glass beads trap kept at $-150°C$. The same trap was heated to 20°C to desorb target compounds without desorbing water vapor. Helium was used to transfer the desorbed compounds to the Tenax TA trap kept at $-20°C$, a temperature high enough to prevent CO_2 trapping, but low enough to collect target compounds. The compounds where then thermally desorbed (at 180°C) and transferred to the capillary kept at $-180°C$ for cryofocusing.

A very efficient automated GC-MS apparatus for routine atmospheric field measurements of CFCs, HFCs, and HCFCs was developed by Simmonds and coworkers [67]. A schematic diagram of the adsorption/desorption system (ADS), and the various valve positions (helium purging, air sampling and GC-MS analysis) is shown in Fig. 6. The microtrap is a stainless-steel tubing (0.068 cm I.D. × 0.109 cm O.D.) filled with three adsorbents: Carbotrap (5 mg), Carboxen 1003 (5 mg), and Carboxen 1000 (4 mg), with 425–500 μm glass beads to separate the individual adsorbents. The trap is housed in a small insulated aluminum block kept at the temperature of approximately $-50°C$ by means of two two-stage cascaded thermoelectric Peltier devices. The small dimensions of the trap ensure that chromatographic peak shape and resolution are retained during the thermal desorption stage, when the trap is ohmically heated to a temperature of approximately 220°C. This apparatus allows the successful detection of trace amounts of ubiquitous HCs in the presence of much larger concentration of atmospheric hydrocarbons in 2-L air samples. Before entering the analytical apparatus, ambient air is dried by passage through a Nafion® permeation dryer that is continuously counterpurged with approximately 150 mL min^{-1} dry nitrogen gas.

The apparatus is in use for remote field monitoring of the hy-

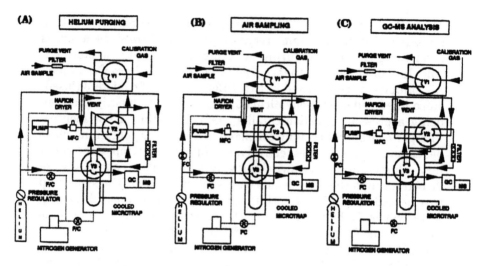

Fig. 6 Schematic diagram of the adsorption/desorption system. (From Ref. 67.)

drohalocarbons to determine the growth rates and lifetimes for these compounds in the frame of the Global Atmospheric Gases Experiment/Advanced GAGE (GAGE/AGAGE) network.

A modified version of this apparatus is equipped with two ECDs and designed in order to direct the effluent from the first ECD to the second one [68]. The microtrap design is as in previously described works [53,67]. The system has proved capable of measuring a wide range of HCs at concentrations as low as 0.1 pptv. This apparatus was used in two field campaigns carried in the Arctic [69] for monitoring chlorinated solvents (C_2Cl_4, $CHCCl_3$), which are considered excellent markers of anthropogenically polluted air. The fully automated instrumentation was deployed at Zeppelin station (Ny-Alesund–Arctic Norway) for three months. The instrumentation was also installed for one month on board the RRS Discovery for a cruise in the NE Atlantic. It was observed that while average baseline C_2Cl_4 and $CHCCl_3$ concentrations were 1.77 and 0.12 pptv, respectively, during pollution incidences concentration rose to 5.61 pptv for C_2Cl_4 and to 3.18 pptv for $CHCCl_3$.

The same apparatus was also used to analyze grab samples collected in SUMMA® or Silcosteel® canisters in a study aimed at evaluating biogenic fluxes of halomethanes from Irish peatland ecosystems [70]. Fluxes were measured using static chambers of 25 L volume designed for soil emission studies. Global peatland ecosystem fluxes were evaluated to be 4.7, 5.5, 0.9, 1.4 Gg year^{-1} for $CHCl_3$, CH_3Cl, CH_3Br, CH_3I, respectively, thus demonstrating the importance of terrestrial ecosystems as biogenic sources of these compounds.

Recently, an automated GC-ECD system equipped with a cryotrap made of a stainless tubing (10 cm long $\times$ 0.32 cm O.D.) filled with a 4 cm bed of 120–180 µm glass beads was described [71]. This method allowed the determination of only the most abundant CFCs, together with other chlorinated hydrocarbons (CH_3CCl_3, CCl_4, C_2Cl_4, $CHCl_3$). The apparatus was used for evaluating the variability of concentration of the above-cited compounds over an Industrial Park in Taiwan, sampling 150-mL aliquots of air [72].

B. Water Sample Treatment

Monitoring the presence of volatile halogenated hydrocarbons in water requires several steps before GC-ECD or GC-MS analysis: extraction from the matrix, clean-up, and enrichment procedures. The actual sample must turn into a final sample, dissolved in a suitable solvent, at a concentration compatible to detection limit of the analytical technique chosen. Because these compounds must be determined at sub-ppb level, a suitable pre-concentration technique is needed. However, one should keep in mind that this step is a very delicate one because during pre-concentration, severe sample loss of the analytes may occur that can alter the final results.

Liquid–Liquid Extraction

This well established extraction technique shows some limitations in extracting HCs from water, mainly for the following reasons:

Sample loss may occur, due to the high volatility of the compounds.

The solvent used for the extraction shows analogous volatility and retention times to that of the HCs, making it quite difficult to separate the compounds of interest from the solvent in the GC column.

It is difficult to automate.

It may require large amounts of organic solvent.

However, liquid–liquid extraction (LLE) with pentane has been successfully used by several authors to extract biogenic and anthropogenic halocarbons from water. In particular, samples are extracted with pentane, in which the HCs are enriched. As reported in Ref. 73, for the evaluation of the removal of HCs formed after break-point chlorination in the Arno river, Italy, a 150-mL water sample was extracted with 2 mL of *n*-pentane by manual shaking for 5 mins. The organic layer was allowed to separate and the extracts were kept at 10°C and analyzed within 3 h of being injected directly in the capillary column.

An interesting comparison between LLE and open-loop stripping (see headspace, Section B) is described in Ref. 74 with the aim of investigating the formation of by-products in chlorinated seawater used for drinking-water production on three oil platforms. These compounds were suspected to be present in a wide concentration range (low ng L^{-1} to high µg L^{-1}). Tribromomethane and dibromoacetonitrile were efficiently extracted with *n*-pentane, at a concentration of 19–27 µg L^{-1} and 0.9–1.6 µg L^{-1}, respectively. However, when using this approach, formation of artifacts such as 3-bromo-2-methyl-2-butanol and other halogenated C_5 compounds was observed, due to a possible reaction of excessive halogen with traces of olefin in the solvent. Sodium thiosulfate was used as a quenching agent. Open-loop stripping was more indicated for the extraction of halogenated compounds present at the low nanograms per liter level, such as the iodinated trihalomethanes.

LLE with pentane has been used to extract chloroform, tetrachloroethylene, and bromoform from the seawater of a fjord at the border between Norway and Sweden [75]. These compounds can be used as tracers to study the water mixing in different parts and depths of the Idefjorden, the first two being of anthropogenic origin, whereas bromoform is produced by several species of red algae. A device for extraction and injection of the samples that is totally isolated from the atmosphere has been constructed. On-line extraction of the sample with *n*-pentane was performed in a segmented flow in a glass coil. The glass coil has been internally coated in order to create a hydrophobic surface. After extraction, the phases are separated by a membrane separator and the pentane is directly transferred to the loop of an injector mounted on a gas chromatograph. The extraction efficiency was 83% for chloroform, 41% for tetrachloroethylene, and 80% for bromoform. These values were taken into account for final evaluation of the results.

A similar approach was used by the same author to determine carbon tetrachloride, tetrachloroethylene, 1,1,1-trichloroethane, and bromoform in the Artic seawater in the Svalbard area [76].

The concentration of volatile halocarbons and other compounds into an anoxic fjord environment has been determined using LLE with pentane, containing an internal standard [77].

Headspace

When one performs quantitative trace analysis of organics, uncertainty may occur due to the presence of interferences at the same concentration level. This drawback can be overcome by analyzing a phase, generally a gaseous phase, in thermodynamic equilibrium with the sample. This principle is exploited in the so-called headspace analysis (HS). This technique shows several advantages over classical techniques, such as solvent extraction, steam distillation, or distillation under vacuum. First of all, it eliminates complex preconcentration procedures, it does not overload or contaminate the chromatographic column with water or other compounds, and it lowers unwanted effects due to the matrix [78].

The headspace techniques can be divided in three groups. The first one, also known as static or equilibrium headspace, is based on the principle that the analytes dissolved in water and contained in a closed vial are in thermodynamic equilibrium with the vapor phase. Their vapor pressure is constant at constant concentration and temperature; thus, injecting in a gas chromatograph a known amount of the supernatant gas phase, the peak area of the analytes is directly related to their concentration.

The concentration (mol fraction) x_i of a given substance can be determined according to the following equation:

$$x_i = \frac{A_i}{k_i P_i^{\circ} \alpha_i}$$

where P_i° is the vapor pressure of pure substance at temperature T α_i is the activity coefficient of the substance in solution A_i is the peak area of the substance.

Static headspace requires rigid control of the sample temperature, sample withdrawal, and other parameters. Once the equilibrium between the two phases is reached, it does not change with time, however it is disturbed after sampling. As a consequence, sampling method and volume should be chosen carefully.

Using this approach, Mohnke and Buijten [79] were able to determine several halogenated hydrocarbons in water at the low ppb level.

In the late 1970s Kolb and coauthors [80] were able to determine seven simple HCs in water using a FID and, for the separation, a packed GC column. The limit of detection was lower than several μg L^{-1}, demonstrating the high sensitivity of static headspace analysis. Similar results were obtained years later by Vitenberg et al. [81] using a capillary column. Again, Kolb and coauthors [82] successfully combined static headspace gas chromatography with a dual capillary column and a dual detector (ECD and photo-ionization detector, PID) arrangement for the extraction of volatile aromatic and halogenated hydrocarbons in water and soil. ECD was used for the determination of HCs, and PID was used for aromatic hydrocarbons. A schematic diagram of the apparatus is shown in Fig. 7; in Fig. 8, an ECD chromatogram is shown.

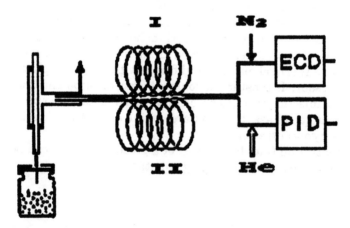

Fig. 7 Instrumentation for dual-channel ECD/PID headspace analysis of volatile halogenated and aromatic hydrocarbons in water. Channel I: ECD for halogenated hydrocarbons 50 m × 0.32 mm fused silica capillary column, methyl-5%-phenyl silicone, bonded phase, 2 μm film. Channel II: PID for aromatic hydrocarbons 50 m × 0.32 mm fused silica capillary column, methyl-5%-phenyl silicone, bonded phase, 5 μm film. (From Ref. 82.)

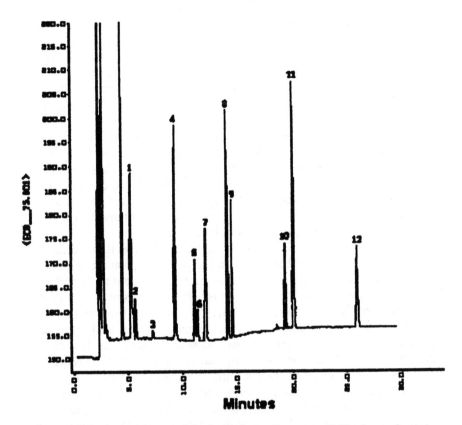

Fig. 8 Volatile halogenated hydrocarbons in water, ECD channel, as in the previous figure. Peak identification: 1 = 1,1-dichloroethene (5.9 µg L^{-1}); 2 = dichloromethane (9.0 µg L^{-1}); 3 = 1,1-dichloroethane (8.0 µg L^{-1}); 4 = chloroform (1.4 µg L^{-1}); 5 = 1,1,1-trichloroethane (0.1 µg L^{-1}); 6 = 1,2-dichloroethane (11.9 µg L^{-1}); 7 = carbon tetrachloride (0.07 µg L^{-1}); 8 = trichloroethene (0.7 µg L^{-1}); 9' = dichlorobromomethane (0.2 µg L^{-1}); 10 = dibromochloromethane (0.2 µg L^{-1}); 11 = tetrachloroethene (0.16 µg L^{-1}); 12 = bromoform (1.4 µg L^{-1}). (From Ref. 82.)

Static headspace requires a very simple apparatus that can be easily automated; however, strict control of the temperature is necessary as well as the knowledge of the equilibrium coefficients between gas and liquid phases. It is particularly suitable for highly volatile and medium volatile compounds, however, it fails when trace components or components with very low vapor pressure are analyzed. To overcome these limitations, the concentrations of the analytes in the gas phase can be increased by raising the temperature, which increases the vapor pressure of the trace components. As an alternative, the values of the activity coefficients can be increased by adding an electrolyte or a nonelectrolyte ("salting out" effect).

In the second HS approach, the sample is stripped from the matrix by a gas stream and addressed toward a suitable trap, from which it is eluted either thermally or with a solvent. This strip-trap technique can be done in either of two systems: in an open system, where the stripping gas passes through the sample and the trap and is vented to the atmosphere; or in a closed system, where the gaseous phase is recycled through the sample and trap.

The third approach is called dynamic headspace; it is similar to strip-trap procedure and is based on the principle that a gas stream that continuously flows over the supernatant is able to "strip" the compounds present in the liquid phase. If this process is carried on long enough, the analytes are extracted completely from water. The equilibrium between gas and liquid phase is continuously reestablished, leading to an exponential decrease of the concentration of the analytes in the liquid phase. Temperature control may be required for those substances that show a limited vapor pressure or are highly soluble in water, although it is important to keep in mind that water itself can be "stripped" to a certain extent. The stripping gas is directed to an adsorption trap in which the analytes are concentrated prior to gas chromatographic analysis. Dynamic headspace has been described by Pellizzari et al. and this technique is considered more convenient than others when foaming fluids have to be extracted for VOC analysis [83].

In an article by Kaiser and Oliver [84], a modified headspace technique is described in which the water sample was placed into a laboratory funnel, leaving airspace of approximately 2 mL. The funnel was inverted and evacuated at $7.5 \ 10^{-2}$ Pa, then submersed in a water bath at 30°, 50°, 70°, or 90°C. This procedure generated small gas bubbles that would eventually reach the surface. After 30 min the headspace was returned to atmospheric pressure and a 5 µL-sample of this headspace was withdrawn and injected into a GC-ECD system. If the procedure was repeated, multiple determination of the same water sample was possible. This method allowed the determination of several halocarbons such as $CHBrCl_2$ and $CHBr_2Cl$, as well as CCl_4 and $CHBr_3$.

Purge-and-Trap

Similar to dynamic headspace in principle, purge-and-trap (P&T) is more efficient because the gas stream directly bubbles into the sample by means of a gas dispersion tube equipped with a frit. This technique is also known under the name of gas stripping. After stripping, the analytes are addressed to an adsorption trap.

When capillary GC-MS detection is performed, further preconcentration of the analytes might be required, due to the small injection volumes typical of this technique. In order to accomplish this goal, several modes have been described. Thermal desorption and cryofocusing in a small volume of liquid nitrogen, followed by flash heating of the trap and subsequent injection into the capillary column has been described in Ref. 85. Another approach consists of substituting the trap with a condenser (for water vapor) and a capillary coated with a thick film of stationary phase [86]. An empty stainless-steel capillary has been used as a dryer for the effluent from a trap [87].

EPA Method N° 601/SW-846 Method 8010/8021 for the determination of volatile halocarbons in water and soil is a purge-and-trap gas chromatographic method [88]. An inert gas is bubbled through a 5-mL water sample contained in a specially designed purging chamber. The volatile halocarbons are efficiently transferred from the aqueous phase to the vapor phase. The vapor is swept though an adsorption tube where the halocarbons are trapped. After purging is completed, the adsorption tube is back-flushed with the inert gas for analytes. Desorption onto a gas chromatographic column takes place. The compounds are detected with an electrolytic conductivity detector (ELCD), a halide-specific detector. This method allows the detection of the halocarbons at $1-10 \ \mu g \ L^{-1}$.

Vapor stripping has also been described in a HP Application Note [89]; in this procedure, volatile halocarbons are extracted from water by a stream of helium gas bubbled through the water sample and then carried to the adsorption trap. The trap, packed with Tenax, is then connected to a GC-MS system for analytes separation.

Narang and Bush [90] described a stripping procedure that is based on Grob's closed-system stripping [91], in which the organics are stripped from water at varying temperatures in closed systems, transferred to a charcoal trap, and eluted with carbon disulfide. Grob's procedure works very well for high-boiling compounds, but recoveries are poor for volatiles. In Narang and Bush's work, the graduated impinger is kept at constant temperature (45°C) and connected to the circulating system. Helium was used as the stripping gas and let bubble for 15 min after the sample had reached thermal equilibrium. The adsorption tube was filled with Porapack, and after the stripping it was rinsed with 1–1.5 mL of distilled methanol. This volume, collected by gravity, was analyzed by GC-ECD. The recoveries were above 80% for the selected compounds with small relative standard deviation values. This method is compatible with a conventional automated injector of a gas chromatograph.

Closed-loop stripping with Tenax traps was used to collect HCs released to seawater by marine microalgae [92]. The stripping procedure was unsuccessful for the recovery of the extremely volatile compounds, such as CH_3Cl and CH_3Br, or relatively nonvolatile compounds, such as CBr_4 or CHI_3. Identification and quantification was achieved using a GC-MS system.

An automatic system for the analysis of HCs, based on purge-and-trap, followed by GC-ECD detection is described in Ref. 93. The trace level analysis of chloroform, 1,1,1-trichloroethane, carbon tetrachloride, 1,1,2-trichloroethylene, and 1,1,2,2-tetrachloroethylene in rainwater and ambient air is discussed.

A new sampling device [94] for HCs has been developed for the collection of samples without contact with the atmosphere. The device enables direct introduction of the sample into the analytical system, thus eliminating the risk of any contamination. It is made out of polytetrafluoroethere (PTFE), a chemically inert material, and it allows the storage of the samples for up to 16 h in the case of CFCs and carbon tetrachloride and several days in the case of other HCs. After sampling, liquid-gas extraction, followed by cold trapping, is performed and the separation of the HCs is achieved by GC-ECD.

Solid-Phase Micro-Extraction (SPME)

Solid-phase micro-extraction (SPME), first introduced at the end of the 1980s by Pawliszyn and coworkers, represents one of the most valid examples of "solvent free" extraction technique. It can be considered the ultimate evolution of the well-established solid-phase extraction (SPE), and it consists a fused silica fiber coated with a nonvolatile polymeric or a graphitized carbon black coating. The fiber is exposed to the sample (direct-SPME) or to its headspace (headspace-SPME) for the adsorption of the analytes, then thermally desorbed in the injector of a gas chromatograph for separation and quantitation. Due to its geometric characteristics, the fiber can be mounted into a syringe-like holder that will eventually act as a classical syringe for sample injection into the chromatograph.

The operating principle is based on the partition of the analytes between sample and fiber. At the equilibrium, the amount of analyte adsorbed on the fiber can be described by the following equation:

$$n = \frac{K_{fs}V_f C_0 V_s}{K_{fs}V_f + V_s}$$

where

n is the mass of analyte adsorbed by the coating; V_f and V_s are the volumes of the coating and of the sample, respectively; K_{fs} is the partition coefficient of the analyte between the coating and the sample matrix; C_0 is the initial concentration of the analyte in the sample.

This equation shows that the relationship between the initial concentration of the analyte in the sample and the amount adsorbed by the fiber is linear.

If the V_s value is very large, then the amount of analyte extracted by the fiber is

$$n = K_{fs}V_f C_0$$

meaning that n is independent from the sample volume and that it is directly proportional to C_0. If K_{fs} values are quite large, organic compounds show very strong affinity for the adsorbent used. For such compounds, SPME is a quite efficient pre-concentration technique that leads to good sensitivity. When K_{fs} values are small, it means that the analytes show a weak affinity for the fiber used and that they are not completely extracted from the matrix. However, SPME is considered an equilibrium sampling method because in many cases K_{fs} values are not sufficiently large to extract the analytes from the matrix completely.

Due to its intrinsic simplicity, SPME can be easily used for field analysis, because it combines sampling, extraction, concentration, and injection in one single device. The amount extracted from the matrix is dependent on the mass transfer of analyte to the coating, and in direct-SPME it is maximized when the solution is homogeneously stirred.

Direct-SPME is more suitable when organic compounds are extracted from gaseous samples and relatively clean water samples. In all other cases, such as solid matrices or wastewater samples, headspace SPME can be successfully used, depending on the volatility of the target compounds. Volatile organics tend to vaporize quite easily into the headspace, whereas semivolatile compounds may require longer extraction times. Extraction can be limited by two different aspects: One is thermodynamic, when the affinity of the analytes for the matrix is stronger than their affinity for the fiber coating. In this case K_{fs} values are small and the extraction efficiency is poor. The other aspect is kinetic, and it is mainly due to the amount of time required to extract the analytes from the matrix. In this case, simply increasing the sampling temperature will solve the shortcoming. The extracted analytes can be desorbed either in a GC or in a HPLC injection system.

SPME has been successfully used in many applications for the analysis of different classes of compounds from various matrices. Several authors applied SPME to extract HCs from water. In Ref. 95, volatile halocarbons were extracted both from air and water and quantified using an electron-capture detector. The aqueous standard was spiked into a 40-mL vial, and 0.3-mL headspace was exposed to the SPME needle. During extraction the sample was stirred continuously. The sensitivity of the fiber was compared to that of valved, gaseous injection (Fig. 9) for a 400 ng L^{-1} standard solution, demonstrating the high performance of this technique.

When using SPME, especially when sampling highly volatile compounds, one should be aware that some interferences, due to the presence of solvents in the laboratory, may occur. In Ref. 96, SPME was used for the determination of 60 volatile organic compounds (VOC) in drinking water. Immersion and headspace (HS)-SPME were compared, demonstrating that the two techniques show similar sensitivities, except for the very light VOCs, although HS-SPME showed shorter equilibration times. Water samples from the contaminated drinking water in the low µg L^{-1} range with 1,1,1-trichloroethane, trichloroethene, and tetrachloroethene were extracted and analyzed. The results obtained were compared to those obtained by purge-and-trap, and a reasonable agreement between the two sets of data was demonstrated (Table 7). SPME was extensively used in

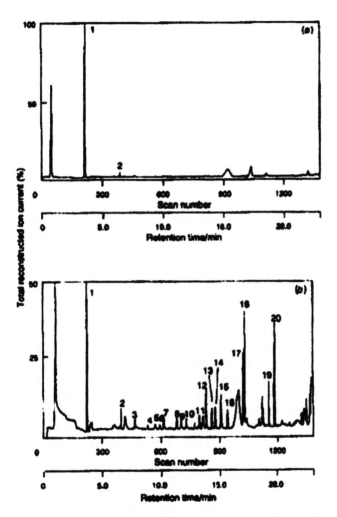

Fig. 9 Total reconstructed ion current for a 400 ng L^{-1} standard sampled with (*a*) a 0.5-mL gas-tight syringe and (*b*) a 3-min fiber exposure. 1 = fluoroethane; 2 = dichloromethane; 3 = *trans*-1,2-dichloroethene; 4 = chloroform; 5 = 1,1,1-trichloroethane; 6 = carbon tetrachloride; 7 = benzene; 8 = trichloroethene; 9 = 1,2-dichloropropane; 10 = bromodichloromethane; 11 = *cis*-1,3-dichloropropene; 12 = toluene; 13 = *trans*-1,3-dichloropropene; 14 = 1,1,2-trichloroethane; 15 = perchloroethylene; 16 = dibromochloromethane; 17 = chlorobenzene; 18 = ethylbenzene; 19 = bromoform; 20 = 1,1,2,2,-tetrachloroethane. (From Ref. 95.)

Table 7 Measurements by SPME and Purge-and-Trap of Trace Concentrations (μg L^{-1}) of VOCs in Drinking Water Samples in a Purification Plant

Sampling site	1,1,1-Trichloroethane		Trichloroethene		Tetrachloroethene	
	P&T	SPME	P&T	SPME	P&T	SPME
Well	3.0	4.7	1.9	2.7	24	24
Filter 1	3.1	3.8	2.0	2.6	25	22
Filter 2	3.2	3.9	2.1	2.7	25	24

Source: from Ref. 95.

monitoring the presence of HCs in ground and drinking water [97–99] as well as in the air and water of an indoor swimming pool [100].

In order to validate a standard method that employs SPME for the quantitative analysis of VOCs, including several HCs, in aqueous samples, an interlaboratory comparison was organized [101]. Twenty laboratories were involved in this two-round study; the majority participated to both rounds, and many of them used purge-and-trap and headspace, besides SPME, for comparison. HS-SPME and immersion SPME, using a 100-μm polydimethylsiloxane fiber, were compared, and detection limits were examined. Separation and quantification was obtained with a GC-MS system only in one-third of laboratories; in the others, GC-FID and/or GC-ECD detection was acquired. The statistical data treatment was performed in accordance with ISO standard 5725, which is based on the analysis of variance (ANOVA) technique. In Table 8 the comparison between HS-SPME and SPME is shown.

In Table 9 the comparison between the four techniques is shown. From the data obtained, the linearity of SPME is very good, with LODs in the low ng L^{-1} with the MS detection. The accuracy is comparable to that of the reference methods, purge-and-trap and headspace, and the precision is satisfactory for most quantitative routine analysis of HCs in water. Better precision is achieved by HS-SPME than by SPME, while the accuracy of the two approaches is equal.

Purge-and-Membrane

The classic extraction methods (LLE; static and dynamic HS; P&T) suffer from some disadvantages: sensitivity of solvent extraction and static HS is often not satisfying; dynamic HS and P&T are more

Table 8 Repeatability Standard Deviation (S_r), Inter-Laboratory Standard Deviation (S_L), Reproducibility Standard Deviation (S_R), Repeatability Limit (r), Reproducibility Limit (R), Gross Average ($\bar{\bar{x}}$) and Confidence Interval (CI) Determined in Accordance with the ISO Standard 5725 and Expressed in µg L^{-1} for HS-SPME and SPME.

Compound	Technique	S_r	S_r	S_R	r	R	$\bar{\bar{x}}$	CI
Chloroform	HS-SPME	0.28	1.13	1.16	0.78	3.25	3.96	3.96 ± 0.78
	SPME	0.44	1.59	1.65	1.25	4.63	4.52	4.52 ± 1.25
1,1,1-Trichloroethane	HS-SPME	0.32	1.55	1.58	0.90	4.42	3.21	3.21 ± 0.90
	SPME	0.36	1.02	1.08	1.01	3.04	3.04	3.04 ± 1.01
Carbon tetrachloride	HS-SPME	0.28	1.23	1.27	0.80	3.55	3.02	3.02 ± 0.80
	SPME	0.25	1.48	1.50	0.70	4.20	3.14	3.14 ± 0.70
Benzene	HS-SPME	0.30	0.55	0.63	0.85	1.76	3.10	3.10 ± 0.85
	SPME	0.47	1.15	1.24	1.32	3.47	3.47	3.47 ± 1.32
Trichloethane	HS-SPME	0.21	0.80	0.83	0.59	2.31	2.98	2.98 ± 0.59
	SPME	0.32	1.13	1.17	0.89	3.27	2.88	2.88 ± 0.89
Bromodichloromethane	HS-SPME	0.17	0.96	0.97	0.47	3.73	2.93	2.93 ± 0.47
	SPME	0.27	1.29	1.32	0.75	3.70	3.23	3.23 ± 0.75
Toluene	HS-SPME	0.50	0.00	0.48	1.41	1.34	2.71	2.71 ± 1.41
	SPME	0.34	1.22	1.27	0.96	3.54	2.61	2.61 ± 0.96
Tetrachloroethene	HS-SPME	0.35	1.13	1.18	0.99	3.30	3.32	3.32 ± 0.99
	SPME	0.37	1.12	1.18	1.05	3.29	3.05	3.05 ± 1.05
Ethylbenzene	HS-SPME	0.29	0.74	0.80	0.82	2.23	3.26	3.26 ± 0.82
	SPME	0.47	1.67	1.73	1.32	4.86	3.09	3.09 ± 1.32
p-Xylene	HS-SPME	0.25	0.80	0.84	0.70	2.34	3.09	3.09 ± 0.70
	SPME	0.37	1.51	1.55	1.05	4.35	2.69	2.69 ± 1.05
Bromoform	HS-SPME	0.31	0.32	0.45	0.87	1.26	2.95	2.95 ± 0.87
	SPME	0.30	0.92	0.97	0.85	2.71	2.95	2.95 ± 0.85
1,2-Dichlobenzene	HS-SPME	0.16	0.57	0.59	0.44	1.66	2.70	2.70 ± 0.44
	SPME	0.28	0.96	1.00	0.78	2.80	2.57	2.57 ± 0.78

Source: From Ref 101.

Table 9 Comparison of HS-SPME and SPME with the Reference Methods Purge-and-Trap and HS

Technique	P&T	HS	HS-SPME	SPME
Average value	0.983	0.993	0.993	0.994
Stragglers [%]	9	8	2	6
Outliers [%]	8	12	10	8
Average relative s_r [%]	9	4	9	11
Average relative s_R [%]	27	29	29	42
Lowest $\bar{\bar{x}}$	2.17 (bromoform)	1.99 (toluene)	2.70 (1,2-dichlorobenzene)	2.57 (1,2-dichlorobenzene)
Highest $\bar{\bar{x}}$	5.39 (chloroform)	4.38 (benzene)	3.96 (chloroform)	4.52 (chloroform)
Average $\bar{x}$	2.96	2.95	3.10	3.10

Source: From Ref. 101.

sensitive, but recovery of highly volatile compounds is not always reliable, due to their low breakthrough volumes with Tenax adsorbent. A new method that couples dynamic headspace and membrane extraction was developed and applied to the screening of HCs in water samples [102]. The water sample is purged by a stream of helium and the purged compounds are collected from the gas phase through a silicone hollow fiber membrane and addressed to a GC-ECD system.

The entire process is the result of two steps: the purge step and the membrane step. The sensitivity of the method is dependent on the recovery (R) of the analyte from water in the first step [Eq. (1) and (2)] and on its diffusion through the silicon membrane [Eq. (3)]:

$$R = (n_{i,w}^{\circ} - n_{i,w}(t))/n_{i,w}^{\circ} = 1 - \exp\left(\frac{-Ft}{V_g + n_{i,w}^{RT/H_{i,w}}}\right) \tag{1}$$

where $n_{i,w}^{\circ}$ and $n_{i,w}(t)$ are the number of moles of component I in the sample before and after purging; F is the gas flow-rate; t is the purge time; V_g is the volume gaseous phase; R is the gas constant (8314.4 mL kPa mol^{-1} K^{-1}); T is the absolute temperature; $H_{i,w}$ is the Henry coefficient (kPa).

$$H_{i,w} = p_i/x_{i,w} = p_i^{0/x_{i,w}}(sat.) = p_i^0 y_{i,w} \tag{2}$$

where p_i is the vapor pressure; $x_{i,w}$ is the mole fraction i in the aqueous sample; p_i^0 is the vapor pressure of pure i; $x_{i,w}$ is the mole fractional solubility of i in water; $y_{i,w}$ is the activity coefficient.

$$I_{ss} = ADS(P_s/l) \tag{3} \text{ (Fick's equation)}$$

where I_{ss} is the steady state flow through the membrane; A is the membrane surface area; D is the diffusion constant; S is the solubility constant; P_s is the vapor pressure of the analyte on the sample side of the membrane; l is the thickness of the membrane.

According to Eq. (1), nonpolar compounds such as HCs can be efficiently purged from water samples due to their large Henry's coefficients, as demonstrated also by this study in which the analytes were 100% extracted in few minutes. Their selective extraction through the membrane is assured by their good solubility in the silicone membrane.

Purge-and-membrane cannot be used for quantification, unless a mass spectrometer is used for the detection of the analytes. In GC-

Table 10 Rise times (RT) and Limits of Detection (LODs); (signal-to-noise (S/N) = 5/1 measured by purge-and-membrane-ECD and membrane-ECD)

Compound	Membrane-ECD LOD μgL^{-1}	Membrane-ECD RT min	Purge-and-Membrane-ECD LOD μgL^{-1}	Purge-and-Membrane-ECD RT min
Vinylchloride	2.20	3.00	1.30	0.20
1,2-Dichloroethene	1.20	4.00	0.70	0.15
1,2-Dichloroethane	1.50	5.00	1.00	0.20
Trichloroethane	0.07	5.50	0.08	0.70
1,1,1-Trichloroethane	0.06	4.50	0.15	0.70
Carbon tetrachloride	0.02	4.00	0.03	0.60
Tetrachloroethene	0.01	5.00	0.01	0.80
1,1,2,2-tetrachloroethane	0.08	7.00	0.20	1.50
Chlorobenzene	4.00	3.50	3.00	0.80
1,4-Dichlorobenzene	1.20	5.00	0.80	1.50
1,2,4-Trichorobenzene	0.02	7.00	0.05	2.50

Source: From Ref. 102.

ECD mode, it can be successfully used for rapid screening of HCs in water samples.

Comparison with direct membrane showed similar LODs; however, purge-and-membrane showed shorter response times, as displayed in Table 10.

In order to extend selectivity, the same authors introduced purge-and-membrane mass spectrometry (PAM-MS) for the analysis of individual VOCs in water and soil samples [103]. Water or soil samples were purged with nitrogen gas directed to the membrane module. The membrane was made of dimethylpolysiloxane and was interfaced with the electron ionization ion source of a quadrupole mass spectrometer. Excellent linearity and repeatability was achieved together with fast analysis of the samples.

Other Approaches

Hollow fibers have been used in a module connected to a multiplex gas chromatograph, for routine analysis and continuous monitoring of VOCs, including 1,1,1-trichloroethane and trichloroethene in contaminated groundwater and raw sewage water [104]. In the multiplex approach, multiple sample injections are made at high frequency in the GC column. As a result, a multiplex chromatogram is obtained that cannot be interpreted by the chromatographer, requiring a computer interpretation of the results. The extraction with the hollow fiber was performed at two different flow rates: (1) low flow rates were used for quantitative extraction, eliminating temperature and matrix effects; (2) high flow rates improved the sensitivity dramatically, because of the increase in extraction rates. At the optimum flow rate, a detection limit of 4 µg L^{-1} was obtained for trichloroethene, using GC-FID detection. Increasing the length of the hollow fiber or using a more sensitive detector can improve the sensitivity.

C. Chromatographic Separation

Because of the high volatility of the compounds considered, gas chromatography is the technique of choice for separation. Both packed columns employing extremely selective phases and capillary columns can be used, depending upon the total number of compounds to be determined, the complexity of the sample (presence of interferences), and the detector used.

Initially, when only a few compounds were identified and capillary column and instrumentation technology were not sufficiently developed, packed columns were the adequate and the only possible choice for this kind of analysis, being compatible with large volumes of samples to be injected into the gas chromatograph.

Then, following the need for extremely high resolution and the improvements in fused silica and instrumentation technology, capillary columns have been preferentially used.

Packed columns

Table 11 shows a quite comprehensive, though not exhaustive, list of the various packed columns developed through the years for HCs analysis [30,31,33,43,105–122]. The type of detector used and traceable reference data are also given.

Gas-solid chromatography employing different stationary phases, such as Al_2O_3, Porapack Q and W, Porasil C, D and F, and Chromosil 310, has been basically applied to the determination of CFC 11 and CFC 12, and occasionally extended to the simultaneous determination of the other main target compounds (i.e. CH_3CCl_3, CCl_4, $CHCl_3$, C_2HCl_3, C_2Cl_3, C_2Cl_4). Both constant (i) or pro-

Table 11 Packed Columns Used for HCs Analysis

Stationary phase	Halocarbons separated	Ref.	Det	Year
3% SE-52 on 80–100 mesh Chromosorb W	CF_2Cl_2, CH_3CCl_3, CCl_4, $CHCl_3$, C_2HCl_3, C_2Cl_4	105	ECD	1973
20% DC-200 on 30–60 mesh Chromosorb W	CF_2 Cl_2, CF_2Cl CF_2Cl, $CFCl_3$, $CFCl_2$ CF_2Cl, CH_3CCl_3, CCl_4, $CHCl_3$, CH_3I, C_2HCl_3, C_2Cl_4, CH_3 Br, CH_3Cl, C_2H_5Cl	106, 107	ECD	1977 i
20% DC-200 on 80–100 mesh Supelcoport			ECD	1983
10% SF-96 on 100–120 mesh Chromosorb W	$CHClF_2$, CF_2Cl_2, $CClF_2$ $CClF_2$, CH_2CHCl, CH_3Cl, CH_3Br, $CHCl_2F$, CH_3CH_2Cl, CCl_3F, CH_3I, CCl_2FCF_2Cl, CH_2Cl_2, cis-$CHCl$: $CHCl$, $CHCl_3$, CH_2ClCH_2Cl, CH_3CCl_3, CCl_4, $CHCl$:CCl_2, CCl_2: CCl_2, $CHBr_3$	108	ECD	1979 p
0.4% Carbowax 1500 on 80–100 mesh Chromosorb W	CF_2 Cl_2, $CFCl_3$, CH_3CCl_3, CCl_4, $CHCl_3$, C_2HCl_3, C_2Cl_4	106	ECD	1977
Al_2O_3 80–100 mesh	CF_2 Cl_2, $CFCl_3$	106	ECD	1977
Chromosil 310 80–100 mesh	CF_2 Cl_2, $CFCl_3$, $CFCl_2$ CF_2Cl, CH_3CCl_3, CCl_4, CH_3I, C_2HCl_3, C_2Cl_4, CH_3 Br, CH_3Cl	106	ECD	1977
Durapack n-octane/Porasil C 100–120 mesh	CF_2 Cl_2, $CFCl_3$, $CFCl_2$ CF_2Cl, $CHFCl_2$, CH_2Cl_2, CH_3CCl_3, CCl_4, C_2HCl_3, CH_3Br, CH_3I	30	MS	1975 p
		109	MS	1979 p

Column	Compounds	Ref.	Detector	Year	Mode
Durapack Carbowax 400/Porasil F (low k')	CH_3Cl, CH_3CCl_3, C_2Cl_4	30	MS	1975	p
0.5% SP-1000 on Carpopack B 80–100 mesh	CH_3Cl, $CF_2 Cl_2$, CF_2Cl-CF_2Cl, $CFCl_3$, $CHFCl_2$, $CFCl_2 CF_2Cl$, CH_3CCl_3, CCl_4, C_2HCl_3, C_2Cl_4, CH_2Br_2, CH_2Cl_2	110, 112, 113	ECD / MS	1978 / 1981	p / p
10% Na_2SO_4 on Porasil A, 80–100 mesh	$CF_2 Cl_2$, $CFCl_3$, C_2HCl_3, CCl_4, C_2Cl_4	114, 115	ECD	1974	p
30% Didecylphtalate on 100–120 mesh Chromosorb P	$CF_2 Cl_2$, $CFCl_3$, $COCl_2$	116			
Porapack Q, 80–100 mesh	$CF_2 Cl_2$, $CFCl_3$	117	ECD	1981	a/i
Porasil C, 80–100 mesh	$CF_2 Cl_2$, $CFCl_3$	118	ECD	1973	a/i
Porasil D, 80–100 mesh	$CF_2 Cl_2$, $CFCl_3$	43	ECD	1980	a/i
10% OV-101 on Chromosorb W 80–100 mesh	$CF_2 Cl_2$, $CFCl_3$, $CFCl_2 CF_2Cl$, CH_3CCl_3, CCl_4, C_2HCl_3, C_2Cl_4	33, 43	ECD	1978	a/i
10% SP2100 on Supelcoport 100–120 mesh	$CFCl_3$, $CFCl_2 CF_2Cl$, CH_3CCl_3, CCl_4, $CBrClF_2$, $CH_3 Br$, CH_2BrCl, $CHBr_2Cl$, $C_2H_4Br_2$, $CHBr_3$, CF_3Br, $CBrF_3$, $CBrClF_2$	43, 119, 120	ECD / ECD / ECD	1978 / 1984 / 1984	p/s.a
1% PEG modified with nitroterephtalic acid on 60–80 mesh GBC	$CFCl_3$, $CHCl_3$, CH_3CHCl_2, CH_3CCl_3, CCl_4, C_2HCl_3, C_2Cl_4	121	FID	1990	p
5% Fluorcol on Carbopack B 60–80 mesh	$CClF_2CClF_2$, CCl_2FCF_3	122	ECD	1994	i

i = isothermal; p = programmed temperature; a = ambient temperature; s.a. = subambient temperature.

grammed (p) temperature modes were used and good results were obtained as shown by the example in Fig. 10.

Gas-liquid chromatography using silicon fluids as stationary phase [30,33,43,105–109,116,119–122], has been used for general purpose separation of CFCs. Separation was essentially in order of the different halocarbon boiling points. However, using this kind of

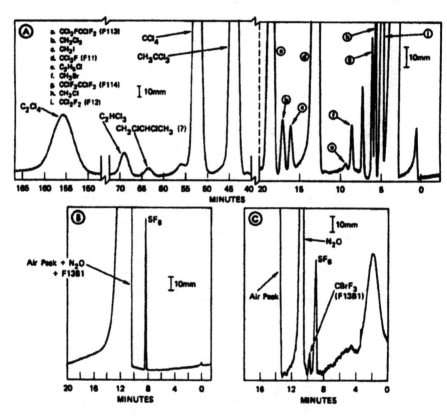

Fig. 10 Separation of selected halogenated species from a Pacific air sample: (A) A chromatogram based on the analysis of a 200-mL air sample onto a 10 m × 3.175 mm 20% DC 200 column at 40°C (isothermal). (B) The SF_6 separation, which used a 300-mL air sample and a 4 m × 3.175 mm molecular sieve 5A column at 50°C. (C) A molecular sieve 5A column of 8-m length shows the separation of $CBrF_3$ and SF_6 at 110°C. Only $CBrF_3$ was measured this way. In its shorter configuration, both $CBrF_3$ peak and the large N_2O peak merge with the large air peak. (From Ref. 107.)

column, some compounds with similar boiling points were difficult to resolve isothermally and subambient temperature programming was required, or, as an alternative, multiple injection of the sample was necessary. A chromatogram obtained analyzing 500 mL of rural air on a 6 mm × 3 m stainless steel column packed with 10% SF-96 on 100/120 mesh Chromosorb W is shown in Fig. 11.

During the transfer of the sample onto the head of the column, this was held at a temperature of −10°C in order to retain the sample on the head of the column before the temperature-programmed analysis is initiated.

Such a problem was solved using gas-liquid-solid chromatography (GLSC) [110,113], employing graphitized carbon black as solid matrix and a polar liquid phase containing free carboxyl groups, such as SP1000.

The column chosen for this purpose was a 3 m long × 2 mm I.D. glass packed with a graphitized carbon black whose surface area was

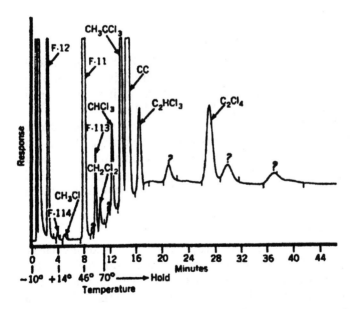

Fig. 11 ECD chromatogram of 500 mL of rural air. Column 6mm × 3m, stainless steel, packed with 10% SF-96 onto 100/120 mesh Chromosorb W. Temperature program: −10°C for 1 minute, then 8°C/min to 70°C. (From Ref. 31.)

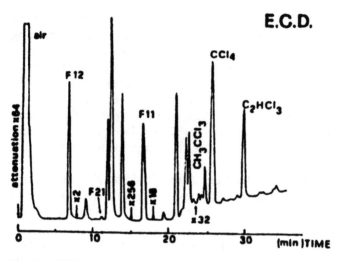

Fig. 12   EC detection of an actual air sample collected in a rural area. (From Ref. 113.)

about 90 m^2 g^{-1} (Carbopack B) coated with 0.5% SP 1000. Thus, the outstanding selectivity of GLSC, in which the driving force of the chromatographic process is molecular polarizability, has been exploited. A chromatogram is reported in Fig. 12.

It should be noted that the elution order is, in this case, strictly related to the molecular dimensions and not to boiling points. This leads to a wide range of retention times and good separation factors for the compounds considered. Until a few years ago, the GAGE-AGAGE team was still using packed columns to measure halocarbons in unattended monitoring stations.

Capillary Columns

Problems encountered when using capillary columns for halocarbon analysis are in that on capillaries with common film thickness (i.e., ~0.2 μm), the most volatile halocarbons show very short retention time at room temperature. Therefore, time-consuming sub-ambient chromatography is required for adequate retention and resolution. For this reason, thick film (2–5 μm) capillary columns are preferred. Capillary columns [36,38,49,50,52,65,67,123–129] developed for CFCs analysis are summarized in Table 12.

Table 12 Capillary Columns Used for HC Analysis

Column type	Halocarbons separated	Ref.	Det.	Year	
5μ PS-255, 25 m × 0.32 mm I.D.	CH$_2$Cl$_2$, C$_2$H$_4$Cl$_3$, CHCl$_3$, CH$_3$CCl$_3$, CCl$_4$, CHCl$_2$Br, C$_2$Cl$_4$	123	ECD	1983	i
PLOT Al$_2$O$_3$, 50m × 0.32 mm I.D.	CF$_2$Cl$_2$, CH$_2$Cl$_2$, CFCl$_3$, CHCl$_3$, CH$_3$CCl$_3$, CCl$_4$, C$_2$HCl$_3$, C$_2$Cl$_4$	49	ECD	1985	p
PLOT Al$_2$O$_3$ + KCl, 50m × 0.32 mm I.D.	CClF$_3$, CClF$_3$ (B1), CF$_3$CClF$_2$, CF$_2$Cl$_2$, CH$_3$Cl, CHClF$_2$, CF$_2$ClC F$_2$Cl, CFCl$_3$, CHCl$_2$F, CCl$_2$FCClF$_2$	50 124 38	ECD FID FID	1987 1987 1994	p p p
5μ CPSil 5 CB, 50 m × 0.32 mm I.D. 2μ SE-54 immobilised with DCUP, 28 m × 0.30 mm I.D.	CF$_2$Cl$_2$, CFCl$_3$, CCl$_2$FCClF$_2$, CHCl$_3$, CH$_3$CCl$_3$, CCl$_4$, C$_2$HCl$_3$, C$_2$Cl$_4$	36 125	ECD ECD	1990 1990	p p
Poraplot Q, 10m × 0.32 mm I.D. DB624, 75 m × 0.53 mm I.D. DB1, 60 m × 0.25 mm I.D.	CH$_2$Br$_2$, C$_2$HCl$_3$, CHBr$_3$, CF$_2$Cl$_2$, CH$_3$Cl, CClF$_2$ClF$_2$, CBrClF$_2$, CFCl$_3$, CCl$_2$FCClF$_2$, CHCl$_3$, CH$_3$CCl$_3$, CCl$_4$, CHClCCl$_2$, C$_2$Cl$_4$	65 38 126	MS ECD ECD	1993 1994 1996	p p p
5μ CPSil 5, 50 m × 0.32 mm I.D.	CF$_2$ClCF$_3$, CF$_3$Br, CHClF$_2$, CFClCF$_2$, CF$_2$Cl$_2$, CH$_3$CF$_2$Cl, CH$_3$Cl, CF$_2$ClC F$_2$Cl, CF$_3$CFCl$_2$, CF$_2$ClBr, CH$_3$Br, CHCl$_2$F, CFCl$_3$, CH$_3$CFCl$_2$, CF$_2$ClCFCl$_2$, CHCl$_3$, CH$_3$CCl$_3$, CCl$_4$, C$_2$HCl$_3$,	67	MS	1995	p
3μ DB 624, 70 m × 0.53 mm I.D. GLOT, GBC and AT1000, 30m × 0.25 mm I.D.	CF$_2$Cl$_2$, CFCl$_3$, CF$_2$ClCFCl$_2$, CCl$_4$ CF$_2$Cl$_2$, CF$_2$ClCF$_2$Cl, CFCl$_3$, CF$_2$ClCFCl$_2$, CH$_3$CCl$_3$, CCl$_4$, C$_2$HCl$_3$, C$_2$Cl$_4$	127 52	ECD MS	1996 1994	i p
Ciclodextrin 30m × 0.32 mm I.D.	CHF$_3$, CBrF$_3$, CH$_2$F$_2$, CF$_3$CClF$_2$, CCl$_2$F, CHF$_3$CF$_3$, CH$_3$CF$_3$, CHClF$_2$, CH$_3$Cl, CH$_2$FCF$_3$, CBrClF$_2$, CHF$_2$CH$_3$, CClF$_2$CClF$_2$, CH$_3$Br, CH$_3$CClF$_2$, CHClFCF$_3$, CCl$_3$F, CH$_2$Cl$_2$, CH$_3$CCl$_2$F, CHCl$_2$CF$_3$, CHCl$_3$	128, 129	MS FID	1996 1999	p p

i = isothermal; p = programmed termperature.

As early as 1983, K. Grob and G. Grob [123] showed in fact that for film thickness up to 5 μm several advantages can be obtained over traditional wall-coated capillary columns (0.1 ÷ 0.5 μm). These can be summarized in an at least equivalent loading capacity combined with higher separation efficiency, better stability characteristic, and increased retention, which is essential in the analysis of very volatile substances. Following these observations, several authors proposed capillary columns for the analysis of the main $C_1 - C_2$ halocarbons in ambient air, and the simultaneous separation of eight compounds (Table 12; Refs. 36 and 125) in less than 20 min was obtained, with detection limits ranging between 1 and 16 ppt. Later on [38], Blake and Rowland showed the simultaneous separation of 11 HCs, including $CBrClF_2$, in ambient air collected at 1320 m altitude (Fig. 13) using a 3 μm DB 625 Megabore, 0.53 mm I.D. capillary column.

Alternatively, the need for subambient temperature may be avoided by using gas-solid chromatography (GSC). For this reason, alumina Porous Layer Open Tubular (Al_2O_3 PLOT) columns were introduced in halocarbon analysis in 1985 [49]. In these columns, the porous adsorbent (Al_2O_3) is deposited on the inside tube wall which, in itself, acts as a stationary phase in a gas-solid chromatographic process. As the activity of the alumina surface causes peak tailing and the retention mechanism is very sensitive to the water

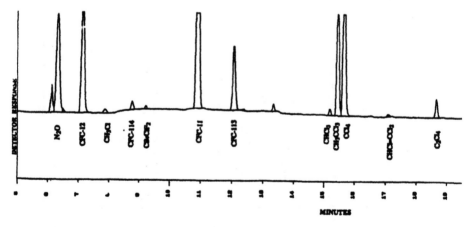

Fig. 13 Chromatogram of background ambient air collected at 1320 m altitude using a 60 m × 0.53 mm DB-1 column. (From Ref. 38.)

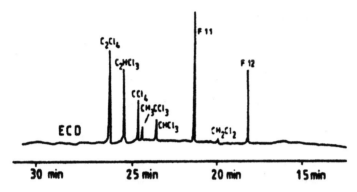

Fig. 14 ECD chromatogram of an ambient air sample, 50 m alumina PLOT column. (From Ref. 49.)

content of the carrier gas, deactivation with KCl or Na_2SO_4 is necessary. In Fig. 14 the ECD trace showing the separation of light halogenated hydrocarbons obtained by using a 50-m Al_2O_3 PLOT column is reported.

However, it was observed [124] that at elevated temperature some halogenated compounds reacted during the chromatographic process. In fact, the inherent basicity of the alumina stationary phase causes dehydrochlorination of hydrogenated halocarbons, probably leading to the formation of the corresponding unsaturated material. This phenomenon is enhanced by the presence of water or chlorine. The more polar compounds were shown to have the highest tendency to undergo dehydrochlorination. The measure for the inertness of a compound with respect to the reactivity of the alumina stationary phase is given by the stability factor K_s. Irreversible adsorption or catalytic decomposition is indicated by $K_s = 0$; no catalytic reactivity is indicated by values close to unity. From Table 13 it can be seen that, as expected, fully halogenated compounds (except CCl_4) are unaffected, whereas partly halogenated compounds interact strongly with the alumina stationary phase.

Sturrock et al. [130], in a detailed investigation regarding the evaluation of different capillary columns for the separation of CFC replacement compounds, observed the same phenomenon. Further, these authors demonstrated that fluorine-substituted HFCs (CF_3CH_3, CF_2HCF_3, and CH_2FCF_3) had K values close to unity, as well as the fully halogenated CBrClF2, and confirmed that com-

Table 13 The Influence of the Column Temperature on K_s

Compound	Temperature °C					
	75	100	125	150	175	200
CCl_3F	—	—	0.8	1.0	0.9	0.9
$C_2Cl_2F_4$	1.0	0.9	0.9	1.0	1.0	1.0
$C_2Cl_3F_3$	—	—	—	1.1	1.0	1.1
C_2Cl_4	—	—	1.1	1.0	1.0	1.1
CCl_4	—	—	0.5	0.3	0.4	0.2
CH_3CCl_3	—	—	0	0	0	0
$CHClF_2$	0.6	0.3	0.1	0	0	0
$CHCl_2F$	—	—	0	0	0	0

Source: From Ref. 124.

pounds containing chlorine and hydrogen atoms (namely, $CHClF_2$, CH_3CF_2Cl, CF_3CHCl_2, and CH_3Cl) were affected by dehydrochlorination.

To overcome the above-mentioned problems, porous polymers can be used as stationary phase. Three different fused silica capillary columns packed with porous copolymers (i.e., PoraPLOT Q, PoraPLOT S, and PoraPLOT U) with different surface areas and polarities were tested.

These columns provide high selectivity, short analysis times, good reproducibility, nonsensitivity to oxygen and water, and maximum operating temperature of 250°C. Principal disadvantages are their relatively long retention times for higher boiling compounds and bleed at elevated temperature, which complicate their use in GC-MS. The separations of a hydrogenated halocarbons standard mixture on the three different PoraPLOT columns are depicted in Fig. 15.

The separation of halocarbons and hydrogenated halocarbons on PoraPLOT columns is governed both by molecular mass and boiling point. Furthermore, the polarity of the sorbate and sorbent appear to be of some influence. Therefore, the elution order and retention time depend on the compound boiling point, its molecular size, and possible surface interaction with the stationary phase. This is clearly shown by the different retention times of CF_3Br on the three columns.

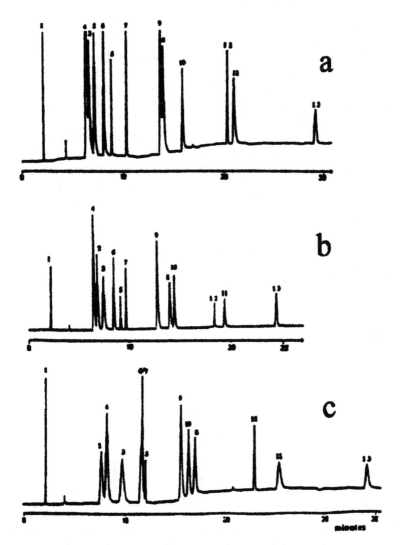

Fig. 15 Separation of HFC standard mixture on PoraPLOT columns. Temperature program: 40°C (3 min) the to 85°C at 9°C min^{-1}, 85°C (3 min) then to 140°C at 10°C min^{-1}. a) PoraPLOT Q, carrier gas 65 kPa; b) PoraPLOT S, carrier gas 25 kPa; c) PoraPLOT U, carrier gas 65 kPa. Peak identification: 1) methane, 2) CF_3Br, 3) HFC-125, 4) HFC-143a, 5) HCFC-22, 6) HFC-134a, 7) CH_3Cl, 8) HCFC-124, 9) HCFC-142b, 10) $CBrClF_2$, 11) HCFC-123, 12) CH_2Cl_2, 13) $CHCl_3$. (From Ref. 130.)

Graphite layer open tubular (GLOT) columns were used by Mangani and coworkers [52]. The main advantages of graphitized carbon blacks are the high separation factors of adsorption chromatography (further enhanced by modifying their surfaces with liquids or solids), the linear elution capability, and the low HEPTs that can be obtained. A chromatogram showing the separation of a halocarbons mixture can be seen in Fig. 16.

The above-reported chromatogram is relative to a standard mixture. Obviously, in actual sample analysis, difficulties encountered in separating nearby eluting compounds present at very different concentration levels (i.e., HFC-134a and CFC-12) are solved by using the individual specific ion traces of the single compounds.

In an above-mentioned paper, Simmonds et al. [67] tackled the problem of setting up an automated station for monitoring the main CFCs and the relative replacement compounds, with the goal of detecting as many compounds as possible. Taking into account that MS detection and trap enrichment of the air samples were required,

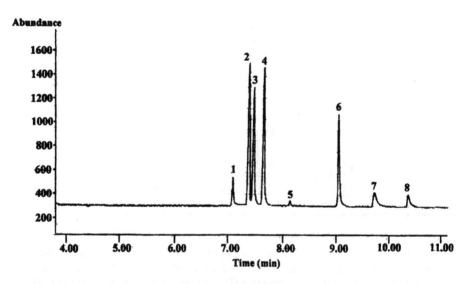

Fig. 16 Reconstructed total ion chromatogram of a standard mixture. Peak identification: (1) HFC-134a; (2) CFC-12; (3) HCFC-22; (4) HFC-142b; (5) CFC-114; (6) CFC-11; (7) HCFC-141b; (8) CFC-113. GLOT column 90 m × 0.25 mm I.D. Stationary phase Carbograph 1 coated with SP1000. Temperature program: 10 min at 35°C, then 10°C min^{-1} to 150°C. (From Ref. 52.)

they reviewed the different capillary columns discussed above. The authors concluded that the thick-film CPSil-5 silicone column represents the best compromise, because the compounds of interest can be resolved at modest temperature, while alumina/KCl PLOT columns and PoraPLOT Q columns either induce dehydrohalogenation process or require higher elution temperatures. Fig. 17 shows the separation of 26 ubiquitous compounds detected in a 2-L air sample collected at Mace Head, Ireland, including halocarbons, hydrohalocarbons, and hydrocarbons on a 5-μm CPSil-5 methyl silicone capillary column (100 m × 0.32 mm I.D.)

Selected ion monitoring (SIM) and a temperature program from 30°C up to 100°C were used. An alternative route proposed by other authors is the simultaneous use of multiple columns and/or multiple conventional detectors for the detection of halocarbons and hydrocarbons. A capillary column equipped with a three detector system, (i.e., ECD, FID, and flame photometric detector [FPD] [49] has been used for the analysis of air samples, while two capillary columns, each one equipped with an ECD, have been used for the shipboard simultaneous analysis of halocarbons in seawater and air [127]. Excellent results in this direction have been obtained by Blake and Rowland [59] with regard to the simultaneous determination of halocarbons and NMHCs (non-methane hydrocarbons) in background ambient air. The air sample is split into three chromatographic systems, (1) Al_2O_3 PLOT capillary columns plus FID for C_2–C_7 analysis, (2) a DB-1 capillary column plus FID for C_4–C_{10} analysis, and (3) a DB-624 column for CFCs and other halocarbons. About 40 compounds were detected at ppt levels in 27 minutes by temperature programming, including 11 halocarbons.

GSC columns with cyclodextrin as stationary phase were used in HC analysis by different authors [128,129]. In the study described in Ref. 129, a 30 m × 0.32 mm cyclodextrin column was tested for the separation of a number of HCs. This column was able to strongly retain also the more volatile HCs. Furthermore, all the other HCs show excellent resolution, with the exception of three pairs of compounds (i.e., CFC-12 and HFC-125, HCFC-142b and HCFC-124, CH_3Cl and HFC-134a). Also, in this case, the use of mass spectrometry by selecting ions characteristic to each compound helps in solving this problem. Also, the retention times of the less volatile compounds is not too excessively long. The elution order is governed by a combination of factors: boiling point, hydrogen bonding, and number of

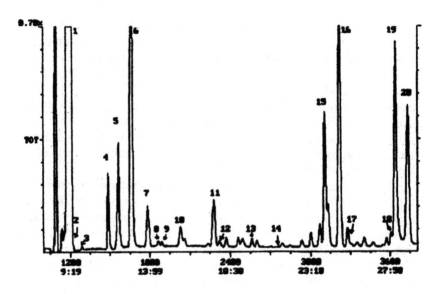

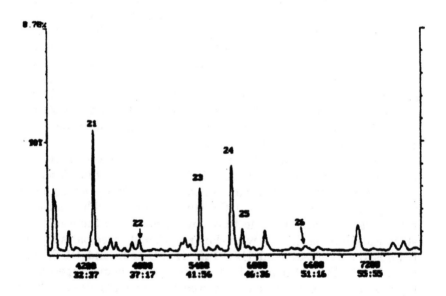

chlorine atoms. Molecules that can form hydrogen bonds are particularly strongly adsorbed, i.e., those compounds with adjacent hydrogen and fluorine atoms. An example is CH_3CHF_2, whose boiling point is $-24°C$, which is more retained than $CBrClF_2$, whose boiling point is $-4°C$. A chromatogram obtained analyzing a standard mixture of HCs is reported in Fig. 18.

D. Mass Spectrometric Detection

Detection of HCs at higher concentrations (>0.2% w/w) can be performed using the most common FID. Other modes of detection, such as thermal conductivity, atomic emission, and infrared (IR) spectroscopy have been used. However, the detection and routine observation of HCs at unprecedented low levels of concentration (ppt) were possible after the invention of the ECD by Lovelock. Actually, the unexpected results obtained by Lovelock during a halocarbon measurement campaign over the Atlantic [131,132] were the starting point for the Rowland-Molina theory. Even in recent years, developments of the ECD have been proposed [68,133] aimed at optimizing detection of HCs. However, a deep discussion of these methods is beyond the scope of this review.

Mass spectrometric detection (MSD) has been used because it shows an adequate sensitivity and a higher selectivity than ECD. Actually, GC-MS analysis is imperative when ECD response is low and/or a very high number of components have to be determined, whereas some of them may be present at low ppt level.

Mass spectrometry is a destructive analytical technique that requires a very little amount of sample (a few picograms) for the identification of a molecule and, in certain conditions, for its structural elucidation [134]. A wide variety of mass spectrometers is available to satisfy the analyst's demand, but all of them are based on the capability to assign mass-to-charge (m/z) values to ions, although

---

Fig. 17 GC-MS analysis of a 2-L air sample collected at Mace Head (Ireland). Peak identification: (1) Xe, (2) CFC-115, (3) CF_3Br, (4) HCFC-22; (5) CFC 1113, (6) CFC-12, (7) unknown, (8) HCFC-142b, (9) CH_3Cl, (10) CFC-114 and CFC-1214a, (11) butene, (12) CF_2ClBr, (13) CH_3Br, (14) CFC-21, (15) butane, (16) CFC-11, (17) 3-methyl-1-butene and HCFC-141b, (18) CH_2Cl_2, (19) unknown, (20) CFC-113, (21) pentane, (22) $CHCl_3$, (23) CH_3CCl_3, (24) benzene, (25) CCl_4, (26) $CHClCCl_2$. (From Ref. 67.)

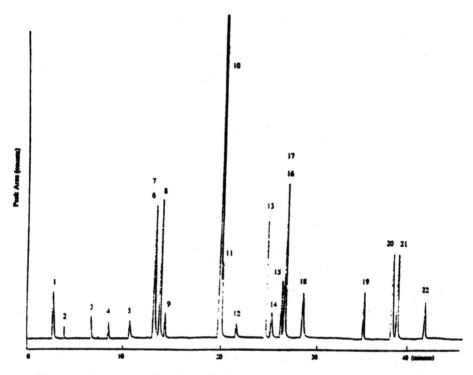

Fig. 18 Separation of a HC mixture on a 30 m cyclodextrin column. Detector FID. Temperature program: 35°C hold 15 min, then 5°C min^{-1} to 120°C, hold 4 min, then 5°C min^{-1} to 180°C, hold 5 min. Peak identification: (1) N$_2$O, (2) HFC-23, (3) Halon-1301, (4) HFC-32, (5) CFC-115, (6) CFC-12, (7) HFC-125, (8) HFC-143a, (9) HCFC-22, (10) CH$_3$Cl, (11) HFC-134a, (12) Halon-1211, (13) HFC-152a, (14) CFC-114, (15) CH$_3$Br, (16) HCFC-142b, (17) HCFC-124, (18) CFC-11, (19) CH$_2$Cl$_2$, (20) HCFC-141b, (21) HCFC-123, (22) CHCl$_3$. (From Ref. 128.)

operating principles and fields of application can differ greatly. A mass spectrometer is capable of producing ions from a neutral molecule and separating them in the gas phase according to their mass-to-charge ratio (m/z).

A mass spectrometric analysis involves different steps:

Sample introduction—samples can be gaseous, liquid or solid; liquid or solid samples must be brought to gas phase prior or during ionization.

Ionization
Mass analysis
Ion detection data analysis

The most common ionization technique used in the analysis of GC amenable compounds is electron ionization (EI), also known as electron impact ionization. It is the oldest and best characterized of all the ionization methods. The ion source of the MS operates at reduced pressure in order to prevent collisions of ions with residual gas molecules in the analyzer during the flight from the ion source to the detector. The analyte is introduced into the evacuated ion source ($<10^{-4}$ Pa) and subsequently ionized by collisions with energetic electrons, generated by a heated filament in the ion source. The electron energy is defined by the potential difference between the filament and the source housing and is usually set to 70 eV ($\sim 1.12 \times 10^{-17}$J). As a consequence, the molecule loses one of its electrons and becomes a positively charged radical ion:

$$M + e^- \rightarrow M^{+\bullet} + 2\ e^-$$

where $M^{+\bullet}$ is called the molecular ion.

The unpaired electron of the molecular ion can occupy various excited electronic and vibrational states. If these excited states contain enough energy, bonds will break and fragment ions and neutral particles will be formed. With an electron energy of 70 eV, enough energy is transferred to most molecules to cause extensive fragmentation. Decreasing the electron energy can reduce fragmentation, but it also reduces the number of ions formed. All ions are subsequently accelerated out of the ion source toward the analyzer (Fig. 19).

The mass spectrum obtained from recording all of these ions contains signals of varying m/z and intensities, depending on the numbers of ions that reach the detector. The fragmentation pathways of the molecular ion depend on the structure of the molecule, such that similar structures give similar mass spectra. Mass spectral libraries are available today that contain over 160,000 spectra that can be used to help identify unknowns and search for common substructures.

Chemical ionization (CI) relies on the interaction of the molecule of interest with a reactive ionized reagent species. Many such reactants are gaseous Brönsted acids. For example, some of the most

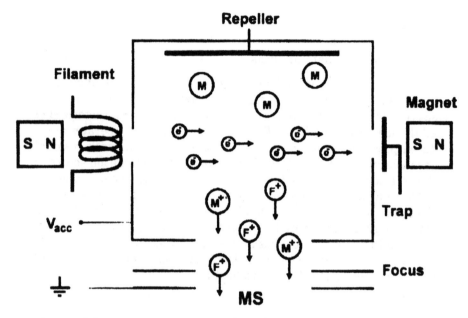

Fig. 19 Schematic representation of an electron ionization ion source. M represents neutral molecules; e⁻, electrons; M⁺ the molecular ion; F⁺, fragment ions; V_{acc}, accelerating voltage; and MS, the mass spectrometer analyzer. (From Ref. 134.)

widely used reactant species are generated by EI from methane. The first ion formed is CH_4^+ which then reacts to give the Brönsted acid CH_5^+ according to the following reaction:

$$CH_4^{+\cdot} + CH_4 \rightarrow CH_5^+ + CH_3^\cdot$$

If a neutral molecule M in the source has a higher proton affinity than CH_4, the protonated species MH^+ will be formed in an exothermic reaction.

$$M + CH_5^+ \rightarrow MH^+ + CH_4$$

The instrumentation used for CI is very similar to that used for EI. The major difference in the design of the source is that it is more gas-tight so that the reactant gas is retained at higher pressures in order to favor ion/molecule reactions. The pressure inside the CI ion source is typically of the order of 13.3–133 Pa. A wide variety of reagents have been reported in the literature, some of which lead to fragmentation (e.g., hydrogen [H_2], for which the reactant ion is H_3^+). Other reagents give rise to mass spectra that, depending on the analyte, mainly show the protonated molecule (e.g., i-butane (C_4H_{10}), for which the reactant ion is $C_4H_9^+$). Negative ions can be generated in a similar manner by a reaction involving an election capture process.

Quadrupole Mass Filter and Quadrupole Ion Trap Instruments

The basic principles of the quadrupole mass filter were published in the early 1950s by Paul and Steinwedel [135]. It has now become one of the most widely used types of mass spectrometers because of its ease of use, small size and relatively low cost. Mass separation in a quadrupole mass filter is based on achieving a stable trajectory for ions of specific m/z values in a hyperbolic electrostatic field. An idealized quadrupole mass spectrometer consists of four parallel hyperbolic rods (Fig. 20).

To one pair of diagonally opposite rods a potential is applied consisting of a DC voltage and a rf voltage. To the other pair of rods, a DC voltage of opposite polarity and a rf voltage with a 180° phase

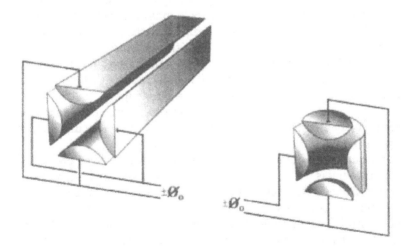

Fig. 20 Schematic representation of a quadrupole mass filter and an ion trap, where f_0 is the potential applied to opposite pairs of rods and/or caps [see Eq. 1]. (From Ref. 134.)

shift is applied. The potential f_0 applied to opposite pairs of rods is given by;

$$\pm f_0 = U + V \cos wt \tag{1}$$

where U is a DC voltage; V cos wt, the time-dependant rf voltage in which V is the rf amplitude; w is the rf frequency.

At given values of U, V and w, only certain ions will have stable trajectories through the quadrupole. The range of ions of different m/z values, capable of passing through the mass filter, depends on the ratio of U to V. All other ions will have trajectories which are unstable (i.e., they have large amplitudes in x- or y-direction) and will be lost. The equation of motion for a singly charged particle can be expressed as a Mathieu equation from which one can define expressions for the Mathieu parameters a_u, and q_u,

$$a_u = a_x = -a_y = 4zU/mw^2r_0^2 \quad q_u = q_x = -q_y = 2zV/mw^2r_0^2 \tag{2}$$

with m/z, the mass-to-charge ratio of the ion; r_0 is half the distance between two opposite rods.

There is no z parameter, because the ac field only acts in the x/y-plane (z is the main axis of the linear quadrupole). Scanning the mass range on a quadrupole means changing the values U and V at a constant ratio a/q = 2U/V, while keeping the rf frequency w fixed.

The resolving power of a quadrupole mass filter also depends on the number of cycles experienced by an ion within the rf field, which in turn depends on its velocity. Thus, the resolution will increase with increasing mass, as ions of higher mass have lower velocity. However, the transmission efficiency will decrease, due to the longer time ions of higher masses spend in the quadrupole.

The ruggedness and high reproducibility of this analyzer make it particularly amenable for field analysis and in remote unattended apparati [66,52,136].

The quadrupole ion trap is based on the same principle as the quadrupole mass filter, except that the quadrupole field is generated within a three-dimensional trap. This trap consists of a ring electrode and two end caps as shown in Fig. 20. In the original design, the potential f_0 was applied to the ring electrode and $-f_0$ to the end caps. With this arrangement, ions were detected with resonance techniques. As is the case with quadrupole mass filters, the quadrupole field is closest to the theoretical ideal in the center of the trap. For this reason, a moderator gas like helium is often introduced into the trap in addition to the sample, to dampen the oscillations of the ions and hence concentrate them in the center of the trap. As indicated by the name, the ion trap can store ions over a long period of time, making it possible to study gas phase reactions. In particular, the ion trap has excellent MS/MS capabilities. The mass range of commercial instruments is 650 Da, scanned at over 5000 Da s^{-1}. By reducing the scan speed to 0.015 Da s^{-1}, a resolving power of 1.2 × 10^7 (full width at half maximum, FWHM) can be achieved. Also, under special conditions, (i.e., resonance with external field to cause ejection) a mass range of up to 45,000 Da and sensitivities in the attomole range have been obtained. The very high sensitivity of the ion trap is a consequence of the fact that all ions formed can in theory be detected. However, space charge effects (ion–ion coulombic interactions) reduce the accuracy of mass assignment for an ion trap. Even though ion/molecule reactions take place within the trap, EI spectra generally compare well to EI spectra acquired on quadrupole mass filters. The applications potential of the ion trap is very great. Size, speed, sensitivity, MS/MS capabilities and compatibility with most ionization techniques favor further development of this mass analyzer.

The analyzer used in the above-cited work by Simmonds et al. [67] was an ion trap that had been optimized to increase the overall sensitivity for the specific analysis of hydrohalocarbons. The ion trap storage parameters were modified in order to "cut off" mass from m/z 20 to 32. The automatic gain control and tune factors were optimized so that an almost twofold gain in sensitivity for most of the halocarbons was achieved.

Membrane Introduction Mass Spectrometry (MIMS)

On-site analysis of environmental threatening chemicals is becoming more and more important. Continuous release of rules and regulation of environmental legislation require that any industry or laboratory must demonstrate, on-site, that their emissions are within the limits regulated by law. Environmental samples are often quite complex, thus demanding sophisticated methods, in the mean time transportable, sensitive and reliable. Traditional methods, such as headspace and purge-and-trap are difficult to adapt to an on-line procedure. Sample introduction restrictions can be overcome thanks to membrane introduction mass spectrometry (MIMS). This technique was first described in 1963 for the study of reaction kinetics in water [137]; however, a large number of studies have been devoted to the development of this direct interfacing technique between air and water samples and a mass spectrometer. The membrane can take the shape of a sheet across which the sample passes or of a capillary tube through which the sample flows (hollow fiber). In the latter form it shows the advantage over planar membrane because the surface of the hollow fiber is entirely self-supporting, with no need for mechanical support. A polymer membrane acts as the interface between the aqueous or gaseous sample and the ion source of a mass spectrometer. In this way, permeation of the organic compounds through the membrane occurs, providing part per million (ppm) to part per billion (ppb) sensitivity. As described in two exhaustive reviews [138,139], the transport mechanisms through the membrane is the result of three different steps:

Adsorption to the outer surface of the membrane

Diffusion through the membrane

Evaporation from the inner surface into vacuum

Diffusion through the membrane determines the rate of the process, while adsorption and evaporation are considered instantaneous. The permeation rate depends on the solubility of the analyte in the membrane, the membrane thickness, and the temperature, as described in Eq. (1):

$$I_{ss} = ADS(P_s/l) \qquad \text{(Eq. 1, Fick's equation)}$$

where I_{ss} is the steady-state flow of analyte through the membrane (mol/s); A is the membrane surface area (cm^2); D is the diffusion constant (cm^2 s^{-1}); S is the solubility constant (mol Pa^{-1} cm^{-3}) of the analyte in the membrane matrix; P_s is the vapor pressure of the analyte on the sample side of the membrane (Pa); l is the membrane thickness (cm).

A thorough discussion on the effect of the experimental parameters that influence the permeation rate, such as fiber dimension, flow rate, temperature, inlet configuration, is presented in Ref. 140. Kinetics of the extraction process is described in Ref. 141 through two mathematical descriptions. The solutions obtained were compared to each other and to experimental data, with the purpose of investigating the optimal conditions to obtain quantitative recoveries.

Evaluation of the influence of various parameters (geometry of the fiber, flow rates, Henry's constant, etc.) were taken into account. Spiraling of the fiber allowed the use of longer ones and improved recoveries at higher sample velocities.

The membranes commonly used are made of organic polymers, such as polyethylene and Teflon for gas monitoring, and silicone-based polymers for organics in aqueous solutions and in air.

A hydrophobic polypropylene hollow fiber and a silicone membrane were used to extract 1,1,1-trichloroethane, trichloroethene, and tetrachloroethene from water [142]. The polluted sample was pumped through the hollow fiber, while an inert gas was flowing counter currently around the exterior of the fiber. Cryofocusing of the analytes was necessary prior to injection into the GC column. Two different instrumental configurations were investigated: in the first one, a switching valve permitted the introduction either of sample or carrier gas. In the second one, the module was connected directly to the injector of the GC. In this case the stripping gas was constituted by the carrier gas. While the two membranes performed similarly, quantitative data were obtained in the valve configuration, that ensure better linearity.

For MIMS experiments, two different solutions can be adopted (see Fig. 21).

The first one, called flow-by (Fig. 21a) [138] employs a device that is mounted in the fluid vessel. It has been successfully employed in several studies, due to its simplicity that makes it easy to adapt to most reactors. However it suffers from memory effects, poor response times and condensation of analytes may occur along the lines. The second device, called flow-through (Fig. 21b), resembles a direct insertion probe, placed into the ion source of the MS. The sample flows across the membrane, which is from one side exposed to the vacuum of the ion source. This configuration totally eliminates memory effects, improves reproducibility and sensitivity. It was successfully used in the determination of volatile organic compounds in water [143,144] at trace level. Compatibility with continuous monitoring, especially in flow injection analysis (FIA) is another advantage of this operating mode. A considerable increase in sensitivity can be achieved when a stream of helium continuously sweeps the permeant out of the hollow fiber into the ion source. This minimizes the formation of a gas-phase analyte-enriched region in the hollow fiber interior [145].

(a)

(b)

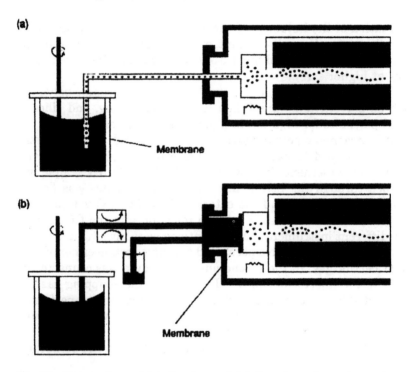

Fig. 21 Comparison of (a) flow-by and (b) flow-through configurations of membrane introduction devices. (a) The membrane is mounted inside the reactor. (b) The sample is continuously withdrawn from the reactor and transported to a membrane inlet mounted inside or near the ion source. (From Ref. 138.)

All kinds of analyzers have been used with MIMS. In early works, quadrupole analyzers were more frequently used; however, ion traps are certainly increasing in popularity due to their overall simplicity and their mode of operation. Ions coming from the matrix can be suppressed, boosting sensitivity at ppt level [146], and MS/MS can be performed.

The 59 compounds listed in EPA method 524.2 were detected at ppt level, directly from water, with sampling rates of 6 min or less. The official method is based on purge-and-trap extraction followed by GC-MS detection. Using MIMS, the authors are able to detect the selected compounds from water at ppt level without interference from the sample matrix [147]. Table 14 lists the detection limits obtained. As it can be seen, most of the compounds were detected at or below 500 ppt.

An on-line device for multiple gas/liquid stream monitoring technique has been developed and used for the extraction of organics, including several HCs, from complex and dirty matrices with little or no sample preparation [148]. This method has been used in a wastewater treatment process, monitoring the contaminants in the influent and effluent water stream as well as in the influent air stream. Minimization of errors and simplification of calibration are important qualities of this approach, thanks to the single technique and single analyzer, a quadrupole.

A set of associated papers describe a number of advances in MIMS. In answer to the more and more imperative demand of on-site chemical analysis, an advanced mobile analytical laboratory (AMAL) has been developed in the EUREKA/EUROENVIRON subproject EU674. The AMAL consists of four vehicles, one equipped with the newest sample collection methods, three with the most recent techniques for sample handling and analysis for inorganic and organic compounds. A membrane inlet mass spectrometric method for on-site analysis in AMAL is based on a helium purge type [149]. Several VOCs, including a bunch of HCs, can be determined at subppb level with response time of 1–2 min, proving the good sensitivity of MIMS. Linearity is also very good with a linear dynamic range of at least four orders of magnitude. The type of membrane chosen for this work can be easily used in any commercial GC-MS instrument without modifications. The aqueous sample continuously flows over the membrane, which is continuously purged from the inside by a helium stream directed towards the ion source of a quadrupole mass spectrometer. A diagram of the apparatus is shown in Fig. 22.

The use of tailored wave form methods to achieve, in favorable cases, parts-per-quadrillion (ppq) detection limits for organic compounds, such as 1,2-dichloroethene, in aqueous solutions [150] is described. Tailored wave form resonance methods of mass selective ion storage allows the limited ion storage capacity of an ion trap to be fully utilized for analyte ions. In this work, the stored wave form inverse Fourier transform (SWIFT) method was applied to selectively eject all but the analyte ions during ionization [151].

In another article, MIMS was employed for the on-line monitoring of biological reactions at ppt level [152]. Chloroethylenes repre-

Table 14 MIMS Volatile Organic Detection Limits

Compound name	Quantitation ion (m/z)	EPA Quantitation ion (m/z)	Detection limit (ppb)	MW
Toluene	91	91	0.01	92
Chloroform	83	83	0.025	118
o-Xylene	91	106	0.025	106
p-Xylene	91	106	0.025	106
m-Xylene	106	106	0.05	106
o-Chlorotoluene	91	91	0.05	126
p-Chlorotoluene	91	91	0.05	126
1,2-Dichloroethane	62	62	0.1	98
1,3-Dichloropropene (1:1 cis/trans)	75	75	0.1	110
Benzene	78	78	0.1	78
Carbon tetrachloride	117	117	0.1	152
Ethylbenzene	91	91	0.1	106
Styrene	104	104	0.1	104
n-Butylbenzene	91, 92	91	0.1	134
n-Propylbenzene	91	91	0.1	120
1,1-Dichloroethane	63, 83	63	0.2	98
1,1-Dichloroethene	96, 60	96	0.2	96
1,1-Dichloropropene	75	n/a[a]	0.2	110
1,2,3-Trichloropropane	75	75	0.2	146
1,2,4-Trimethylbenzene	105	105	0.2	120
1,2-Dibromo-3-chloropropane	75	75	0.2	234
1,2-Dibromoethane	107	107	0.2	186
1,3-Dichloropropane	76	76	0.2	112
Bromodichloromethane	83	83	0.2	162
Bromoform	173	173	0.2	250
Chlorobenzene	112	112	0.2	112
Dibromochloromethane	129	129	0.2	206
Isopropylbenzene	105	105	0.2	120

Compound				
Tetrachloroethene	166	166	0.2	164
Trichlorofluoromethane	101, 103	101	0.2	136
Vinyl chloride	62	62	0.2	62
cis-1,2-Dichloroethene	105	105	0.2	96
sec-Butylbenzene	105	105	0.2	134
trans-1,2-Dicloroethene	96	96	0.2	96
1,2-Dichlorobenzene	146	146	0.3	146
1,4-Dichlorobenzene	146	146	0.3	146
2,2-Dichloropropane	77	77	0.3	112
Bromobenzene	156, 158	156	0.3	156
1,3,5-Trimethylbenzene	105	105	0.4	120
1,3-Dichlorobenzene	146	146	0.4	146
Naphtalene	128	128	0.4	128
Trichloroethene	130	96	0.4	130
p-Isopropyltoluene	119	119	0.4	134
1,1,1-Trichloroethane	96	97	0.5	132
1,2,4-Trichlorobenzene	180, 182	180	0.5	180
Dibromomethane	174, 176	93	0.5	172
Methylene chloride	49	84	0.5	84
tert-Butylbenzene	119	119	0.5	134
1,1,2-Trichloroethane	96	83	0.6	132
1,2,3-Trichlorobenzene	180, 182	180	0.7	180
1,1,1,2-Trichloroethane	131	131	0.8	166
1,1,2,2-Trichloroethane	83	83	0.8	166
1,2-Dichloropropane	76	63	0.8	112
Bromochloromethane	49	128	1	128
Bromomethane	94, 96	94	1	94
Dichlorodifluoromethane	85, 87	85	1	120
Hexachlorobutadiene	224	225	1	258
Chloroethane	49	64	5	64
Chloromethane	49	50	10	50

^a Not applicable.

Source: From Ref. 147.

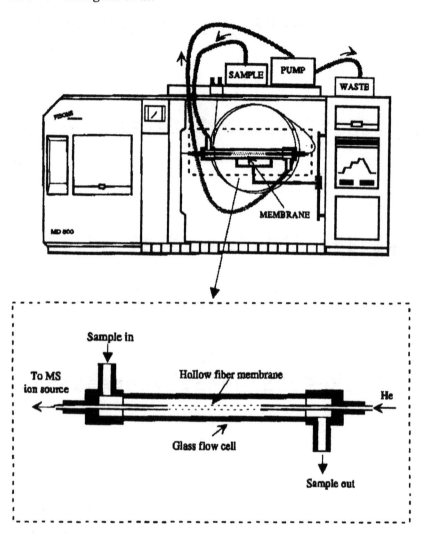

Fig. 22 Schematic presentation of the measurement system showing a helium purge type of membrane inlet mounted into the gas chromatograph oven of a Fisons MD-800 mass spectrometer and an expanded view of the membrane inlet. (From Ref. 149.)

sent a serious threat for human health even at very low concentration (1 μg L⁻¹). They are usually not degraded in the environment and their monitoring in the groundwater reservoirs has become an imperative. In this work, the oxygenation of such compounds by cultures of methanotroph *Methylococcus capsulatus* (Bath) has been monitored on-line at the low ppt level using a flow-through membrane inlet coupled to a single quadrupole mass spectrometer.

In another application, a two-stage hollow fiber membrane has been used for rapid and sensitive determination of VOCs in air by ion-trap mass spectrometry [153]. Although the compounds of interest were in air, the same membrane can be used for sequential analysis of water and soil samples without modifications of the instrument. The two-stage membrane worked well with an ion-trap mass spectrometer, since the helium flow rate used was perfectly compatible with the ion-trap buffer gas. Fig. 23 shows the detection of toluene in air (1 ppbv), water, and soil (in each, 1 ppb by weight).

Direct detection of VOCs in aqueous samples at ppt level is accomplished using jet separator/MIMS [154].

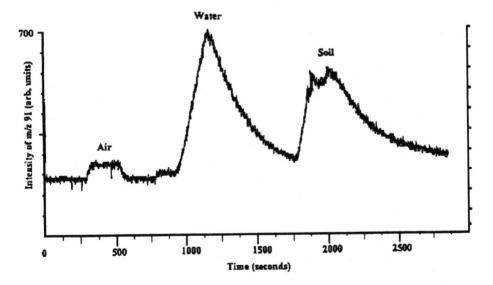

Fig. 23 Ion chromatogram showing the signal for m/z 91 and 92 from 1 ppb toluene in air, water, and soil obtained by simply moving the inlet to the membrane separator from one sample to another. (From Ref. 153.)

Sample enrichment is performed in two consecutive stages: one by means of a semi-permeable membrane interface, the other one by a jet separator. The aqueous sample flows coaxially over a semi-permeable capillary membrane, the interior of which is continuously purged by helium. The jet separator conveys the sample towards the mass spectrometer, removing much of the helium purge gas, as well as water vapor. This system is tested with two instruments: a single quadrupole and an ion trap and its performance is compared to that of membrane probes. The quadrupole was fitted with a membrane/jet separator, while the ion trap was fitted with (1) a capillary direct-insertion membrane probe (C-DIMP), (2) a membrane/jet separator system, and (3) both interfaces.

For the ion-trap membrane probe experiments, a C-DIMP was fitted with a 1.5-cm Silastic, hollow fiber membrane (0.635 mm I.D. × 1.19 mm O.D.). The sample solution was pumped through the probe at a flow rate of 2 mL min^{-1} at 30°C, using a peristaltic pump. For the ion trap membrane/jet separator experiments a 15-cm Silastic hollow fiber membrane was used (0.635 mm I.D. × 1.19 mm O.D.). The fiber was coaxially encased in a 2 mm I.D. Pyrex tube. This assembly is connected to a quartz jet separator and a vacuum pump is connected in order to remove helium and water from the analyte stream. The jet separator, operating at room temperature, was connected to the ion trap via a stainless steel tubing.

For the quadrupole experiments, a 15 cm silastic hollow fiber membrane (0.635 mm I.D. × 1.19 mm O.D.) was encased in a glass tube and connected to a custom built stainless-steel jet separator. As opposed to ion trap, the experiments were carried out at elevated temperatures.

The performance of the two MIMS/jet separator system is equivalent. Comparison between C-DIMP and jet separator experiments show that the latter permits achievement of lower detection limits that, for most of the tested compounds, span from 30 ppt to a few ppb, with a response linearity of 3 orders of magnitude.

E. Calibration

The absolute calibration of HCs measurements in the ppt range is probably the most challenging problem associated with the analysis of these compounds, especially for the most volatile ones. In fact, common volumetric methods, involving volumetric measurement by means of a microliter syringe of the standard components introduced in a container in which a diluting gas is contained, are often affected by scarce measure accuracy, further magnified by the high dilution factor that has to be achieved in a single or multiple step dilution process. Therefore, gravimetric and/or "ad hoc" volumetric methods should be used. Calibration in HCs analysis is usually carried out by using secondary ("working") standards (i.e., samples of air are periodically calibrated against primary standards).

Permeation Devices

Permeation tubes, first introduced by O'Keefe and Ortman [155], proved to be a simple and reliable method to create primary standards of several gaseous pollutants at very low concentration levels. Some years after, a method for the generation of accurate primary standards with permeation tubes especially for halocarbons was introduced [156,157]. Permeation tube method is based on a very simple principle and is amenable to the generation of standards for almost any condensable substance with a vapor pressure above a few thousands of Pa under ambient conditions. The substance of interest (normally liquid at room temperature or liquefied) is sealed in a inert plastic tube immersed in a diluent gas (e.g., N_2 or He). Permeation of the substance through the plastic tube walls occurs at a constant rate readily measured by means of an analytical balance.

In HC analysis, fluorinated ethylene propylene resin (FEP Teflon) proved to be the most appropriate, and consequently the most widely used, tube material, being chemically inert, sufficiently elastic for sealing, commercially available in different diameters and wall thickness. Furthermore, its mutual solubility with the compounds studied is nearly zero.

PTFE plugs, heat seals, tapered plugs, and mechanical clamps were tried as tube sealing device; however, simple stainless-steel spheres, having a diameter 1.5 times the internal diameter of the tube, proved very effective, forming a leak-proof seal against even high pressures.

The permeation process occurs, since inside the tube a liquid-vapor equilibrium is established regulated by the Clausius Clayperon equation, which states for any state or phase change, gas permeates the tube in accordance with the following equation:

$$F = \frac{2\,\pi\,DS(p_2 - p_1)}{\ln(b/a)}$$

where F is the rate of flow per unit length of the cylinder; D is the diffusion constant; S is the solubility coefficient; p_1 and p_2 are the partial pressures outside and inside, respectively; b and a are outer and inner radii, respectively.

Therefore, the main driving force is the difference in partial pressure between the inner and outer walls of the tube.

The permeation rate P is affected by the temperature, T, according to:

$$P = P_0 \exp(-E/RT)$$

where E is the activation energy, which changes according to the different compounds. A constant permeation rate in the steady state, therefore requiring a rigorous control of temperature; consequently tubes have to be kept at a constant temperature, controlled within $\pm 0.1°C$. This is obtained by placing the tubes in the inner space of a double-walled glass holder fed by water maintained at the desired temperature by means of a thermostatic bath [156,157]. A schematic diagram of such apparatus is shown in Fig. 24.

The permeation process actually involves three steps: the dissolution of the compound vapors in Teflon, the diffusion through the Teflon wall, and the evaporation from the outer surface. Permeation rate depends on various factors, namely the nature of the compound considered, the nature of the tube, wall thickness, tube length, and, as stated above, temperature at which the tube is kept. A constant permeation rate is reached within 1–6 weeks after the tube has been prepared, according to the nature of the compound. Permeation continues as long as some liquid is still present in the tube (up to several months).

The major merit of this device is that it can be considered as a true primary standard as it is based on gravimetric measurements. The whole tube and its content can be periodically weighed by means of an analytical balance, and permeation rate is calculated dividing weight loss by time (minutes) passed between two subsequent weighings.

Calibration curves of selected halocarbons are reported in Fig. 25 (from Ref. 157).

For actual calibration purposes the permeation rate is generally calculated by linear regression of four data points preceding the moment of quantitative analysis and two data points after it.

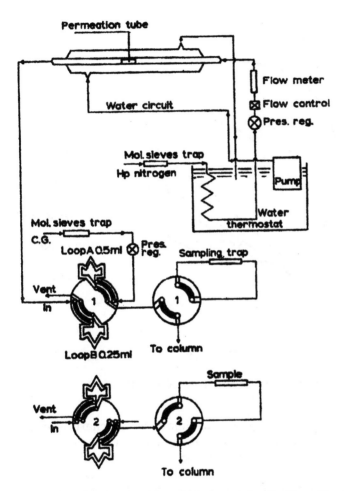

Fig. 24 Schematic diagram of the apparatus for the calibration and analysis of halocarbons in atmospheric samples. (From Ref. 157.)

However, for weighing the tube on the analytical balance, it must be removed from the thermostat. For this reason, an interesting apparatus to avoid removal of the tube from the thermostat was developed [158], in which the permeation tube to be calibrated is suspended from an electromicrobalance in a water-jacketed hangdown tube, held at a fixed temperature by thermostated water. The weight loss is recorded continuously on a 1mV recorder. Using this

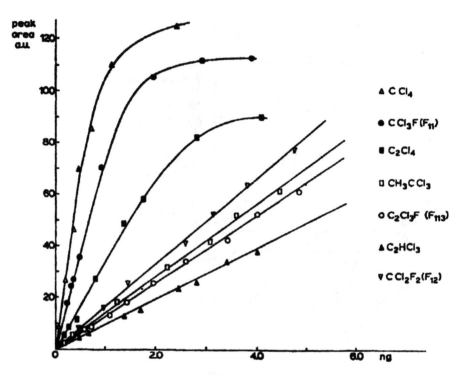

Fig. 25 Calibration graphs for several HCs. (From Ref. 157.)

continuous recording of weight versus time on a strip chart recorder, the permeation rate can be determined in hours or days instead of weeks or months. Thus, times required for the calibration are remarkably shortened. The apparatus is illustrated in Fig. 26.

By means of permeation tubes, a wide range of known concentrations at levels comparable with those found in the actual air sample can be created, just varying the flow rate of the diluent gas, and measuring it accurately.

Alternatively, when a slower permeation rate is required, or the vapor pressures of the compounds considered are too high, permeation vials can be used instead of permeation tubes. The permeation vial, described by Teckentrup and Klockow [159], is made of a glass vial whose cap is substituted by a PTFE membrane of different thickness according to the vapor pressure of the compound considered. Advantages of these devices are that they are refillable, they

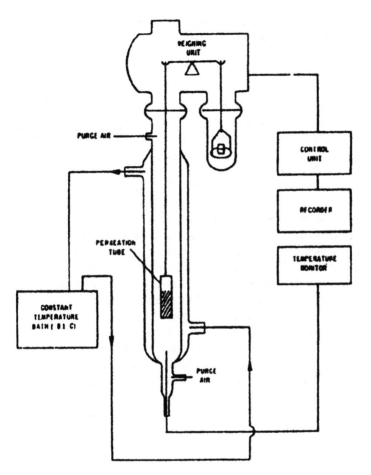

Fig. 26 Apparatus for gravimetric calibration of permeation devices. (From Ref. 158.)

can contain a large volume of liquid, and the permeation area can be kept to a minimum; thus, permeation vials can last for several years.

The permeation process occurs regulated by the same parameters, which state for permeation tubes, being proportional to the compound's partial pressure difference in- and outside the device (p_i vs p_0). Obviously, permeation rate (r) is also proportional to the permeation surface area (A_p) and inversely proportional to the membrane thickness (d_p), as can be seen from the following equation [160], where B is the permeation coefficient:

$$r = B(p_i - p_0) \frac{A_p}{d_p}$$

The use of permeation vials to calibrate an ECD for stratosphere halocarbons trace analysis was described by Noij et al. [161]: five permeation vials containing different CFCs (CFC-12, CFC-114, CFC-11, CFC-113, and CFC-40) were made by modifying GC autosampler vials (1.5 mL) provided with a screw cap, whose rubber septum was replaced by a PTFE disk (thickness 1–2 mm) closely fitting in the neck of the vial. The vials were placed in a glass permeation holder immersed in a thermostated water bath kept at the temperature of 34.90 ± 0.05°C. The glass holder was flushed with helium that passed a 1m × 4 mm I.D. glass coil before entering the main compartment. The apparatus, shown in Fig. 27, provided a well-controlled gas flow at flow rates between 10 and 200 ml min⁻¹, and with a temperature stability better than 0.01°C.

A known volume of the permeation gas was quantitatively transferred to the GC-ECD, yielding the absolute detector response for each compound: in this respect permeation vials were used as primary calibrants of the ECD. Additionally, concentrated standard mixtures were prepared statically, containing other compounds of interest (reported in Table 15 together with the ECD response factors relative to CFC-12) as well as the primary calibrants. Split injections of the standard mixtures yielded the response factors, and in

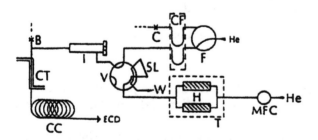

Fig. 27 Gas delivery system incorporated in the sample introduction system. T = thermostated permeation system; H = permeation tube holder; MFC = mass flow controller; V = six-port valve; SL = sample loop; W = waste; I = injector; CT = cold trap; CC = capillary column. (From Ref. 161.)

Table 15 ECD Response Factors Relative to CFC-12

Compound	CFC trade no.	Response factor	rsd (%)	Q_0 (pg)	Source*
$CBrF_3$	13BI	2.00	0.2	0.2	b
C_2ClF_5	115	0.0171	1.1	25	b
CCl_2F_2	12	**1.00**	—	0.5	—
CH_3Cl	40	0.0092	1.4	50	d
$C_2Cl_2F_4$	114	0.256	1.1	2	b
CCl_3F	11	13.8	1.3	0.08	c
CH_2Cl_2	30	0.0205	2.1	40	c
$C_2Cl_3F_3$	113	1.69	2.8	0.5	a
$CHCl_3$	20	1.10	1.2	0.8	c
CH_3CCl_3	140a	3.12	2.2	0.3	c
CCl_4	10	9.4	5	0.1	c
C_2HCl_3	1120	1.39	3.2	0.8	c
C_2Cl_4	1100	6.2	6	0.2	c

Q_0 = minimum detectable amount. *a = split injection permeation gas.
b = split injection gas standard. c = split injection liquid standard.
d = splitless injection permeation gas.
Source: From Ref. 50.

combination with the absolute response data obtained from the permeation gases experiment, absolute detector sensitivities were calculated for all the compounds studied.

Inconveniences of this approach lie in that the response of the ECD is very sensitive to variations of operational parameters like detector temperature, standing current, pulse voltage, gas flow, and composition of carrier gas and makeup gas. Therefore, frequent calibrations are required to minimize the lack of reliability.

The major disadvantages of calibration with permeation devices are the following: (1) for establishing a constant permeation rate several weeks (or months) are required; (2) extreme care should be taken to vent the effluents which are a possible major contaminant source for the environment in which analysis are performed; (3) the use of permeation devices is unsatisfactory for field use, because tubes and vials cannot be held at a constant temperature during transportation to and from field operation. Therefore, several days or weeks of calibration time are required to gravimetrically recheck the permeation rate.

For this reason, most researchers, especially those involved in field measurements in remote areas, make use of calibrated air as secondary standard (i.e., compressed air cylinders containing ambient air accurately calibrated by means of gravimetric standards or any other primary standard, like those that will be described in the following sections).

Gravimetric Dilution

This method, described by Novelli et al. [162] for preparation of CO primary standards at actual atmospheric concentration level, involves the preparation of gravimetric standards using a two-step dilution process: the first step implies the preparation of high concentration standard, the second one implies the dilution of these high concentration "parents" to lower concentration levels. The cylinders to be used for standard preparation were first evacuated to less than 5 10^{-1} Pa and accurately weighed on a Voland balance. A 5 cm^3 stainless steel tube fitted with a brass valve was evacuated and then filled to a specific pressure with the high-purity compound considered. The tube has been weighed empty five times on an analytical balance and then weighed again five times after being filled to the pressure which was previously calculated to yield the target concentration of the standard. The mole fraction of the compound of interest was calculated from the difference in weight due to the addition of the compound and the molecular weight of the compound considered. As diluent gas, natural air purified by high-temperature catalytic combustion was used. Furthermore, a stainless-steel trap containing molecular sieve 13X was placed in line between the air tank containing diluent air and the valve to which the weighing tube was attached to the manifold. This trap was used to remove any trace organic compounds or water which may have remained in the air after combustion. To prepare the parent standard, the valves of the air cylinder and the weighing tube were connected to the manifold, and with both valves closed, the manifold and the standard cylinder were evacuated to less than 5 10^{-1} Pa. Once this pressure was reached, the valve of the HC-filled weighing tube was opened to the evacuated manifold. Thus, the HCs disperse throughout the manifold lines and into the standard cylinder. With the valve of the weighing tube still open, the lines were flushed with the diluent air; thus, the HCs in the manifold were repeatedly diluted and transferred to the cylinder. The cylinder was then pressurized such that the final weight of diluent air (determined by a platform balance) yielded the target HC concentration. Finally, the weight of the pressurized cylinder was accurately determined on the Voland balance.

The procedure used to prepare the tropospheric concentration level standards starting from the obtained parents is essentially the same as that described above. However, instead of filling the weighing tube with pure HC, the tube was filled with the high concentration parent, using all the precautions that are necessary to avoid any contamination. A schematic diagram of the high pressure manifold used for the preparation of gravimetric standards is shown in Fig. 28.

A further approach for preparation of gravimetric primary standards was described [118,163,164]. It is the two-step "bootstrap" technique used at the Scripps Institution of Oceanography (SIO) (La Jolla, Ca, USA) to calibrate measurements of N_2O-halocarbons mix-

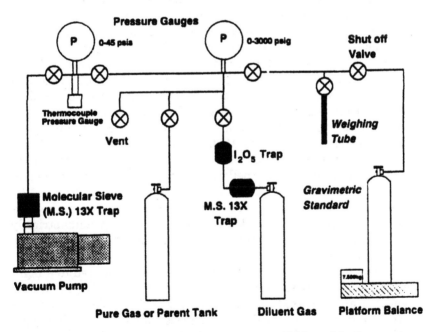

Fig. 28 Schematic of the high-pressure gas manifold used in the preparation of primary standards. The platform balance was used to gain a rough estimate of the amount of diluent air added to a cylinder during standard preparation. (From Ref. 162.)

ture. This method emphasizes the use of large volume of gas to minimize measurement and handling errors.

Gravimetric halocarbons-N_2O mixtures were prepared using a stainless-steel high vacuum line fitted with a digital vacuum gauge, used to fill a fixed volume capillary (about 100 cm^3) with a known amount of gas. The capillary, sealed on one end, was connected to a canister containing the pure halocarbon (>99.9% purity). The known amount of gas was transferred into the capillary by freezing with liquid nitrogen. The evacuated capillary was sealed with a micro torch while halocarbon is still frozen in liquid nitrogen. The mass of halocarbon sealed in the capillary was determined by weighing on a Mettler micro-balance. Before filling, the capillary was sealed on one end and weighed. After the capillary was filled and sealed again, both pieces were weighed together. It was demonstrated that less than one microgram was lost by the sealing process. As approximate masses halocarbon range from 20 to 100 milligrams, and the filled capillaries weighed between 100 and 200 mg, their content can be determined with an uncertainty ranging from 4 to 10 micrograms that negligibly contributes to the overall error of the calibration method. The sealed and weighed capillaries, one for each halocarbon, were placed in stainless-steel bellows and returned to the vacuum line to which an evacuated 850 cm^3 high pressure stainless steel canister was connected as well. Two capillaries of liquid degassed water were also added in order to obtain a water vapor pressure in the canister of about 1330 Pa, to inhibit the degradation of CH_3CCl_3 and of CCl_4, which was observed in dry containers [34]. By bending the stainless steel bellows, the capillaries were broken and their contents vacuum-transferred to the canister. About 25 g (~12.7 L) of N_2O were then transferred to the canister to complete the mixture. The amount of N_2O added was determined both volumetrically and gravimetrically (by weighing the canister before and after the mixture was added).

A small aliquot of this mixture was then transferred to a high-pressure stainless-steel tank (volume 36 L) subsequently filled with synthetic air (prepared from ultra-pure O_2 and N_2). The zero air was further purified of all the detectable traces of HCs and N_2O using a molecular sieve 13X trap kept at −78°C. By using this spiking system, small aliquots of the halocarbon-N_2O mixture were transferred at pressures above 10^3 hPa and without exposure to large surface area or dead volumes.

The gas tank is then allowed to mix and from the initial ratios of the gases and from accurate measurements of the resulting N_2O concentration, the HC concentrations can be calculated accurately.

In turn, the N_2O calibration scale [165] is based on the preparation of accurate mixture of N_2O in CO_2 at ambient ratios then diluted to ambient levels, and on accurate measurements of CO_2. Thus, as CO_2 is the "bootstrap" gas for N_2O, N_2O is the "bootstrap" gas for halocarbons.

Primary standards prepared using the above described gravimetric HC/N_2O mixtures are at the base of the SIO 1993 calibration scale, whose accuracy ranges from slightly better than 1% for the more volatile and more abundant compounds, to about 2% for less volatile and less abundant compounds.

Calibration curves of CFC-12 and CFC-11 are reported by Bullister and Weiss [164]. These were generated by multiple injections of known volumes of standard gas using two gas sample loops whose volumes were 3 and 0.5 cm^3. CFC-11 sensitivity tends to increase with increasing concentration, while CFC-12 sensitivity tends to decrease with increasing concentration.

However, employing the above-described procedure, some problems might arise when standard solutions of less volatile compounds (e.g. CFC-113, CH_3CCl_3, and CCl_4) must be prepared. The scarce precision attained is due to the absorption of small amounts of the compounds on the walls of the high vacuum system during the introduction of the halocarbons/N_2O mixtures. To solve such problem, Cunnold et al. [6] constructed a separate system to introduce the halocarbons/N_2O mixture into the primary standard cylinder using a high-pressure gas chromatography sampling valve with no unflushed "dead volumes" and never exposed to pure CFCs, thus reducing the surface area exposed to the injected aliquots by about two order of magnitude.

Static Dilution

This approach was initially described by Rasmussen and Lovelock [42]: an accurately measured amount of a pure compound (~1 mg) was dispersed into a sealed empty room, whose volume (50–100 m^3) was large enough to give in a single step a concentration level in the range of ppb or less. As walls of this room were made of a material inert toward the vapor of the compounds of interest, vapor concentration would persist indefinitely and constitute a primary standard.

Obviously, the construction of such a room was really expensive. Therefore, as an alternative, a more conventional room was used with positive pressure clean air ventilation at a known rate and with a fan for stirring the air, so that the vapor concentration in the room was uniform but was undergoing dilution at a constant rate. By observing the rate of decay of the vapor together with that of a reference gas (i.e., hydrogen) the concentration of the time zero can be found. Furthermore, since the vapor concentration decays exponentially, the logarithm of its concentration is a linear function of the elapsed time from the start of the dilution. In this way, an accurate standard with a concentration down to a tenth of the initial one was obtained. The decay of concentration (C) with time (t) is:

$$\frac{dC}{dt} = -\frac{U}{V}C$$

where V is the chamber volume and U is the flow rate of ventilating air.

To prepare the text mixture, the same authors used the following methods: pure samples of halocarbons were degassed (by freezing and thawing on a vacuum line) and used to fill an exactly known volume with vapor at a known pressure. The sample was then frozen in an evacuated glass ampoule kept at $-180°C$ by means of liquid nitrogen, and immediately sealed by gas flame. The ampoule, after being introduced in the dilution chamber, was broken and the text mixture released. In the case of less volatile halocarbons, the text mixture was prepared by adding them, using a 10-µL syringe, to a weighed volume of heptane in a small flask. In this case, the quantity added was found by reweighing.

Primary standards obtained by using the above-described procedure were used in the first phase of the ALE/GAGE project (1978–1983) to calibrate the secondary standard to be used in the different ALE monitoring stations. However, due to a strong evidence that a 2% shift in secondary calibration occurred, the Advanced Global Atmospheric Gases Experiment (AGAGE) team developed a new primary calibration scale based on the SIO primary standards above described [6].

Secondary Standards

The use of a secondary standard, which consists of calibrated clean air, is compulsory when performing a comparative study in which several laboratories are involved. This is the case of the Atmospheric Lifetime Experiment (ALE)/Global Atmospheric Gases Experiment (GAGE) carried out since 1977, and the Advanced Global Atmospheric Gases Experiment (AGAGE) which started in 1993. Measurements are carried out four times daily in five globally distributed sites. Therefore, the instrumentation located in the different sites must be daily calibrated by using reference mixtures.

In the first phase of the ALE/GAGE experiment [42] secondary standards were prepared by cryogenically liquefying clean Oregon marine air into stainless-steel passivated SUMMA® tanks (static volume 35 L) fitted with stainless-steel Nurpo® valves. For this purpose, tanks are floated in liquid nitrogen; and when the bottom of the tank has reached equilibrium with the liquid nitrogen ($-196°C$), the side valve is opened and clean air is drawn into the tank by the vacuum created as liquid air forms on the bottom surface. The filling was accelerated by pumping air into the tank by means of an ultra-clean metal bellow pump. The total amount of compressed air was generally ~1,000 L, resulting from a pressure of $3\ 10^4$hPa. The water condensed was removed by draining and the concentration of the halocarbons were determined against the primary standards. The stability of halocarbons and N_2O was followed for five years, rechecking the tanks every few months. Tests demonstrated that HCs and N_2O were unchanged in tanks down to pressure of $<7\ 10^3$ hPa, with a deviation for CFC-11 equivalent to 0.8 pptv/v, which was within the overall accuracy required of the measurements. As concentration levels of halocarbons in the tanks were in the order of 100–200 pptv, heterogeneous reactions that could take place on the tank walls could be important. Therefore, clean air samples were collected cryogenically into 1.6-L stainless-steel SUMMA® canisters. Samples were equivalent in concentration range and pressure to the 35-L tanks, the only difference being a smaller surface/volume ratio of 0.99 and 2.46 for the 1.6-canister and the 35-L tank, respectively. For all the species, except CCl_4, no concentration decrease relative to 35-L tanks was observed. Furthermore, in small canisters any volume-dependent effect would be magnified by a factor of 22 over the effect in the 35-L tanks.

The AGAGE group prepared secondary standards using compressed clean marine air taken at La Jolla. To avoid the presence of liquid water in the secondary tanks, air was pumped to a pressure higher than $4\ 10^4$ hPa (that is, the pressure generally used in the tanks), and was then expanded from a water vapor trap down to 4 10^4 hPa. In this way, the water vapor pressure in the tank was below saturation.

These secondary standards were calibrated at La Jolla, using the SIO primary calibration scale described in the foregoing section, both before and after use at the different AGAGE stations. When a standard tank was replaced at a station, the new tank was carefully measured against the old tank, using the station instrument. This was done automatically by reducing, but not suspending, the frequency of atmospheric measurements. Errors occurred during the GAGE experiment were mainly associated with changes in the secondary standard tanks. By carrying on such comparison both at the stations and at La Jolla, such errors have been reduced significantly.

Coulometry

To determine the absolute concentration of strongly electron-absorbing species, such as CFC-11 and CCl_4, coulometry has also been used [166,167]. Halocarbons are ionized irreversibly by the reaction with free electrons in the ECD, and for each molecule ionized, one electron is lost. This ionizing reaction can be used as a gas phase coulometer in which absolute measurements of trace quantities of halocarbons is possible by calculating the fraction f of a compound that is ionized at known analytical condition. Using the ECD in the constant pulse frequency, the chromatographic peak area in coulombs is obtained, and after dividing by f, an estimate of the number of molecules present can be found by applying Faraday's law. However, this method, even though it gives a good analytical precision for replicate analysis, is prone to serious systematic inaccuracies, and can be used only under very specific operational conditions [168].

REFERENCES

1. M. J. Molina and F. S. Rowland, *Nature, 249*: 810 (1974).
2. F. S. Rowland and M. J. Molina, *Rev. Geophys. Space Phys, 13*: 1 (1975).
3. *Montreal Protocol to Reduce Substances that Deplete the Ozone Layer, Final Report*, U.N. Environmental Programme, New York, 1987.

4. J. Elkins, T. Thompson, T. Swanson, J. Butler, B. Hall, S. Cummings, D. Fisher, and A. Raffo, *Nature*, *364*: 780 (1993).

5. P. G. Simmonds, D. M. Cunnold, G. J. Dollard, T. J. Davies, A. McCulloch, and R. G. Derewent, *Atmos. Environ.*, *27A*: 1397 (1993).

6. D. M. Cunnold, P. J. Fraser, R. F. Weiss, R. G. Prinn, P. G. Simmonds, B. R. Miller, F. N. Alyea, and A. J. Crawford, *J. Geophys. Res.*, *99*: D1, 1107 (1994).

7. P. D. Fraser, D. Cunnold, F. Alyea, R. Weiss, R. Prinn, P. Simmonds, B. Miller, and R. Langenfelds, *J. Geophys. Res.*, *101*: 12,585 (1996).

8. D. M. Cunnold, R. F. Weiss, R. G. Prinn, D. Hartley, P. G. Simmonds, P. J. Fraser, B. R. Miller, F. N. Alyea, and L. Porter, *J. Geophys. Res.*, *102*: D1, 1259 (1997).

9. R. G. Derewent, P. G. Simmonds, S. O'Doherty, and D. B. Ryall, *Atmos. Environ.*, *32* (21): 3689 (1998).

10. T. J. Wallington, W. F. Schneider, D. R. Worsnop, O. J. Nielsen, J. Sehested, W. J. Debruyn, and J. A. Shorter, *Environ. Sci. Technol.*, *28*: 320 (1994).

11. S. Pinnock, M. H. Hurley, K. P. Shine, T. J. Wallington, and T. J. Smyth, *J. Geophys. Res.*, *100*: D11, 23,227 (1995).

12. V. Ramaswamy, M. D. Schwarzkopf, and K. P. Shine, *Nature*, *355*: 810 (1992).

13. R. Toumi, S. Bekki, and K. S. Law, *Nature*, *372*: 348 (1994).

14. Copenhagen Amendment to the Montreal Protocol. UNEP (United Nations Environment Programme) 1992. Report of the Fourth Meeting of the Parties to the Montreal Protocol on Substances that Deplete the Ozone Layer. United Nations, New York (1994).

15. J. H. Butler, J. W. Elkins, B. D. Hall, S. O. Cummings, and S. A. Montzka, *Nature*, *359*: 403 (1992).

16. K. R. Redeker, N.-Y. Wang, J. C. Low, A. McMillan, S. C. Tyler, and R. J. Cicerone, *Science*, *290*: 966 (2000).

17. *EPA Federal Register*, June 25, 2001 (Volume 66, Number 122), pp. 33680–33681.

18. K. P. Shine, Y. Fouquart, V. Ramaswamy, S. Solomon, and J. Srinivasan, in *Climate Change 1994: Radiative Forcing of Climate Change*, Cambridge, UK: Cambridge University Press, 1995, p. 177.

19. Intergovernmental Panel on Climate Change, *Climate Change 1994: The Science of Climate Change*, Cambridge, UK, Cambridge University Press, 1995.
20. P. M. Gschwend, J. K. Mac Farlane, and K. A. Newman, *Science, 227*: 1033 (1985).
21. T. Class and K. Ballschmitter, *J. Atmos. Chem., 6*: 35 (1988).
22. W. T. Sturges, G. F. Cota, and P. T. Buckley, *Nature, 358*: 660, 1992.
23. Y. Yokouchi, H. Mukai, H. Yamamoto, A. Otsuki, C. Saitoh, and Y. Nojiri, *J. Geophys. Res., 102*: 8805 (1997).
24. M. J. Molina, in C. S. Sloane and T. W. Thoms, Eds., *Atmospheric Chemistry* Lewis, Chelsea, Mich., 1991, pp. 1–8.
25. Global Ozone Research and Monitoring Report No. 37: *Scientific Assessment of Ozone Depletion: 1994*, World Meteorological Organization, Geneva, Switzerland, 1995.
26. C. Farman, B. G. Gardiner, and J. D. Shanklin, *Nature, 315*: 207 (1985).
27. Intergovernmental Panel on Climate Change, *Climate Change 1995: The Science of Climate Change*, Cambridge, UK Cambridge University Press, 1996.
28. J. J. Rook, *J. Soc. for Water Treatment and Examination, 23*: 234 (1974).
29. ATSDR (Agency for Toxic Substances and Disease Registry), *Toxicological Profile for Bromoform and Chlorodibromomethane*, ATSDR/TP-90-05, 1990.
30. E. P. Grimsrud and R. A. Rasmussen, *Atmos. Environ., 9*: 1014 (1975).
31. R. A. Rasmussen, D. E. Harsch, P. H. Sweany, J. P. Kransec, and D. R. Cronn, *J. Air Pollut. Contr. Assoc., 27*: 579 (1977).
32. D. E. Harsch, D. R. Cronn, and W. R. Slater, *J. Air Pollut. Contr. Assoc., 29*: 975 (1979).
33. Y. Makide, Y. Kanai, and T. Tominaga, *Bull. Chem. Soc. Jpn., 53*: 2681 (1980).
34. A. Yokohata, Y. Makide, and T. Tominaga, *Bull. Chem. Soc. Jpn., 58*: 1308 (1985).
35. J. Rudolph, F. J. Johen, and A. Khedin, *Int. J. Environ. Anal. Chem., 27*: 97 (1986).
36. S. Muller and M. Oheme, *JHRC & CC 13*: 34 (1990).
37. S. A. Motzka, R. C. Myers, J. H. Butler, and J. W. Elkins, *Geophys. Res. Lett., 21*: 2483 (1994).

38. D. R. Blake, T. W. Smith Jr., T. Y. Chen, W. J. Whipple, and F. S, Rowland, *J. Geophys. Res.*, *99*: *D1*, 1699 (1994).
39. D. E. Oram, C. E. Reeves, S. A. Penkett, and P. J. Fraser, *Geophys. Res. Lett.*, *22*: 2741 (1995).
40. S. M. Shauffer, W. M. Pollock, E. L. Atles, L. E. Heidt, and J. S. Daniel, *Geophys. Res. Lett.*, *22*: 819 (1995).
41. F. Mangani, M. Maione, L. Lattanzi, and J. Arduini, *Int. J. Environ. Anal. Chem.*, *74* (9): 273 (2001).
42. R. A. Rasmussen and J. E. Lovelock, *J. Geophys. Res.*, *88*: C13, 8369 (1983).
43. R. G. Prinn, P. G. Simmonds, R. A. Rasmussen, R. D. Rosen, F. N. Alyea, C. A. Cardelino, A. J. Crawford, D. M. Cunnold, P. J. Fraser, and J. E. Lovelock, *J. Geophys. Res.*, *88*: C13, 8353 (1983).
44. J. Rudolph, D. M. Ehalt, A. Khedin, and C. Jebsen, *J. Chromatogr.*, *217*: 301 (1981).
45. L. E. Heidt, J. F. Vedder, W. H. Pollock, R. A. Lueb, and B. E. Henry, *J. Geophys. Res.*, *94*: D9, 11599 (1989).
46. F. Bruner, G. Bertoni, and G. Crescentini, *J. Chromatogr.*, *167*: 399 (1978).
47. G. Crescentini, F. Mangani, A. R. Mastrogiacomo, A. Cappiello, and F. Bruner, *J. Chromatogr.*, *280*: 143 (1983).
48. H. Frank and W. Frank, *Frezenius Z Anal. Chem.*, *332*: 115 (1988).
49. F. J. Reineke and K. Bachmann, *J. Chromatogr.*, *323*: 323 (1985).
50. T. Noy, P. Fabian, R. Borchers, F. Janssen, C. Cramers, and J. Rijsk, *J. Chromatogr.*, *393*: 343 (1987).
51. W. T. Sturges and J. W. Elkins, *J. Chromatogr.*, *642*: 123 (1993).
52. F. Mangani, M. Maione, L. Lattanzi, and J. Arduini, *Chromatographia*, *51* (5/6): 325 (2000).
53. S. J. O'Doherty, P. G. Simmonds, and G. Nickless, W. R. Betz, *J. Chromatogr.*, *630*: 265 (1993).
54. S. J. O'Doherty, P. G. Simmonds, and G. Nickless, *J. Chromatogr.*, *A*, *657*: 123 (1993).
55. A. Jordan, J. Harnisch, R. Borchers, F. Le Guern, and H. Shinohara, *Environ. Sci. Technol.*, *34*: 1122 (2000).
56. F. Bruner, M. Maione, and F. Mangani, *Intern. J. Environ. Anal. Chem.*, *62*: 255 (1996).

57. P. V. Doskey, *J. High Res. Chromatography*, *14*: 724 (1991).
58. Y. Makide, A. Yokohata and T. Tominaga, *J. Trace and Microprobe Techn.*, *1* (3): 265 (1983).
59. D. R. Blake, T-Y. Chen, T. W. Smith Jr., C. J.-L. Wang, O. W. Wingenter, N. J. Blake, and F. S. Rowland, *J. Geophys. Res.* *101*: D1, 1763 (1996).
60. N. J. Blake, D. R. Blake, O. W. Wingenter, B. C. Sive, C. H. Kang, D. C. Thornton, A. R. Bandy, E. Atlas, F. Flocke, J. M. Harris, and F. S. Rowland, *J. Geophys. Res.* *104*: D17, 21803 (1999).
61. S. A. Montzka, R. C. Myers, J. H. Butler, J. W. Elkins, and S. O. Cummings, *Geophys. Res. Lett.*, *20*: 703 (1993).
62. G. J. Sharp, Y. Yokouchi, and H. Akimoto, *Environ. Sci. Technol.*, *26*: 815 (1992).
63. H. Frank, W. Frank, H. J. C. Neves, and R. Englert, *Fresenius J. Anal. Chem.*, *340*: 678 (1991).
64. N. Kirshen and E. Almasi, *J. High Resolut. Chromatogr.*, *14*: 484 (1991).
65. Y. Yokouchi, H. Bandow, and H. Akimoto, *J. Chromatogr.*, *642*: 401 (1993).
66. Y. Yokouchi, H-J Li, T. Machida, S. Aoki, and H. Akimoto, *J. Geophys. Res.*, *104*: D7, 8067 (1999).
67. P. G. Simmonds, S. O'Doherty, G. Nickless, G. A. Sturrock, R. Swaby, P. Knight, J. Ricketts, G. Woffendin, and R. Smith, *Anal. Chem.*, *67*: 717 (1995).
68. M. R. Bassford, P. G. Simmonds, and G. Nickless, *Anal Chem.*, *70*: 958 (1998).
69. C. H. Dimmer, A. McCulloch, P. G. Simmonds, G. Nickless, M.R. Bassford, and D. Smythe-Wright, *Atmos. Environ.*, *35*: 1171 (2001).
70. C. H. Dimmer, P. G. Simmonds, G. Nickless, and M. R. Bassford, *Atmos. Environ.*, *35*: 321 (2001).
71. J.-L. Wang, C.-J. Chang, W.-D. Chang, C. Chew, and S.-W. Chen, *J. Chromatogr. A*, *844*: 259 (1999).
72. C.-C. Chang, G.-G. Lo, C.-H. Tsai, and J.-L. Wang, *Environ. Sci. Technol.*, *35*: 3273 (2001).
73. P. Peruzzi, D. Cursi, and O. Griffino, *J. High Resolut. Chromatogr. & CC*, *8*: 450 (1985).
74. N. K. Kristiansen, M. Frøshaug, K. T. Aune, and G. Becher, *Environ. Sci. Technol.*, *28*: 1669 (1994).

75. E. Fogelqvist and M. Krysell, *Marine Poll. Bull.*, *17*: 378 (1986).
76. E. Fogelqvist, *J. Geophys. Res.*, *90*: 9181 (1985).
77. K. Abrahamsson, S. *Klick Chemosphere*, *18*: 2247 (1989).
78. A. J. Nunez, L. F. Gonzales, and J. Janak, *J. Chromatogr. Chromatography Reviews*, *300*: 127 (1984).
79. M. Mohnke, and J. Buijten, *Chromatographia*, *37*: 51 (1993).
80. B. Kolb, H. Kraus, M. Auer, *Application of Gas Chromatographic Headspace Analysis*, *Vol. 21*, Bodenseewerk: Perkin-Elmer, 1978.
81. A. G. Vitenberg and M. I. Kostikina, *Zh. Anal. Khim.*, *88*: 341 (1988).
82. B. Kolb, C. Bichler M. Auer, T. C. Voice, *J. High Resolut. Chromatogr.* *17*: 299 (1994).
83. E. D. Pellizzari, R. A. Zwedlinger, and M. D. Erickson, *Environmental Monitoring Near Industrial Sites: Brominated Chemicals, Part I and II*, EPA-560/6-78-002, U.S. EPA, Office of Pesticides and Toxic Substances, Washington, DC, 1978, p. 325 and references therein.
84. K. L. E. Kaiser and B. G. Oliver, *Anal. Chem.*, *48*: 2207 (1976).
85. L. C. Michael, E. D. Pellizzari, and D. L. Norwood, *Environ. Sci. Technol.*, *25*: 150 (1991).
86. S. Blomberg and J. Roeraade, *J. Chromatogr.*, *394*: 443 (1987).
87. A. Zlatkis, R. P. J. Ranatunga, and B. S. Middledirch, *Chromatographia*, *29*: 523 (1990).
88. EPA Method N° 601.
89. D. P. Beggs, Hewett-Packard Co., GC-MS Application Note AN 176-24.
90. R. S. Narang and B. Bush, *Anal. Chem. 52*: 2076 (1980).
91. K. Grob and J. Grob, *J. Chromatogr.*, *90*: 303 (1974).
92. P. M. Gschwend, J. K. MacFarlane, and K. A. Newman, *Science*, *227*: 1033 (1985).
93. M. W. T. Jung, H. Tatematu, D. H. Sohn, and T. Maeda, *J. High Resolut. Chromatogr.*, *14*:83 (1991).
94. J. Pruvost, O. Connan, Y. Marty, and P. Le Corre, *Analyst*, *124*: 1389 (1999).
95. M. Chai, C. L. Arthur, J. Pawliszyn, R. P. Belardi, and K. F. Pratt, *Analyst*, *118*: 7005 (1993).
96. T. Nilsson, F. Pelusio, L. Montanarella, B. Larsen, S. Fac-

chetti, and J. O. Madsen, *J. High Resolut. Chromatogr.*, *18*: 617 (1995).

97. A. A. M. Hassan, E. Benfenati, G. Facchini, and R. Fanelli, *Toxicological and Environmental Chemistry*, *55*: 73 (1996).

98. A. A. M. Hassan, E. Benfenati, and R. Fanelli, *Bull. Environ. Contam. Toxicol.*, *56*: 397 (1996).

99. F. J. Santos, M. T. Galceran, and D. Fraisse, *J. Chromatogr. A*, *742*: 181 (1996).

100. J. Czerwiński, B. Zygmunt, and J. Namieśnik, *Fresenius Envir. Bull.*, *5*: 55 (1996).

101. T. Nilsson, R. Ferrari, and S. Facchetti, *Analytica Chimica Acta*, *356*: 113 (1997).

102. R. Kostiainen, T. Kotiaho, R. A. Ketola, and V. Virkki, *Chromatographia*, *41*: 34 (1995).

103. R. Kostiainen, T. Kotiaho, I Mattila, T. Mansikka, M. Ojala, and R. A. Ketola, *Anal. Chem.*, *70*: 3028 (1998).

104. M. J. Yang and J. Pawliszyn, *Anal. Chem.*, *65*: 1758 (1993).

105. A. J. Murray and J. P. Riley, *Anal. Chim. Acta*, *65*: 261 (1973).

106. H. B. Singh, L. J. Salas, and L. A. Cavanagh, *Air Pollut. Contr. Assoc. J.*, *27*: 332 (1997).

107. H. B. Singh, L. J. Salas, and R. E. Stiles, *J. Geophys. Res.*, *88*: C6, 3675 (1983).

108. D. E. Harsch, D. R. Cronn, and W. R. Slater, *Air Pollut. Contr. Assoc. J.*, *29*: 975 (1979).

109. D. R. Cronn and D. E. Harsch, *Anal. Lett.*, *12*: 1489 (1979).

110. G. Crescentini and F. Bruner, *Ann. Chim.* (Rome), *68*: 345 (1978).

111. F. Bruner, G. Bertoni, and G. Crescentini, *J. Chromatogr.* *167*: 399 (1978).

112. C. Vidal Madjar, M. F. Gonnord, F. Benchah, and G. Guiochon, *J. Chromatogr. Sci.*, *16*: 190 (1978).

113. F. Bruner, G. Crescentini, F. Mangani, E. Brancaleoni, A. Cappiello, and P. Ciccioli, *Anal. Chem.*, *53*: 798 (1981).

114. N. E. Hester, E. R. Stephens, and O. C. Taylor, *Air Pollut. Contr. Assoc. J.*, *24*: 592 (1974).

115. N. E. Hester, E. R. Stephens, and O. C. Taylor, *Atmos. Environ. 9*: 603 (1975).

116. H. B. Singh, *Nature*, *264*: 428 (1976).

117. C. W. Su and E. D. Goldberg, *Nature*, *245*: 27 (1973).

118. R. F. Weiss, J. L. Bullister, R. H. Gammon, and M. J. Warner, *Nature 314*: 608 (1985).
119. R. A. Rasmussen and M. A. K. Khalil, *Geophys. Res. Lett.*, *11*: 433 (1984).
120. S. Lal, R. Brochers, P. Fabian, and B. C. Kruger, *Nature*, *316*: 135 (1985).
121. G. C. Rhoderick and W. R. Miller, *Anal. Chem.*, *62*: 810 (1990).
122. L. Chen, Y. Makide, and T. Tominaga, *Chem. Lett. (Japan)*: 571 (1994).
123. K. Grob and G. Grob, *JHRC & CR*, *6*: 133 (1983).
124. Th. Noij, J. A. Rijks, and C. A. Cramers, *Chromatographia*, *26*: 139 (1988).
125. W. Frank and H. Frank, *Chromatographia*, *29*: 571 (1990).
126. D. R. Blake, N. J. Blake, T. W. Smith Jr., O. W. Wingenter, and F. S. Rowland, *J. Geophys. Res.*, *101*: D2, 4501 (1996).
127. S. M. Boswell and D. Smythe-Wright, *Analyst*, *121*: 505 (1996).
128. D. W. Armstrong, G. L. Reid III, M. P. Gasper, *J. Microcolumn Separations*, *8*: 2, 83 (1996).
129. S. J. O'Doherty, G. Nickless, M. Bassford, M. Pajot, P. Simmonds, *J. Chromatogr. A*, *832*: 253 (1999).
130. G. A. Sturrock, P. G. Simmonds, G. Nickless, and D. Zwiep, *J. Chromatogr.*, *648*: 423 (1993).
131. J. E. Lovelock, *Atmos. Environ.*, *6*: 915 (1972).
132. J. E. Lovelock, R. J. Maggs, and R. J. Wade, *Nature*, *241*: 194 (1973).
133. G. A. Sturrock, P. G. Simmonds, and G. Nickless, *J. Chromatogr. A*, *707*: 255 (1995).
134. URL: http://ms.mc.vanderbilt.edu/tutorials/ms/ms.htm
135. W. Paul and H. Steinwedel, *Z. Naturforsch.*, *8a*: 448 (1953).
136. S. A. Montzka, R. C. Myers, J. H. Butler, J. W. Elkins, L. T. Lock, A. D. Clarke, A. H. Goldstein, *Geophys. Res. Letters*, *23*: 2, 169 (1996).
137. G. Hoch and B. Kok, *Arch. Biochem. Biophys.*, *101*: 160 (1963).
138. T. Kotiaho, F. R. Lauritsen, T. K. Choudhury, R. G. Cooks, and G. T. Tsao, *Anal. Chem.*, *63*: 875 A (1991).
139. P. S. H. Wong, R. G. Cooks, M. E. Cisper, and P. H. Hemberger, *Environ. Sci. Technol.*, *29*: 215A (1995).

140. M. A. LaPack, J. C. Tou, and C. G. Enke, *Anal. Chem.*, *62*: 1265 (1990).
141. K. F. Pratt and J. Pawliszyn, *Anal. Chem.*, *64*: 2101 (1992).
142. K. F. Pratt and J. Pawliszyn, *Anal. Chem.*, *64*: 2107 (1992).
143. M. E. Bier, T. Kotiaho, and R. G. Cooks, *Anal. Chim. Acta*, *231*: 175 (1990).
144. A. Lister, K. Wood, R. G. Cooks, K. Noon, *Biomed Environ. Mass Spectrom.*, *18*: 1063 (1989).
145. L. E. Silvon, M. R. Bauer, J. S. Ho, and W. L. Budde, *Anal. Chem.*, *63*: 1335 (1991).
146. S. J. Bauer and R. G. Cooks, *Talanta*, *40*: 1031 (1993).
147. S. Bauer and D. Solyom, *Anal. Chem.*, *66*, 4422 (1994).
148. M. A. LaPack, J. C. Tou, and C. G. Enke, *Anal. Chem.*, *63*: 1631 (1991).
149. V. T. Virkki, R. A. Ketola, M. Ojala, T. Kotiaho, V. Komppa, A. Grove, and S. Facchetti, *Anal. Chem.*, *67*: 1421 (1995).
150. M. Soni, S. Bauer, J. W. Amy, P. Wong, and R. G. Cooks, *Anal. Chem.*, *67*: 1409 (1995).
151. M. H. Soni and R. G. Cooks, *Anal. Chem.*, *66*: 2488 (1994).
152. F. R. Lauritsen and S. Gylling, *Anal. Chem.*, *67*: 1418 (1995).
153. M. E. Cisper, C. G. Gill, L. E. Townsend, and P. H. Hemberger, *Anal. Chem.*, *67*, 1413 (1995).
154. L. E. Dejarme, S. J. Bauer R. G. Cooks, F. R. Lauritsen, T. Kotiaho, and T. Graf, *Rapid Commun. Mass Spectrom.*, 7: 935, (1993).
155. A. E. O'Keefe and G. C. Ortman, *Anal. Chem.*, *38*: 760 (1966).
156. H. B. Singh, L. Salas, D. Lillian, R. R. Arnts, and A. Appleby, *Anal. Chem.*, *11*: 511 (1977).
157. G. Crescentini, F. Mangani, A. R. Mastrogiacomo, and F. Bruner, *J. Chromatogr.*, *204*: 445 (1981).
158. L. J. Purdue and R. J. Thompson, *Anal. Chem.*, *44*: 1034 (1972).
159. A. Teckentrup and D. Klockow, *Anal. Chem.*, *50*: 1728 (1978).
160. T. Ibusuki, F. Toyokawa, and K. Imagami, *Bull. Chem. Soc. Jap.*, *52*: 2105 (1979).
161. T. Noij, P. Fabian, R. Borchers, C. Cramers, and J. Rijks, *Chromatographia*, *26*: 149 (1988).
162. P. C. Novelli, J. W. Elkins, L. P. Steele, *J. Geophys. Res.*, *96*: 13109 (1991).

163. J. L. Bullister, *Ph.D. Thesis*, University of California, San Diego, (1984) 172 p.
164. J. L. Bullister and R. F. Weiss, *Deep Sea Res.*, *35*: 839 (1988).
165. R. F. Weiss, C. D. Keeling, and H. Craig, *J. Geophys. Res.*, *86*, 7197 (1981).
166. J. E. Lovelock, R. Maggs, and R. Wade, *Nature*, *241*: 194 (1973).
167. D. Lillian H. B. Singh, *Anal. Chem.*, *46*: 1060 (1974).
168. E. P. Grimsrud and S. W. Warden, *Anal. Chem.*, *52*: 1842 (1980).

5
Microfluidics for Ultrasmall-Volume Biological Analysis

Todd O. Windman, Barb J. Wyatt, and Mark A. Hayes *Arizona State University, Tempe, Arizona, U.S.A.*

Biochemical analysis on a very small scale is beneficial, and there are a variety of reasons for this: small samples lead to minimally invasive clinical/diagnostic procedures; multiple analytes can be measured from single traditional samples; most very small scale procedures lead to faster turn-around and present the possibility of massive parallel analysis; and less waste is generated. A less obvious aspect of small-volume chemical analysis on a fast time scale is that it will soon be possible to monitor biological processes on the same time scale and volume in which the intricate mechanisms of biochemistry occur in living systems. In fact, for microelectrodes, microdialysis, and near-field fiber-optic probes, this capability could be argued as a reality, albeit for a small subset of active biomolecules. This idea of intimately monitoring the chemical activity of living systems has essentially driven the field of ultrasmall volume bioanalysis, although it is unclear if the loose collection of techniques and methods that are utilized for these analyses even constitutes a defined field.

I. INTRODUCTION

Ultrasmall sample volumes bring new problems and opportunities. The relatively new term used to describe the control of ultrasmall fluid volume—microfluidics—defines a class of manipulations in which commonplace macro-scale notions of transport no longer dominate the movement of fluids and molecules within a fluid. Before embarking on an understanding of developments in the field, it is important to define the parameters of microfluidics. For example, flow stream and sample volumes: typical volume flow rates range from a few nanoliters per second to picoliters per second and sample volumes from nanoliter to femtoliter. Just to emphasize how small a femtoliter (10^{-15}L) is, it is instructive to look at the common notation 1/1,000,000,000,000,000 of a liter. On these small scales, gravity and convection are no longer the dominant forces; surface tension, intermolecular and surface interactions, and diffusion typically dominate motions and interactions. Microfluidics and ultrasmall-volume bioanalysis can trace their roots along several paths (Table 1, Fig. 1): From microscopy, histology, and immunohistology; from nano- and micro-chemical probes such as microelectrodes, near-field fiber-optic probes, and ion-selective electrodes; from microdialysis sampling, capillary electrophoresis, and flow cytometry; from ad-

Table 1   Origins of Microfluidics

Microscopy	Microdialysis
Histology	Capillary electrophoresis
Immunohistology	Detection technology
Probes (Nano and micro)	Molecular recognition
Flow cytometry	Fabrication processes

Microfluidics grew out of a number of different fields, each with its own contribution to the state of the art in ultrasmall volume biological analysis.

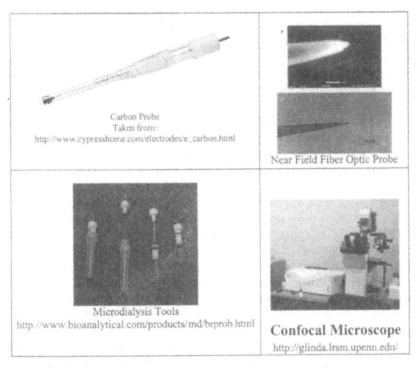

Fig. 1   Types of probes used in the study of biology.

vances in detection technology in spectroscopy, mass spectrometry, and molecular recognition; and from recent developments in microfluidics on microchip devices.

II. MICROSCOPY

Building on this theme of multiple origins, it is instructive to briefly review the various lines of development that have brought about the present technology. It is easiest to start with light microscopy of excised biological samples as the origin of ultrasmall volume samples, where the visualized structures were correlated with healthy and disease states, and eventually with the underlying biochemical processes. Recently, the visual images have been extended to native fluorescence of biological tissues, thus providing some additional chemical differentiation [1]. This correlation with biochemical processes took a large step when immunostaining microscopy techniques were developed. This field has provided a valuable tool that is still quickly evolving—a look at the catalogue of the commercial supply house Molecular Probes (Eugene, OR) provides ample evidence of these advances. Growing out of static staining techniques is dynamic optical imaging, where new information about dynamics and localization of important bioactive agents can be discerned [2,3]. For instance, in one recent report, Qian and Kennedy developed a unique optical measurement of the release of Zn^{2+} based on confocal fluorescence microscopy in pancreatic β-cells in culture. With this novel method, colocalization of exocytotic release sites and Ca^{2+} entry was observed upon stimulation by glucose or K^+. However, there are severe limitations with approaches rooted in traditional techniques [4]. These approaches require incubation of the tissue or cell with fluorescent dye and thus rely on diffusion to transport the dye to the area of interest, which is often not the case. Other techniques require the "fixing" of the tissue, which often destroys cellular viability. There are also location-specific detection limitations defined by the transport of the diffusing dye: the resulting signal is a complex mixture of the concentration of the target and the specific transport mechanisms and kinetics associated with dye transport and localization.

A technique that avoids the delivery of reagents to the cellular matrix is imaging mass spectrometry. From this technique, a rich amount of varied chemical information is available from fixed cellular tissues [5–9]. This class of methods extends far beyond that offered by traditional techniques because the mass spectrum for each sampling spot can conceivably be recorded. The techniques are centered on matrix assisted desorption ionization mass spectrometry (MALDI-MS), with new contributions coming from desorption/ionization on porous silicon (DIOS) and direct imaging secondary ion mass spectrometry (DISIMS). The methods differ in how the appropriate amount of energy is added to the target molecules for desorption and ionization without their destruction. For MALDI-MS, the sample is coprecipitated with an organic matrix that adsorbs the laser energy and literally explodes the sample off the surface relatively intact. DIOS is related to MALDI-MS, but the sample or tissue is deposited or grown on porous silicon, which can perform similar functions as the matrix in MALDI-MS. Both of these methods are well suited for analysis of high-molecular-weight biomolecules, whereas DISIMS uses a direct excitation of the sample with high-energy metal ions and is limited to low-molecular-weight targets because the ion beam destroys other materials [10–12]. With each of these methods, the tissue is either fixed or grown onto an appropriate substrate, and a mass spectrum is recorded from a single spot from the resulting desorbed and ionized materials (Fig. 2). The information from each location can be interpreted or the sampling spot can be rastered across the sample, and an image is reconstructed from the resulting spectra. Even though the amount of chemical information available from these powerful new techniques is somewhat staggering, it is still limited in that no dynamics will be available from these techniques. To analyze the sample, it must be fixed and placed in an appropriate vacuum chamber, extinguishing any biochemical activity.

III. DIRECT PROBES

By fabricating very small direct probes—literally small enough to be inserted into single cells or in very localized space in or about tissue—a variety of advantages for monitoring biochemical activity are realized. These small probes are based on optical or electrochemical interrogation and are distinguished by transducing the concentration of a targeted biomolecule at the site of the probe tip to interpretable changes in the light or electric properties. These probes include microelectrodes, nanoprobes, nanosensors, or biosensors. The electrochemical transduction mechanism has been used to monitor fast changing localized concentrations of biomolecules that can

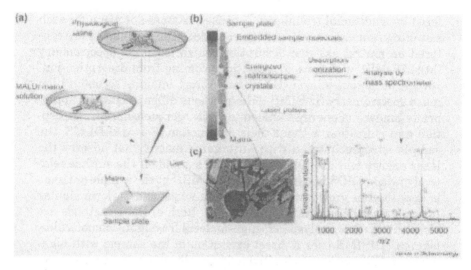

Fig. 2 Overview of single-cell MALDI-MS. (a) The strategy of cell isolation and sample preparation. After animal dissection, the tissue or ganglion of interest is pinned down in a dissection dish. Individual cells are then isolated either in the presence of physiological saline (freshwater specimens) or within a solution of MALDI matrix (marine specimens). Most cells are subsequently placed onto a MALDI sample plate containing a drop of matrix and allowed to air dry. (b) The MALDI-ionization process. In MALDI-MS, the analyte is dilutely embedded in a solid matrix and bombarded with ultraviolet or infrared laser pulses. The laser causes an explosion-like ion plume, which desorbs and ionizes the compounds in the sample into the gas phase, and the ions are mass analyzed with a time-of-flight mass analyzer. Shown is a typical mass spectrum from a single bag cell neuron from *Aplysia californica*. (c) A cultured neuron with cellular processes and matrix crystals visible. (From Ref. 63.)

easily undergo electron transfer reactions at modest electrochemical potentials, termed electroactive species. These probes can be extremely well localized (<1 µm to ~30 µm) and are a good match for high transient concentrations associated with cellular release or intracellular containment. The high-fidelity measurements (both in terms of spatial and temporal resolution) available from microelectrodes has allowed monitoring of individual exocytic events where the changing concentration of neurotransmitter, hormone, or neuropeptide can be measured at the site of release and enable one to track the microsecond–time scale events [13]. The electrochemical probes can be operated in a potentiometric or amperometric form, depending on the specificity of the required measurement and the preexisting knowledge of the tissue or cell being investigated. The chemical resolution of electrochemical measurements is limited to naturally electroactive species and generally no more than two species for a given potential sweep range.

Along the same lines as microelectrochemical probes is near-field fiber-optic probes—called nanoprobes, nanosensors, and, if combined with some biological recognition, biosensors. These probes are able to generate highly localized light (~20–50 nm in diameter) at the tip of a fiber-optic cable by heating and pulling the tip to a very small diameter and depositing metal on the outer surface except exactly at the tip. This structure allows the penetration of some light even though the orifice is well below the diameter of the diffraction limit. The light penetrates only the near field beyond the tip, so the volume being probed is extremely localized. This limited probe volume carries with it all the advantages and disadvantages of the electrochemical probes, with, however, some important differences in transport and detection issues. First, the biomolecule of interest must possess a unique spectral signature, either in an absorption wavelength with sufficiently small background to monitor altered concentration or fluorescence wavelengths (excitation/emission) that can be sufficiently correlated to a specific species. Alternatively, the probe tip can be modified with molecular/atomic recognition elements that change optical properties when interaction occurs. The other major difference is that most, if not all, of the target molecules are not consumed, so that a depletion layer and the diffusion forces are not initiated. Thus, mass transport is generally not an issue for optical probes. Again, similar to electrochemical probes, there is only a limited subset of important bioactive molecules that possess an appropriate chromophore for selective detection in the complex milieu of living systems.

IV. MICROSAMPLING

From a sampling point of view, all these nanoprobes are dependent on probe positioning and local diffusion for delivery of the target species to be interrogated. Thus, the sampled volumes are defined by the size of the probe and the characteristic diffusion distances, and they are typically in the picoliter to femtoliter range.

In contrast to on-site signal transduction, the removal and analysis of fluids from biological microenvironments has been accomplished with microdialysis probes, albeit with lower spatial resolution than the micro-and nanoprobes. This technique pumps fluid past a semipermeable membrane fabricated at the tip of a tubular probe, allows a limited range of species (based on molecular weight) to be sampled via trans-membrane diffusion, and collects the returning fluid, which is then analyzed for target species. The probe tip and adjoining structures are on the order of a millimeter, and transport to the probe is generally based on diffusion and thus the spatial resolution really is not in the ultrasmall volume regime, but this technique must be addressed for several reasons. It is closely related to newer approaches and is still commonly practiced in a variety of pharmacological and clinical settings. Also, it represents an important developmental step in the continuous sampling of living systems. This method has the distinct advantage of delivering a small sample to a remote location for standard biochemical or chemical analysis techniques. This is in stark contrast to the microscopy, electrochemical, and near-field optical techniques that, in essence, transduce the analyte concentration to a spectral or electrical signal at the probe site. By delivering the sample to a remote site, a variety of analytes can be assessed by a number of methods, including immunoassay, separations, and mass spectrometry. With this combination of sampling and analysis, a wide variety of biomolecules and simultaneous multiple targets can be assessed.

Another very important factor that is unique to microdialysis approaches is the delivery of agents in the perfusion fluid. Pharmacological agents, hormones, and neurotransmitters can be delivered to very localized tissues, and the biochemical response can be monitored directly with the same apparatus. Certainly there are a number of limitations to this approach, of which size is probably the most important: the amount of local tissue damage caused during insertion is significant. It not only decreases the spatial resolution but can cause significant physiological artifacts.

V. DIRECT SAMPLING

Somewhere between microdialysis and nano- and microprobes is direct sampling. This technique removes a minute sample, transports it through a very narrow tube (sub-micrometer to tens of micrometers) to be analyzed at a remote site [14–16]. The sampled volumes can be on the order of picoliter to femtoliter, similar to that of other microprobes, but the fluid is actively transported some distance for remote analysis—consistent with microdialysis. This technique is based on electrokinetic principles, the same ones that make capillary electrophoresis (CE) a high-resolution separation technique: efficient transport through narrow tubes (1–100 micrometers inner diameter), minimal dispersion of the original concentration profiles during transport, and a rigid, but flexible microtubing material. Clearly, using pressure or vacuum on this size scales would not only be impractical but possibly deadly to the tissue or animal under study. In addition, pressure-induced flow generates a significant amount of dispersion, which smears out the original concentration profiles and reduces the concentration through dilution with the surrounding buffer. The detection systems must be compatible with the ultrasmall volumes and modest concentrations of some analytes and have therefore been limited to fluorescence, electrochemistry, and mass spectrometry.

VI. CAPILLARY ELECTROPHORESIS AND ELECTROKINETIC EFFECTS

The fundamental transport mechanisms of CE lend themselves to efficient separation science; CE has been directly applied to many biological problems. These same advantages of the electrokinetic effects have led to their being applied to many microfluidic device transport problems. Because this technique is a key part of microfluidics, some discussion of CE is warranted. Traditionally, CE has been performed in capillaries with inner dimensions of 15–75 μm [17]. Separations in smaller channel dimensions can improve the resolution, and this is one of the current goals for CE [18]. Wei and Shear performed analysis in channels with an inner diameter of approximately 600 nm and a sample volume of 180 fL in 1998 [19]. More recently, the Ewing group was able to achieve separations of dopamine and catechol with sample volumes of 12 fL using 430 nm inner diameter tubing [17]. This work suggests that extremely high resolution in terms of volume, space, and time is available with refinements in existing technologies.

Electroosmosis is the driving fluid movement force in traditional CE. Electroosmosis is the result of the interaction between the solu-

Fig. 3 Electrokinetic transport versus pressure-driven flow. (a) Electrokinetic flow was achieved through a 75-μm I.D. capillary at a 200 V/cm potential field strength. (b) Pressure-driven flow was achieved through a 100-μm I.D. fused-silica capillary (5 cm of H_2O per 60 cm of column length). Images show a caged fluorescein dextran dye moving under both flow mechanisms at $t = 0$ and at $t = 165$ ms.

tion and the surface charge of the wall interface [20]. This fact is well shown by work from the Sandia National Laboratory graphically demonstrating the reduced dispersion when an electroosmotic flow profile is compared with that of a pressure-induced flow (Fig. 3) [21]. Because electroosmosis is an interaction of a surface with a liquid, it has not been limited to capillaries and has been performed in microchannels as well [22].

VII. DIRECT SAMPLING COMBINED WITH MICROFLUIDIC DEVICES

A relatively new approach is to deliver the ultrasmall volume sample to a microchip device for purification, separation, and analysis. This approach allows for various components to be analyzed simultaneously while sampling an inconsequential amount of fluids from the system of interest. Ultrasmall volume bioanalysis can be performed in a variety of fashions on microfluidic devices. This is one method of maximizing the information available from an ultrasmall volume sample that has been directly sampled. The rest of this chapter will focus on the development and application of microfluidic devices for small-volume biological analysis.

VIII. MICROFLUIDICS ON MICROCHIP DEVICES

The field of microfluidics as an analytical focus of study can be traced back to the early 1990s. When Harrison, Ramsey, Manz, and Widmer began fabricating microchips with fluid-filled channels, they launched a new and exciting use for the well-established fabrication methods for microelectronic and microelectromechanical technology [26]. The first chips, which were employed to separate small volume samples by electroosmosis, demonstrated the usefulness of the technique; however, the challenges of this type of system were recognized quickly. There was a need to develop chips with devices incorporated to control flow rates and enhance separation efficiency.

Sample injection was an area that presented a challenge. In initial experiments, the problems of leakage and unpredictable transport were observed and were addressed in subsequent designs. For example, leakage of sample from one channel to another was found to be an extreme problem [27,28]. Studies conducted to determine the extent of the problem showed that leakage could cause an increase of up to 20–30% of background fluorescence signal [29]. Higher applied potentials, used to control the flow and increase flow rate, caused higher leakage effects. So, although methods were developed to increase the rate of analysis and perform more complex functions by using more than one channel, these very methods were creating problems with contamination and signal adulteration.

To combat this problem, pinched injection was introduced and became the method of choice [28,30–32]. This is a method whereby the side channels are held under voltage control to induce a very slight backflow of solution, thus preventing the introduction of side-channel constituents into the main channel (see Fig. 4).

Once the primary problems of separation and injection were explored, the field was primed to move forward with analysis techniques. Capillary electrophoresis, chromatography [33,34], mass spectrometry [35–37], bioassay [38,39], DNA digestion and amplification [40–45], and even combinatorial synthesis reactions have been performed on a microchip system.

Work on improving the technology through increased reaction rates, separation times, and better detection limits continues and has been greatly advanced to this point. Theoretical analysis of the components of microfluidics is also providing insight into new directions in fabrication and analysis [46,47].

Much of the research in the 1990s centered on establishing fabrication methods and demonstrating the capabilities for the new technology. The use of electroosmotic pumps to control the flow rates of fluids provided a means for a valveless method of switching fluid flow through channels [22,48]. The valveless switch has provided a

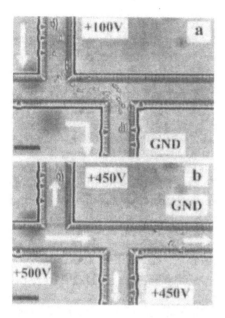

Fig. 4 A microfluidic device pinched injection of *E. coli* cells.

means by which sample manipulation and capture can be performed in volumes at and below microliters (Fig. 5). During this time, continuous separation was accomplished by freeflow electrophoresis integrated on a silicon chip [49,50]. This provided an improved "world-to-chip" interface, which had been and still remains a difficult issue [50]. Moreover, it provided a way for sample injection in a controlled manner at small volumes and low concentrations [51].

The compact size of the microchip systems also allows for optimized space for large-scale analysis. A microfabricated chip with 48 lanes, each with the capacity to handle two samples, demonstrated the use of small volumes and fast analysis. All 96 samples were analyzed in less than 8 minutes [52]. Other means of increasing analysis rates such as increased efficiency and resolution with increased field strengths (for electrophoretic methods) and separation lengths have been successful as well [30,53–56].

The advances in microfluidics on chips cannot be fully described without recognizing those who worked out the principles of preparing the microchip itself. Laser ablation of the microchip medium to

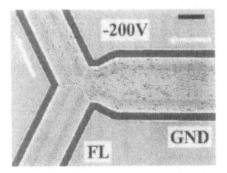

Fig. 5 The path of flow can be changed from the top channel to the bottom with the valveless switch.

produce channels and reservoirs opened the door to more complex reactions and analysis directly on the chip. Injection-molded micro-fabricated electrophoretic separation devises were produced, making another avenue to single-use disposable chips a possibility [57]. Much of the research has focused on the fabrication of these microfluidic devises on planar glass [30]. Glass microchips performed far better than their silicon predecessors and exhibited better fluid control. Although glass has become the popular media in which fluid manipulation is performed, it has suffered from problems in bioanalysis.

IX. POLYMERS

Polymers are being explored as another format for creating microfluidic devices. The growth in research using polymers as a material to fabricate microfluidic devices on chips has grown considerably. The application of a polymer substrate provides several

Table 2 Ideal Polymer Properties

Compatible with photolithographic developers	Optically transparent to detection scheme
Little or no autofluorescence	Inert to experimental condition
Surface can be modified	High dielectric strength
High thermal conductivity	

Table 3 Polymers Used in Microchannels

	Glass transition temperature (K)	Dielectric strength (V/cm)	Thermal conductivity (Wm/K)	Tensile strength (psi)	Effect of strong bases, acids	Effect of organic solvents	Clarity (UV/vis)
Poly(methylmethacrylate)	377	17.7×10^4	0.193	7000–11000	Attacked, acids only	Soluble	Opaque/transparent
Poly(tetrafluoroethylene)	>523	15.7×10^4	0.405	2000–5000	Resistant below 80°C	Very resistant	Opaque/transparent
Poly(carbonate)	430	3800	0.193	8000–9500	Attacked	Soluble	Transparent/transparent
Polybutene	249	NA	0.130	3800–4400	Attacked, acids only	Resistant	Opaque/transparent
Poly(oxydimethylsilylene)	265	NA	0.20	NA	Attacked, bases only	Soluble	Transparent/transparent
Polystryene	381	5000–7000	0.116	5000–12000	Attacked, acids only	Soluble	Transparent/transparent
Glass	>923	11.8×10^4	0.732	4000–8000	Resistant	Resistant	Transparent/transparent
Silicon	>923	30.0×10^4	0.0149	16400	Resistant	Resistant	Opaque/opaque

Values obtained from *Polymer Handbook*, 4th ed., J. Brandup, E. H. Immergut, and E. A. Grulke (Eds.), Wiley-Interscience, New York, 1999.

benefits compared with glass. The physical properties of the polymer must be taken into consideration in fabricating these devices. Table 2 lists a number of characteristics that need to be considered for polymer selection.

Conventional methods of fabricating microfluidic devices in glass have mostly required the fabrication process to be performed in a cleanroom [18]. Polymer devices can be created relatively fast using soft-lithographic techniques and have the benefit, conceivably, of being inexpensive [58]. A number of different polymers have been used for creating the microchannels that are listed in Table 3.

PDMS is an attractive polymer to work with for a number of reasons. PDMS is capable of reproducing surface features at the micrometer level [18]. It is also optically transparent to ~230 nm and exhibits a high electrical volume resistance. Furthermore, it can seal itself if punctured and a variety of methods exist to control its surface chemistry [18].

X. CHANNEL GEOMETRY

Channel designs have evolved alternative geometries, demonstrating progress in designing complex systems for miniaturized total analysis systems (μ-TAS). Initially, turn geometry was deemed a poor choice where it reduced the efficiency of band shape [30]. The reduced efficiency was a consequence of the difference in path length of the inner and outer radius of the turn. Mathies reported on an optimization of turn geometry, trying a variety of different turns shapes; square, rounded, and tapered (Fig. 6). The system was tested by examining the resolution of the 271/281 base pair doublet in the separation of a φX174 HaeIII DNA sizing ladder. The work resulted in two parameters that should be considered in optimizing new geometries. First, that turns with the smallest radius of curvature minimize the effect of tilted lateral diffusion and, second, that the resolution per channel length improves as the radius of curvature is reduced and taper length increases [59].

XI. FABRICATION TECHNIQUES

New fabrication techniques were employed to bring a greater variety of materials control to the microfluidic systems [60,61]. Hot

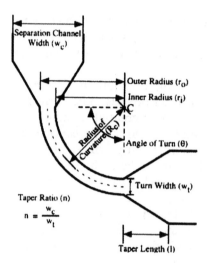

Fig. 6A Defining parameters of a tapered turn in a microfabricated channel. The width of the channel in the turn is denoted w_t and the width in the separation channel is defined as w_c. The turn subtends an arc of θ rad with a radius of curvature R_c. The width, w_t, can also be defined in terms of the inner and outer radii of the channel, r_i and r_o, respectively.

embossing, injection molding, casting, laser ablation, and ion beam etching have all been employed for fabrication of microchannels. With the exception of ion beam etching, all have come about as a fabrication technique for polymer microchannels in the past five years.

Hot embossing is currently the most widely used replication process for fabricating microchannels [58]. A relatively simple technique, hot embossing consists of heating a vacuum chamber to just above the glass transition temperature of the material and applying a stamp with a hydraulic press to the material while it begins to cool. The stamp is then carefully removed, and the cycle is ready to begin again. Hot embossing has been demonstrated to create channels less than a micrometer in width in PMMA [58], and a total fabrication time of less than 5 min [62]. Hot embossing minimizes replication errors due to the controlled nature of the pressure being applied to the stamp [62].

Injection molding is very similar to hot embossing, where its origins can be traced back to the macroscopic world [58]. Polymer pel-

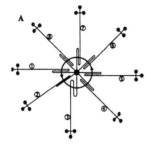

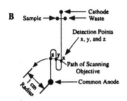

Fig. 6B (A) Layout of the microplate. The 1-cm-radius path of the scanning objective is indicated by the arrow. The parameters of the turns in each channel are summarized in Table 1. (B) Magnified view of a single channel. The sample, waste, and cathode reservoirs were fabricated 3 mm from the injection cross. The 14-μm-deep separation channels were 138 μm wide and the injection cross-channels were 44 μm wide. The scanning objective monitored the separation at points x, y, and z placed 2.75, 3.82, and 4.93 cm, respectively from the injection cross.

lets are fed into a heated syringe to form a flowing polymer [62]. Then polymer is injected by pressure onto the master mold, under vacuum, to aid in filling all spaces. This system differs from hot embossing in that it allows for an automated process, typically about 3 min, and was derived from industry [58].

Another alternative is casting, which is a much slower process for creating microchannels. The amount of time for the channel to be produced depends on the curing time for the polymer in use. Although a much slower process, casting is a much simpler technique that requires minimal capital investment [58], producing channels by pouring the polymer over the master mold and allowing them to cure [18].

Laser ablation allows for rapid fabrication of microchannels on a single device. (See Table 4 for properties of different fabrication techniques.) While removing the need for a master mold, laser abla-

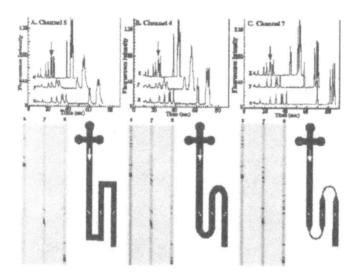

Fig. 6C Electropherograms measured at detection points x, y, and z for channels 5, 4, and 7. An image of the separation and the channel design are given below each set of electropherograms. The square turn, channel 5., the U-shaped turn, channel 4. and the tapered turn with $R_c = 500$ μm, $l = 200$ μm, and $n = 4:1$, channel 7. The fragments of the φX174 *Hae*III digest DNA ladder are of lengths 72, 118, 194, 234, 271, 281, 310, 603, 872, 1078, and 1353 base pairs. The 271- and 218-bp peaks are indicated by the arrow. The channel designs are not drawn to scale.

Table 4 Properties of Different Fabrication Techniques

	Number of masks	Type of mask	Smooth walls, floor	3D structures	Machine in glass, quartz	Machine in polymers
LIGA	2	Optical, x-ray	Yes	Yes	No	Yes
Laser ablation	1	Optical	No	No	No	Yes
Reactive ion etching	0	None	No	No	Yes	Yes
Imprinting	1	Optical	Yes	No	No	Yes
Molding	1 or 2	Optical, x-ray	Yes	No	No	Yes

tion does require a mask, an xy-table with a high degree of precision, and a laser [58]. The polymer is exposed to the laser, which then undergoes decomposition [62]. Depth control is limited to a tenth of a micrometer; however, the surface chemistry is different when compared with the previous techniques due to the interaction of the laser light [58].

XII. DETECTION

There are several detection systems used on microfluidic devices, including laser-induced fluorescence (LIF), electrochemical (ECD), and mass spectrometry (MS). LIF has been argued to be the best-suited detection system [17]. Although chemical derivatization may be necessary, it can be performed post separation in a controlled environment.

Electrochemical detection is useful tool for small-volume detection. ECD is not limited by the reaction time restraints of performing post-separation modification, as with fluorescence. Furthermore, ECD is sensitive to mass, instead of volume. Specifically, amperometric detections are appealing because they are highly sensitive, selective, and relatively low cost to set up and maintain [69].

Existing technologies were adapted to a microchip device with a mass spectrometer. Electrospray created by an electric field at the side of the chip provided the necessary means to transfer effluent to a mass spectrometer [36,37]. With electrospray ionization, the components separated by CE are introduced into the mass spectrometer.

XIII. BIOLOGY

Some would consider DNA and immunoassay arrays to fall within the category of ultrasmall-volume bioanalysis. In fact, by 1997 the lab-on-a-chip concept had its first commercially available products within this category. Substrate separation, immunoassay, DNA digestion and amplification, and analysis of human serum proteins had been accomplished in published systems [40–45]. However, these systems have not been coupled to any ultrasmall-volume or single-cell analysis schemes nor is any dynamic information obtained from these devices, and hence they are somewhat beyond the scope of this chapter [38,39,70,71].

Microfluidic devices can be exploited by either directly growing or transferring cultured or extracted cells, or tissues, to the device for cellular manipulation and analysis. Assessing the properties of cells by flow cytometry is a valuable mechanism for investigating populations of cells either transferred to the microdevice or grown within a portion of the device. Analysis of cells has typically been accomplished in bulk by taking a large homogeneous sample, processing the material, and performing a traditional biochemical or chemical analysis. These methods have been, of course, a valuable paradigm for clinical applications such as finding the average amount of hemoglobin in a cell [23]. Obviously, this method cannot be easily transferred to ultrasmall volumes.

Single-cell analysis can provide more intimate information about the cellular function and its communication with its environment. Rather than looking at the bulk sample of cells and extrapolating information about activity to a single cells, as has been the traditional approach, a large number of single-cell measurements can be performed, which in turn can provide both detailed individual cellular information and a much more detailed statistical basis for cell populations. Single-cell analysis has the potential for generating an early detection system of cell damage or abnormality [24]. However, it is difficult to detect a few cells with irregularities among a large population of healthy cells.

Recently, the sensitivity of bulk cell analysis was compared to that of single cell analysis using human colon adenocarcinoma cells or HT29. Fig. 7A shows the separation of the components of a cell by capillary electrophoresis, whereas Fig. 7B shows a similar analysis of a bulk sample [25]. Fig. 7B illustrates an averaged peak of the total cell population. Does the single cell represent a subpopulation or specific exceptional case or is it within the statistical norm of the distribution of normal or bulk-sampled cells? This question can be adequately addressed with multiple single-cell analyses, whereas traditional techniques fall short.

For application of microfluidic devices to small volume sampling, cell and protein adhesion can become a problem due to debris left on the channel walls [72]. During an initial exposure, the cell generally does not adhere nor does debris obstruct the channel, but subsequent exposure leads to clogging if sufficient cleaning procedures are not followed [73]. Debris on the channel walls can influence the accuracy of transport and analysis. Decreasing the channel dimensions

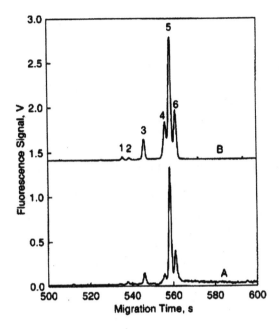

Fig. 7 Electropherograms obtained from (A) single-cell and (B) bulk analysis of the cell extract. The peaks correspond to (1) a tetrasaccharide, Le^y-TMR; (2) a trisaccharide, Le^x-TMR; (3) unreacted substrate, LacNAc-TMR; (4) unidentified product; (5) a monosaccharide, βGlcNAc-TMR; (6) a TMR-aglycone, HO(CH$_2$)$_8$CONH(CH$_2$)$_2$NH-TMR.

has resulted in a better retention efficiency where shear stresses apparently play a large role in cellular transport [72].

Although using small channel sizes can minimize effects of adhered cellular debris, it does not eliminate the problem entirely. As an alternative, coating the channel wall has been one way of minimizing the adhesion of cells. To measure the extent of cell adhesion, *Escherichia coli* has been used because of the tendency of this species to affix to glass. Alternatively, PDMA demonstrated minimal adhesion but generates an unwanted consequence of suppressing electroosmotic flow. Channels coated with PDMA had bovine serum albumin mixed with the buffer to eliminate any remaining adhesion problems [74].

One challenge of performing CE analysis is cell injection. Electrokinetic and hydrodynamic injection systems have been used, depending on the experimental condition. Electrokinetic injections are typically used for cells that adhere to surfaces, and when trying to introduce the sample to a MS system [75]. An injection time of 1 s is enough time to provide an adequate injection when there is a potential of −1 kV [17]. The other traditional method of injecting a cell into a capillary is by hydrodynamics. A hydrodynamic injection works by creating a suction that can hold onto the cell that is suspended in solution. A third technique for performing single-cell injection is optical trapping of a cell [51]. One limitation—or advantage, depending on the specific circumstance with these methods—is that they are limited to a single injection and separation at a time [24].

Once a cell has been introduced into the channel, cell lysis can be induced within the first minute of the separation, typically. Cell debris can and often does create a problem with sticking and/or blocking the channel due to the small channel dimensions [18], and as a consequence the channel must be cleaned in-between each run

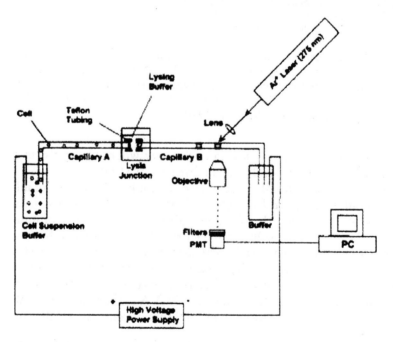

Fig. 8 Schematic of instrumental setup for high-throughput CE analysis of single cells.

or risk affecting further CE runs [25]. Dovichi devised an automated way of performing single-cell analysis while looking at glycosylation and glycolysis from a single human carcinoma cell. Although the technique was somewhat limited, it demonstrated a cell monitoring system during injection. Once introduced to the capillary, the cell could be lysed within 30 seconds of injection, a separation preformed, and then a reconditioning of the separation capillary [25].

Dovichi's work increased the speed in which analysis could be performed, but it still was not an optimized system for two reasons. First, manual observation of cell injection was needed. This proves to be tricky when trying to keep the total injection length below 400

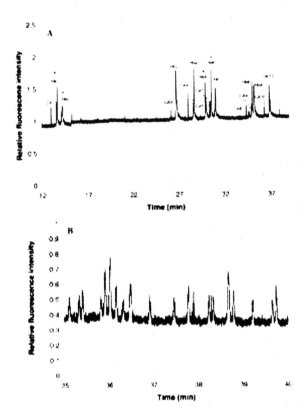

Fig. 9 A high-throughput analysis of single human erythrocytes performed by a continuous single cell injection (A) and with 0.3% SDS in the lysis junction (B).

μm. Second, the system wasn't efficient because of down time while the capillary was rinsed.

Chen and Lillard devised a system that allowed for repeated whole cell injections while at the same time decreasing the amount of cell debris entering into the capillary. They introduced a lysis zone by introduction of a gap junction. The exit of the cell injection capillary is approximately 5 μm away from the injection entrance of the separation capillary. This allows the cell to flow into the junction and then undergo lysis without the use of a detergent, where the cellular contents flow into the separation capillary. The separation of carbonic anhydrase II and hemoglobin from a human erythrocyte sample was demonstrated (Figs. 8 and 9) [24], which illustrates how rapidly multiple cell runs can be performed and the unique nature of each cell.

In addition to performing separations, electrokinetic effects demonstrate that they can effectively control flow dynamics. Work by Harrison demonstrates sample manipulation, on a microchip, of baker's yeast, *E. coli*, and canine erythrocytes (Figs. 4 and 5) [72]. While differing in shape and size, all these cells flow along the electric field lines by electrokinetic effects (electroosmosis and electrophoresis). In addition to controlling the flow of the cells, controlled chemical reactions can be induced. For example, the labeling of proteins and cell lysing can be done on a chip in a controllable manner using a valveless format by changing the electric potential at the junction [72]. In addition, mixing can be performed in these systems [76].

XIV. BIOLOGICAL APPLICATION AND DETECTION

Early work by the Ewing group has led to a variety of interesting applications of CE in monitoring single-cell release [14–16]. A recent and particularly interesting one is monitoring of the cerebral ganglion neurons of *Aplysia californica* [77]. Although the experimental parameters for the study of neural cells is not conducted under conditions encountered in nature, methodology is being developed for future experimentation. Probing *Aplysia californica* with CE/MALDI MS allows for isolation of proteins and peptides in a manner that allows for further study, due to samples being deposited onto a MALDI plate [78]. By coupling CE with MS, two different separation mechanisms (CE charge to drag, and MS charge to mass) are coupled together to obtain highly resolved temporal and spatial data, which is ideal for studying complex biological systems [79]. Such data can begin to give insight into how cellular components are arranged within the cell and how its function varies to adjacent cells.

REFERENCES

1. E. S. Yeung, *Anal. Chem.*, *71*(15): 522A (1990).
2. W. J. Qian and R. T. Kennedy, *Biochem. Biophys. Res. Commun.*, *286*(2): 315 (2001).
3. G. K. Walkup, et al., *J. Am. Chem. Soc.*, *122*(23): 5644 (2000).
4. T. Vo-Dinh, B. M. Cullum, and D. L. Stokes, *Sensors and Actuators B-Chemical*, *74*(1–3): 2–11 (2001).
5. L. J. Li, R. W. Garden, and J. V. Sweedler, *Trends Biotechnology*, *18*(4): 151 (2000).
6. R. A. Kruse, et al., *J. Mass Spectrometry*, *36*(12): 1317 (2001).
7. R. M. Caprioli, T. B. Farmer, and J. Gile, *Anal. Chem.*, *69*(23): 4751 (1997).
8. J. Y. Xu, R. M. Braun, and N. Winograd, Abstracts of Papers of the American Chemical Society, *222*: 180 (2001).
9. D. A. van Elswijk, et al., *Int. J. Mass Spectrometry*, *210*(1–3): 625 (2001).
10. M. L. Pacholski, et al., *J. Am. Chem. Soc.*, *121*(21): 5101 (1999).
11. T. L. Colliver, et al., *Anal. Chem.*, *69*(13): 2225 (1997).
12. D. M. Cannon, N. Winograd, and A. G. Ewing, *Annu. Rev. Biophys. Biomol. Struct.*, *29*: 239 (2000).
13. R. M. Wightman, P. Runnels, and K. Troyer, *Analytica Chimica Acta*, *400*: 5 (1999).
14. T. M. Olefirowicz and A. G. Ewing, *Anal. Chem.*, *62*: 1872 (1990).
15. T. M. Olefirowicz and A. G. Ewing, *J. Neurosci. Meth.*, *34*: 11 (1990).
16. T. M. Olefirowicz and A. G. Ewing, *Chimia*, *45*: 106 (1991).
17. L. A. Woods, et al., *Anal. Chem.*, *73*(15): 3687 (2001).
18. J. C. McDonald, et al., *Electrophoresis*, *21*(1): 27 (2000).
19. J. Wei, et al., *Anal. Chem.*, *70*(16): 3470 (1998).
20. D. C. Grahame, *Chem. Rev.*, *41*: 441 (1947).
21. P. H. Paul, M. G. Garguilo, and D. J. Rakestraw, *Anal. Chem.*, *70*(13): 2469 (1998).
22. D. J. Harrison, A. Manz, and P. G. Glavina, Transducers '91,

Digest of Technical Papers, IEEE 91-CH2817-5((ISBN 0-87942-586-5)): 792 (1991).

23. B. Sterling, et al., *Clin. Chem.*, *38*(9): 1658 (1992).
24. Z. Chen, et al., *J. Chromatogr. A*, *914*(1-2): 293 (2001).
25. S. N. Krylov, et al., *J. Chromatogr. B*, *741*(1): 31 (2000).
26. N. A. Polson and M. A. Hayes, *Anal. Chem.*, *73*(11): 312A (2001).
27. D. J. Harrison, et al., *Anal. Chem.*, *64*: 1926 (1992).
28. Z. H. Fan and D. J. Harrison, *Anal. Chem.*, *66*: 177 (1994).
29. D. J. Harrison, et al., *Science*, *261*: 895 (1993).
30. S. C. Jacobson, et al., *Anal. Chem.*, *66*: 1107 (1994).
31. K. Seiler, et al., *Anal. Chem.*, *66*: 3485 (1994).
32. S. C. Jacobson and J. M. Ramsey, *Electrophoresis*, *16*(4): 481 (1995).
33. A. W. Moore, S. C. Jacobson, and J. M. Ramsey, *Anal. Chem.*, *67*: 4184 (1995).
34. S. C. Jacobson, et al., *Anal. Chem.*, *66*: 2369 (1994).
35. D. Figeys, Y. Ning, and R. Aebersold, *Anal. Chem.*, *69*: 3153 (1997).
36. J. Li, et al., *Anal. Chem.*, *71*: 3036 (1999).
37. R. S. Ramsey and J. M. Ramsey, *Anal. Chem.*, *69*: 1174 (1997).
38. N. Chiem and D. J. Harrison, *Anal. Chem.*, *69*: 373 (1997).
39. L. B. Koutny, et al., *Anal. Chem.*, *68*: 18 (1996).
40. A. T. Woolley, et al., *Anal. Chem.*, *68*: 4081 (1996).
41. E. Delamarche, et al., *Science*, *276*: 779 (1997).
42. A. G. Hadd, et al., *Anal. Chem.*, *69*: 3407 (1997).
43. J. Cheng, et al., *Anal. Biochem.*, *257*: 101 (1998).
44. L. C. Waters, et al., *Anal. Chem.*, *70*: 158 (1998).
45. L. C. Waters, et al., *Anal. Chem.*, *70*: 5172 (1998).
46. S. V. Ermakov, S. C. Jacobson, and J. M. Ramsey, *Anal. Chem.*, *70*: 4494 (1998).
47. N. A. Patankar and H. H. Hu, *Anal. Chem.*, *70*: 1870 (1998).
48. A. Manz, et al., Transducers '91, Digest of Technical Papers, IEEE 91CH2817-5((ISBN 0-87942-586-5)): 939 (1991).
49. D. E. Raymond, A. Manz, and H. M. Widmer, *Anal. Chem.*, *66*: 2858.
50. D. E. Raymond, A. Manz, and H. M. Widmer, *Anal. Chem.*, *68*: 2515 (1996).
51. D. T. Chiu, et al., *Anal. Chem.*, *69*(10): 1801 (1997).
52. P. C. Simpson, et al., *Proc. Natl. Acad. Sci.*, *95*: 2256 (1998).

53. D. J. Harrison, Glavina, and A. Manz, *Sensors Actuators B, 10*: 107 (1993).
54. S. C. Jacobson, et al., *Anal. Chem., 66*: 1114 (1994).
55. C. S. Effenhauser, et al., *Anal. Chem., 66*: 2949 (1994).
56. S. C. Jacobson, et al., *Anal. Chem., 70*: 3476 (1998).
57. R. M. McCormick, et al., *Anal. Chem., 69*: 2626 (1997).
58. H. Becker and C. Gartner, *Electrophoresis, 21*: 12 (2000).
59. B. M. Paegel, et al., *Anal. Chem., 72*(14): 3030 (2000).
60. D. C. Duffy, et al., *Anal. Chem., 70*(23): 4974.
61. C. S. Effenhauser, A. Manz, and H. M. Widmer, *Anal. Chem., 67*: 2284 (1995).
62. S. A. Soper, et al., *Anal. Chem., 72*(19): 642A (2000).
63. S. C. Jacobson, et al., *Anal. Chem., 66*: 4127 (1994).
64. A. T. Woolley, et al., *Anal. Chem., 70*: 684 (1998).
65. A. Manz, et al., *J. Chromatog., 593*: 253 (1992).
66. N. Burggraf, et al., *J. High Res. Chromatogr., 16*: 594 (1993).
67. S. C. Jacobson, et al., *Anal. Chem., 66*: 3472 (1994).
68. K. Fluri, et al., *Anal. Chem., 68*: 4285 (1996).
69. S. Hu, et al., *Anal. Chem., 69*(2): 264 (1997).
70. S. C. Jacobson and J. M. Ramsey, *Anal. Chem., 68*: 720 (1996).
71. S. R. Liu, et al., *Anal. Chem., 71*(3): 566 (1999).
72. P. C. H. Li and D. J. Harrison, *Anal. Chem., 69*: 1564 (1997).
73. T. T. Lee and E. S. Yeung, *Anal. Chem., 64*(23): 3045 (1992).
74. M. A. McClain, et al., *Anal. Chem., 73*(21): 5334 (2001).
75. J. Bergquist, et al., *Anal. Chem., 66*(20): 3512 (1994).
76. G. Blankenstein and U. D. Larsen, *Biosensors & Bioelectronics, 13*(3-4): 427 (1998).
77. Y. M. Liu, T. Moroz, and J. V. Sweedler, *Anal. Chem., 71*(1): 28 (1999).
78. J. S. Page, S. S. Rubakhin, and J. V. Sweedler, *Analyst, 125*(4): 555 (2000).
79. T. F. Hooker and J. W. Jorgenson, *Anal. Chem., 69*(20): 4134 (1997).

6
Recent Trends in Proteome Analysis

Pier Giorgio Righetti and Annalisa Castagna *University of Verona, Verona, Italy*

Mahmoud Hamdan *GlaxoSmithKline Group, Verona, Italy*

I. INTRODUCTION

The goal of proteomics can be defined as the identification of all the proteins encoded in the human (or any other) genome followed by the determination of (1) their range of expression across the different cell types that constitute the various (human) tissues; (2) their subcellular localization; (3) their post-translational modifications; (4) their interaction with other proteins, and (5) their structure–function relationships. Additionally, there exists the need for developing an understanding of the global and temporal expression patterns at different developmental, physiologic, and pathologic states. The implications of such an undertaking for biomedicine are tantalizing, as they range from accelerating drug discovery to the early detection and therapy of cancer and other diseases, just to name a few applications.

It would thus appear that, whereas the genome sequence is fairly unidimensional and finite, the proteome is multidimensional, with extremely large dimensions, stemming from the vast number of cell lineages and subtypes, with the addition of various body fluids, each with its own complement of proteins. Moreover, the proteome is dynamic and constantly changing in response to various environmental factors and other signals, thus giving rise to a mind-boggling number of states. Thus, were we to decide to initiate a single, large international human proteome project, this would be a much more difficult enterprise than the recently terminated human genome project. Much more sophisticated tools for proteome data integration would additionally be required than those devised during the assembling of genomic data. A human proteome project should additionally be composed of various phases, which would ultimately need proper integration:

A. Expression Proteomics

This field involves the identification and quantitative analysis of all proteins encoded in the human genome and a proper assessment of their cellular localization and their post-translational modifications. Hopefully, with very high sensitivity mass spectrometric (MS) tools, such a detailed protein profiling might be obtained by starting with as few as a few thousand cells isolated from a tissue via, for example, laser-capture microdissection. In a future scenario, one could envisage bringing proteomic technologies to a patient's bedside and use such technologies to his or her benefit. Such an approach might require extensive reduction of sample complexity prior to MS analysis, such as obtainable by antibody-based microarray technologies for large-scale, high-throughput, and high-sensitivity protein expression profiling.

B. Functional Proteomics

This field is devoted to the study of protein–protein interaction, to the determination of the role of individual proteins in specific pathways and cellular structures, to the elucidation of their structure, and to the determination of their function. This approach could thus benefit by focusing on particular sets of proteins, such as those that are pathway driven or that are involved in cell signaling, to name just a few examples. Contrary to the aggressive approach that tends to solubilize all proteins in a cell by a strongly denaturant protocol, functional proteomics would require, in most instances, a more gentle approach enabling the harvest of entire classes of proteins under physiological conditions.

C. A Proteome Encyclopedia

A critical issue pertaining to proteome mining efforts would have to deal ultimately with the organization of proteome-related data into a knowledge base. One would have to regroup scattered proteome data, as exist in different databases, into a single, annotated, human protein index. The construction of a human protein encyclopedia would ultimately be the final goal of an international proteome project, a titanic effort vastly superior in breadth and scope to the building of the first encyclopedia attempted in the 18th century by the French encyclopedists, headed by Diderot and D'Alembert.

II. A GLIMPSE AT MODERN TECHNOLOGIES

Although the power of two-dimensional (2-D) electrophoresis as a biochemical separation technique has been well recognized since its introduction, its application, nevertheless, has become particularly significant in the past few years, as a result of a number of developments, outlined below:

> The 2-D technique has been tremendously improved to generate 2-D maps that are superior in terms of resolution and reproducibility. This new technique utilizes a unique first- dimension that replaces the carrier ampholyte-generated pH gradients with immobilized pH gradients (IPG) and replaces the tube gels with gel strips supported by a plastic film backing [1].

> Methods for the rapid analysis of proteins have been improved to the point that single spots eluted or transferred from single 2-D gels can be rapidly identified. Mass spectroscopic techniques have been developed that allow analysis of very small quantities of proteins and peptides [2–5]. Chemical microsequencing and amino acid analysis can be performed on increasingly smaller samples [6]. Immunochemical identification is now possible with a wide assortment of available antibodies.

More-powerful, less-expensive computers and softwares are now available, allowing routine computerized evaluations of the highly complex 2-D patterns.

Data about entire genomes (or substantial fractions thereof) for a number of organisms are now available, allowing rapid identification of the genes encoding a protein separated by 2-D electrophoresis.

The World Wide Web (WWW) provides simple, direct access to spot pattern databases for the comparison of electrophoretic results and to genome sequence databases for assignment of sequence information.

A large and continuously expanding application of 2-D electrophoresis is proteome analysis. The proteome is defined as "the PRO⁻TEin complement expressed by a genOME"; thus, it is a fusion word derived from two different terms [7,8]. This analysis involves the systematic separation, identification, and quantitation of a large number of proteins from a single sample. Two-dimensional PAGE is unique not only for its ability to simultaneously separate thousands of proteins but also for detecting post- and co-translational modifications, which cannot be predicted from genome sequences. Other applications of 2-D PAGE include identification (e.g., taxonomy, forensic work), the study of genetic variation and relationships, the detection of stages in cellular differentiation and studies of growth cycles, the examination of pathological states and diagnosis of disease, cancer research, and monitoring of drug action. Among the books dedicated to this topic, one could recommend those of Wilkins et al. [9], Kellner et al. [10], Rabilloud [11], and Righetti et al. [12]; among book chapters, Westermeier [13] and Hanash [14], to name just a few.

Perhaps one of the key points of the success of present 2-D PAGE was the introduction of the IPG technique, as stated above, and in particular the early recognition that wide, nonlinear pH gradients could be generated, covering the pH 3.5–10 interval. Such a gradient was calculated for a general case involving the separation of proteins in complex mixtures, such as cell lysates, and applied to a number of 2-D separations [15–17]. The idea of this nonlinear gradient came from an earlier study by Gianazza and Righetti [18], computing the statistical distribution of the pIs of water-soluble proteins and showing that as much as two-thirds of them would focus in the acidic region (taking as a discriminant pH 7.0) and only one-third in the alkaline pH scale. Given this relative abundance, it was clear that an optimally resolving pH gradient should have a gentler slope in the acidic portion and a steeper course in the alkaline region. Such a general course was calculated by assigning to each 0.5 pH unit interval in the pH 3.5–10 region a slope inversely proportional to the relative abundance of proteins in that interval. In a separation of a crude lysate of Klebsiella pneumoniae, a great improvement in resolution of the acidic cluster of bands was in fact obtained, without loss of the basic portion of the pattern [15]. This article was also a cornerstone in IPG technology because it demonstrated, for the first time, that the pH gradient and the density gradient stabilizing it did not need to be colinear, because the pH could be adjusted by localized flattening, leaving the density gradient unaltered. This nonlinear pH 3.5–10 gradient formed the basis (with perhaps minor modifications) of most of the wide nonlinear gradients adopted today and sold by commercial companies. The other events that made IPGs so powerful was the recognition that a much wider portion of the pH scale could be explored; thus very acidic pH intervals (down to pH 2.5) were described, for equilibrium fractionation of acidic proteins, such as pepsin [19], as well as very alkaline intervals (e.g., pH 10–12) for analysis of high pI proteases (subtilisins) [20] and even of histones [21]. Already in 1990 our group described 2-D maps in what was, at that time, the most extended pH gradient available, spanning the pH 2.5–11 interval [22]. We went so far as to describe even sigmoidal pH gradients, which have also found applications in 2-D analysis [23,24]. This shows that, already starting from 1990, all the ingredients for success of 2-D maps exploiting IPGs in the first dimension were fully available in the literature.

III. SOME BASIC METHODOLOGY PERTAINING TO 2-D PAGE

Fig. 1 shows what is perhaps the most popular approach to 2-D maps analysis today. The first dimension is preferably performed in individual IPG strips, laid side by side on a cooling platform, with the sample often adsorbed into the entire strip during rehydration. At the end of the IEF step, the strips have to be interfaced with the second dimension, almost exclusively performed by mass discrimi-

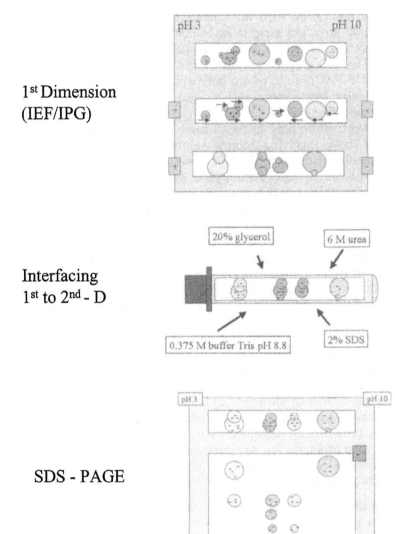

1st Dimension (IEF/IPG)

Interfacing 1st to 2nd - D

SDS - PAGE

Fig. 1 Pictorial representation of the steps required for obtaining a 2-D map, with the first dimension typically done in IPG strips, followed by equilibration of the focused strips in SDS-interfacing solution and, finally, by the second-dimension run in SDS-PAGE.

nation via saturation with the anionic surfactant SDS. After this equilibration step, the strip is embedded on top of an SDS-PAGE slab, where the 2-D run is carried out perpendicular to the 1-D migration. The 2-D map displayed at the end of these steps is the stained SDS-PAGE slab, where polypeptides are seen, after staining, as (ideally) round spots, each characterized by an individual set of pI/Mr coordinates.

It is worthwhile here to recall some important methodologies, especially those involving sample solubilization and preparation prior to the first dimension IEF/IPG step. Appropriate sample preparation is absolutely essential for good 2-D results. Due to the great diversity of protein sample types and origins, only general guidelines for sample preparation are provided here. The optimal procedure should ideally be determined empirically for each sample type. If sample preparation is performed properly, it should result in complete solubilization, disaggregation, denaturation and reduction of the proteins in the sample. We recall here that different treatments and conditions are required to solubilize different types of samples: some proteins are naturally found in complexes with membranes, nucleic acids, or other proteins; some proteins form various nonspecific aggregates; some others precipitate when removed from their physiological environment. The effectiveness of solubilization depends on the choice of cell disruption methods, protein concentration and dissociation methods, and choice of detergents and on the overall composition of the sample solution. Lenstra et al. [25], Molloy et al. [26], and an entire issue of *Methods of Enzymology* [27] can be consulted as general guides to protein purification and/or for specific solubilization problems for, say, membrane proteins and other difficult samples. The following general sample preparation guidelines should always be kept in mind:

Keep the sample preparation strategy as simple as possible, so as to avoid protein losses.

Cells and tissues should be disrupted in such a way as to minimize proteolysis and other modes of protein degradation. Cell disruption should be done at low temperatures and should ideally be carried out directly into strongly denaturing solutions containing protease inhibitors.

Sample preparation solutions should be freshly made or stored frozen as aliquots. High-purity, deionized urea should always be used.

Preserve sample quality by preparing it just prior to IEF or by storing samples in aliquots at −80°C. Samples should not be repeatedly thawed.

Remove all particulate material by appropriate centrifugation steps. Solid particles and lipids should be eliminated because they will block the gel pores.

In the presence of urea, samples should never be heated. Elevated temperatures produce higher levels of cyanate from urea, which in turn can carbamylate proteins. A specific technique has been in fact described for producing "carbamylation trains" as pI markers by boiling specific proteins in urea [28–30].

IV. SOLUBILIZATION COCKTAIL

For decades, the most popular lysis solutions has been the O'Farrell cocktail (9 M urea, 2% Nonidet P-40, 2% β-mercaptoethanol, and 2% carrier empholytes, in any desired pH interval) [31]. Although much in vogue also in present times, over the years new, even more powerful, solubilizing mixtures have been devised. We will discuss below the progress made in sample solubilization because this is perhaps the most important step for success in 2-D map analysis. Great efforts were dedicated to such developments, especially in view of the fact that many authors noted that hydrophobic proteins were largely absent from 2-D maps [32]. These authors noted that, quite strikingly, in three different species analyzed (*Escherichia coli, Bacillus subtilis, Saccharomyces cerevisiae*), all proteins above a given hydrophobicity value were completely missing, independently from the mode of IEF (soluble CAs or IPG). This suggested that the initial sample solubilization was the primary cause for loss of such hydrophobic proteins. The progress made in solubilizing cocktails can be summarized as follows (see also the reviews by Molloy [33] and Rabilloud and Chevallet [34].

A. Chaotropes

Although urea (up to 9.5 M) has been for decades the only chaotrope used in IEF, recently thiourea has been found to further improve solubilization, especially of membrane proteins [35–38]. The inclusion of thiourea is recommended for use with IPGs, which are prone to adsorptive losses of hydrophobic and isoelectrically neutral proteins. Typically, thiourea is added at concentrations of 2 M in conjunction with 5–7 M urea. The high concentration of urea is essential for solvating thiourea, which is poorly water soluble (only approximately 1 M in plain water) [39]. Among all substituted ureas (alkyl ureas, both symmetric and asymmetric) Rabilloud et al. [35] found thiourea to be still the best additive. However, it would appear that not much higher amounts of thiourea can be added to the IPG gel strip, because it seems that at >2 M concentrations, this chaotrope inhibits binding of SDS in the equilibration step between the first and second dimensions, thus leading to poor transfer of proteins into the 2-D gel. It should also be remembered that urea in water exists in equilibrium with ammonium cyanate, whose level increases with increasing pH and temperature [40]. Since cyanate can react with amino groups in proteins, such as the N-terminus α-amino or the ε-amino groups of Lys, these reactions should be avoided since they will produce a false sample heterogeneity and give wrong Mr values upon peptide analysis by MALDI-TOF MS. Thus, fresh solutions of pure-grade urea should be used, concomitant with low temperatures and with the use of scavengers of cyanate (such as the primary amines of carrier ampholytes or other suitable amines). In addition, protein mixtures solubilized in high levels of urea should be subjected to separation in the electric field as soon as possible: in the presence of the high voltage gradients typical of the IEF protocol, cyanate ions are quickly removed and no carbamylation can possibly take place (Herbert and Righetti, unpublished) whereas it will if the protein/urea solution is left standing on the bench.

B. Surfactants

These molecules are always included in solubilizing cocktails to act synergistically with chaotropes. Surfactants are important in preventing hydrophobic interactions due to exposure of protein hydrophobic domains induced by chaotropes. Both the hydrophobic tails and the polar head groups of detergents play an important role in protein solubilization [41]. The surfactant tail binds to hydrophobic residues, allowing dispersal of these domains in an aqueous medium, while the polar head groups of detergents can disrupt ionic and hydrogen bonds, aiding in dispersion. Detergents typically used in the past included Nonidet P-40 or Triton X-100, in concentrations ranging from 0.5% to 4%. More and more, zwitterionic surfactants, such as CHAPS, are replacing those neutral detergents [42–43], often in combination with small amounts (0.5%) of Triton X-100. In addition, minute levels of CAs (<1%) are added, because they appear to reduce protein-matrix hydrophobic interactions and overcome detrimental effects caused by salt boundaries [44]. Linear sulfobetaines are now emerging as perhaps the most powerful surfactants, especially those with at least a 12-carbon tail (SB 3–12). They were in fact already demonstrated to be potent solubilizers of hydrophobic proteins (e.g., plasma membranes), except that they had the serious drawback of being precipitated out of solution due to low urea compatibility (only 4 M urea for SB 3–12) [45]. This drawback of scarce solubility of many nonionic or zwitterionic surfactants with long, linear hydrophobic tails in urea solution seems to be a general problem [46,47] and thus has prompted the synthesis of more soluble variants [48]. The inclusion of an amido group along the hydrophobic tail greatly improved urea tolerance, up to 8.5 M, and ameliorated separation of some proteins of the erythrocyte membranes [49]. Recently, Chevallet et al. [50] embarked on a project for determining the structural requirements for the best possible sulfobetaine. Three major features were found to be fundamental: (1) an alkyl or aryl tail of 14–16 carbons, (2) a sulfobetaine head of 3 carbons, and (3) a 3-carbon spacer between the quaternary ammonium and the amido group along the alkyl chain. The most promising surfactants were found to be ASB 14 (amidosulfobetaine) containing a 14-C linear alkyl tail and C8∅, which possesses a *p*-phenyloctyl tail. Both of these reagents have since been used successfully in combination with urea and thiourea to solubilize integral membrane proteins of both *E. coli* [51] and *Arabidopsis thaliana* [52,53].

C. Reducing Agents

Thiol agents are typically used to break intramolecular and intermolecular disulfide bridges. Cyclic reducing agents, such as dithiothreitol (DTT) or dithioerythritol (DTE) are the most common reagents admixed to solubilizing cocktails. These chemicals are used in large excess (e.g., 20 to 40 mM) so as to shift the equilibrium toward oxidation of the reducing agent with concomitant reduction of the protein disulfides. Because this is an equilibrium reaction, loss of the reducing agent through migration of proteins away from the sample application zone can permit reoxidation of free Cys to disulfides in proteins, which would result not only in horizontal streaking, but also, possibly, in formation of spurious extra bands due to scrambled

—S-S- bridges and their cross-linking different polypeptide chains. Even if the sample is directly reswollen in the dried IPG strip, as customary today, the excess DTT or DTE will not remain in the gel at a constant concentration because, due to their weakly acidic character, both compounds will migrate above pH 7 and be depleted from the alkaline gel region. Thus, this will aggravate focusing of alkaline proteins and be one of the multifactorial factors responsible for poor focusing in the alkaline pH scale. This problem is not trivial at all and deserves further comments. For example, Righetti et al. [54], when reporting on the focusing of recombinant pro-urokinase and urinary urokinase, two proteins with rather alkaline pI values, in IPG gel strips, detected a continuum of bands focusing in the pH 8–10 region, even for the recombinant protein, which exhibited a single, homogeneous band by SDS-PAGE. This protein has an incredible number of Cys residues (no less than 24!), so this extraordinary heterogeneity was attributed to formation of scrambled disulfide bridges, not only within a single polypeptide chain but also among different chains in solution. Curiously, this happened even if the protein was not subjected to reduction of −S-S- bridges prior to the IPG fractionation, but this could also have a logical explanation. According to recent data by Bordini et al. [55], when probing the alkylation by acrylamide of −SH groups in proteins by MALDI-TOF, it was found that the primary site of attack, even in proteins having both disulfide bridges and free −SH groups, was not the free −SH residues, as it should be, but it was systematically one of the −SH engaged in disulfide bridges! This is also an extraordinary result and could only be explained by assuming that, at alkaline pH values (the incubation was carried out at pH approximately 10), disulfide bridges are weakened and probably constantly snapped broken and reformed. The situation would be aggravated when using β-mercaptoethanol, because the latter compound has an even lower pK value; thus, it is more depleted in the alkaline region and will form a concentration gradient toward pH 7, with a distribution in the gel following its degree of ionization at any given pH value along the IEF strip [56]. This is probably the reason for the dramatic loss of any pH gradient above pH 7.5, lamented by most users of conventional IEF in CAs, when generating 2-D maps. The most modern solution to all the above problems appears to be the use of phosphines as alternative reducing agents. Phosphines, which were already described in 1977 by Ruegg and Rüdinger [57], operate in a stoichiometric reaction, thus allowing the use of low concentrations (barely 2 mM). The use of tributyl phosphine (TBP) was recently proposed by Herbert et al. [58], who reported much improved protein solubility for both Chinese hamster ovary cell lysates and intractable, highly disulfide cross-linked wool keratins. TBP thus offers two main advantages: it can be used at much reduced levels as compared to other thiolic reagents (at least one order of magnitude lower concentration) and, additionally, it can be uniformly distributed in the IPG gel strip (when rehydrated with the entire sample solution) because, being uncharged, it will not migrate in the electric field. However, some curious aspects are related to the use of TBP: when this reagent is analyzed by MS, it is shown to be extensively oxidized to Bu_3P0, an event that should occur only when it reacts with disulfide bridges in their reduction process. This occurs not only in old bottles of Bu_3P, but also in fresh preparations and when such bottles are rigorously kept under nitrogen. It is just enough to pipette Bu_3P in the typical buffer used for reduction (10 mM Tris, pH 8.3) and within a few minutes (just the time to bring the solution to the MS instrument and perform the analysis), it can be seen that approximately 95% of it is already present as Bu_3P0, with only approximately 5% remaining in the reduced form [59]. However, such preparations are just as effective in bringing (and maintaining) reduction of $-S-S-$ bridges in proteins, which suggests that the amount needed for reduction is not in the millimolar but instead in the micromolar range. A major drawback of TBP, additionally, is that it is volatile, toxic, and rather flammable in concentrated stocks. As a conclusion of this long excursus, it would appear that, at present, one of the most powerful solubilization cocktails is the one depicted in Fig. 2.

V. SAMPLE REDUCTION AND ALKYLATION: WHEN, WHY, AND HOW?

Reduction and alkylation is also a very important aspect of sample solubilization and pretreatment in 2-D map analysis. Reducing and alkylating the resulting free $-SH$ groups is, of course, a well-known procedure and highly recommended in any biochemistry textbook dealing with protein analysis. Alkylation will prevent all the noxious phenomena reported above, like reformation of disulfide bridges, producing smears and even spurious bands due to inter-chain cross-linking, even among unrelated polypeptide chains. However, as

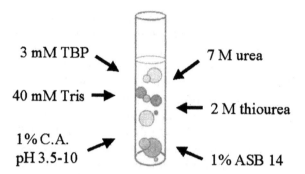

Fig. 2 Composition of the cocktail today preferred for best sample solubilization in preparation of the first-dimension run. In some cases, the level of urea is reduced from 7 to 5 M. The wide range carrier ampholyte solution is added to facilitate sample entry in the IPG strip.

strange as it sounds, although reduction and alkylation is performed by anyone working with 2-D maps, this step is not done at the very beginning of sample treatment, just prior to the first dimension IEF/IPG run, but in between the first and second dimension, during the interfacing of the IEF/IPG strip with SDS-denaturing solution, in preparation for the SDS-PAGE final step. This probably stems from earlier reports by Goerg et al. [60,61] who recommended alkylation of proteins with iodoacetamide during the interfacing between the first and second dimension, on the grounds that this treatment would prevent point streaking and other silver-staining artifacts. As luck goes, this recommendation was followed without any questioning by everyone working with 2-D maps. Clearly, in the light of the above discussion, it appears to be a much smarter move to reduce and alkylate the sample just to start with (i.e., prior to even the first IEF/IPG dimension). The drawbacks for alkylation with iodoacetamide in between the two-dimension runs have also been highlighted, recently, by Yan et al. [62], although on different grounds than those discussed above. These authors noted that, in most 2-D protocols, the discontinuous Laemmli's [63] buffer is used, which calls for the stacking and sample gels to be equilibrated in a pH 6.8 buffer. In order to prevent pH alterations, most people use a modified stacking buffer with a reducing agent (DTT or DTE) and alkylating agent (iodoacetamide) at pH 6.8, so that the strips can be loaded as such after these two treatments, avoiding any further pH manipulations [64]. However, at this low pH both reduction and al-kylation are not so efficient, because the optimal pH for these reac-tions is usually at pH 8.5–8.9 [65]. As a result of this poor protocol, Yan et al. [62] have reported additional alkylation by free acrylamide during the SDS-PAGE run, with the same protein exhibiting a major peak of Cys-carboxyamidomethyl and a minor one of Cys-propionam-ide. It should be borne in mind that whereas the risk of acrylamide adduct formation is much reduced in IPG gels (but not in unwashed IEF gels!), it is quite real in SDS-PAGE gels, due to the fact that these gels are not washed and that surfactants, in general, hamper incorporation of monomers into the growing polymer chain [66]. Dif-ferent alkylating residues on a protein will complicate their recogni-tion by MALDI-TOF analysis, a tool much in use today in proteo-mics. Here too, however, although Yan et al. [62] clearly identified the problem, they did not give the solution we are proposing—namely, to reduce and aklylate the sample prior to any electropho-retic step. They suggest retaining still the original protocol of alkyl-ation in between the two dimensions, but increasing the pH of the equilibration to pH 8.0, adding a much larger amount of alkylating agent (125 mM iodoacetamide), and increasing the time of incuba-tion to 15 min. We again stress that although this new protocol is an improvement over previous ones, it still does not cure the prob-lems of the first IEF/IPG dimension, namely smears and formation of spurious bands, both due to reoxidation of reduced but nonalkyl-ated Cys residues.

Alkylation with iodoacetamide in the presence of thiourea is a conflicting situation. We have seen above that reduction and alkyl-ation prior to any electrophoretic step is a must if one wants to avoid a number of artifacts typically encountered in the alkaline pH re-gion. For example, recently, Herbert et al. [67] have demonstrated, by MALDI-TOF MS, that failure to alkylate prior to the IEF/IPG step would result in a large number of spurious spots in the alkaline pH region, due to "scrambled" disulfide bridges among like and un-like polypeptide chains. This series of artifactual spots comprises not only dimers but also an impressive series of oligomers (up to nonamers) in the case of simple polypeptides such as the human α- and β-globin chains, which possess only one (α-) or two (β-) −SH groups (Fig. 3A). As a result, misplaced spots are to be found in the resulting 2-D map, if performed with the wrong protocol. Conversely, if the protein is reduced and immediately alkylated prior to the first

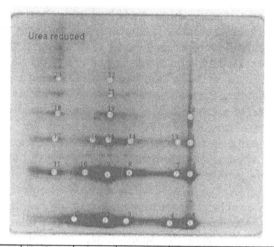

Spot	Identification	Score (second)	Hits/total/ coverage (%)	Comments	2.Search (hits/total/ coverage %)
1	β-chain	2.32	7/8/63		Nothing
2	β-chain		2/19/29	Hit to non-human species Findmod: PALM (41-69), D to N (1,83-104) or H to R (0, 121-144)	Nothing
3	α-chain	0.42	5/16/49	Findmod: bit of amidation (NOT CONFIRMED), E to K (1,17-31)	Nothing
4	α-chain	0.71	4/13/35	As above	Nothing
5	α-chain	0.25	4/15/35	As spot 3	
6	β-chain	1.65 (0.35)	6/22/48		α-chain 4/16/35
7	α-chain	1.8 (0.95)	6/19/55		β-chain 4/13/35
8	β-chain	1.31 (1.19)	6/22/48		α-chain 5/16/55
9	β-chain	2.23 (0.42)	6/14/48		α-chain 3/8/42
10	β-chain	2.09	6/20/48		Nothing
11	β-chain	0.69	3/7/32		Nothing
12	β-chain	0.85 (1.62)	6/17/48		α-chain 6/11/55
13	α-chain			Poor spectrum	
14	?				
15	β-chain	1.68	6/18/48		
16	β-chain	1.61	5/12/49		Nothing
18	β-chain		4/32/35	Keratin contamination	
19	β-chain	1.71	7/16/56		Nothing
20	α-chain	0.13	3/14/43		
21	β-chain	1.97	6/17/48		Nothing
22	?				
23	β-chain	1.10	4/12/38		Nothing

(A)

Fig. 3 (A) Two-dimensional map of reduced but non-alkylated globin chains. (B) Two-dimensional map of reduced and alkylated globin chains. For spot identification, the spots were cut out, digested *in situ* with trypsin, and the peptides eluted and analyzed by MALDI-TOF MS. The tables under panels (a) and (b) give the spot identification via database searching. A, sample solubilized in plain urea, reduced but not alkylated; B, solubilization in urea/thiourea, followed by alkylation with acrylamide. (From Ref. 67.)

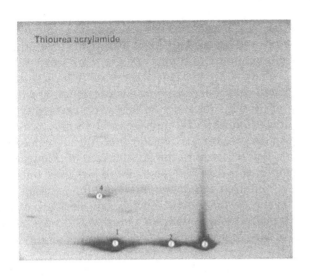

Spot	Identification	Score (second)	Hits/total/ coverage (%)	Comments	2.Search (hits/total/ coverage %)
1	β-chain	2.36	11/26/60		Nothing
2	β-chain	2.04 (2.08)	11/31/71		α-chain 6/20/55
3	α-chain	2.32	7/23/75		Nothing
4	Carbonic anhydrase	1.0	6/25/35		

(B)

dimension, all oligomers simply disappear from the 2-D map (Fig. 3B). Subsequently, the same group [68], *via* the same MALDI-TOF MS procedure, has monitored the kinetics of protein alkylation by iodoacetamide over the period 0–24 h at pH=9. Alkylation reached ~70% in the first 2 min, yet the remaining 30% never quite reached 100%, even when allowing up to 6 h of incubation (the reaction somehow plateaued at 80%). The use of SDS during the alkylation step resulted in a strong quenching of this reaction (thus further corroborating the notion that alkylation in between the first and second dimension is a useless procedure), whereas 2% CHAPS exerted a much reduced inhibition. It was during these investigations that the same authors found a disturbing phenomenon: during alkylation of α- and β-globin chains, substantially different results were obtained when the sample was dissolved in plain 8 M urea or in the mixture

5 M urea + 2 M thiorurea. In the former case, almost complete disappearance of all homo- and hetero-oligomers was achieved; in the latter case, appreciable amounts of dimers and trimers were still left over, even upon prolonged incubation. A thorough investigation by ESI-MS demonstrated that indeed thiourea was competing with the free −SH groups of proteins for reaction and was scavenging iodoacetamide at an incredible rate (in approximately 5 min of incubation all iodoacetamide present in solution had fully disappeared). Iodoacetamide was found to add to the sulfur atom of thiourea; the reaction was driven in this direction also by the fact that, once this adduct was formed, thiourea was deamidated and the reaction product generated a cyclic compound (a thiazolinone derivative) [69]. Thus, those using urea/thiourea solutions should be aware of this side reaction and of the potential risk of not achieving full alkylation of the free, reduced −SH groups in Cys residues. Fortunately, alkylation of proteins is still substantial if iodoacetamide is added as a powder to the protein solution treated with the reducing agent, instead of as a solution. In addition, due to the fact that all iodoacetamide is scavenged by thiourea, the final concentration of thiourea in the solubilizing solution will be reduced by the same molar amounts of iodoacetamide added.

In conclusion, reduction and alkylation at the very start of any 2-D fractionation would have the following immediate advantages [67–69]:

1. Reduce considerably smears in the alkaline region, above pH 8.

2. Prevent formation of spurious bands due to mixed disulfide bridges.

3. Abolish formation of a mixed population of Cys-propionamide and Cys-carboxyamidomethyl species, due to alkylation with acrylamide and iodoacetamide, respectively.

There remain the aspects of how to perform reduction; if it is done with DTT or DTE, it must be a two-step operation. The sample, brought to pH 8.5 to 8.9 (with a weak base, like free Tris and/or basic CAs!) is first incubated for approximately 1 h to allow for full reduction of −S-S- bridges. After that, a two-fold molar excess of alkylating agent (iodoacetamide) is added and the sample left to incubate for an additional hour (or more) to allow for full alkylation (of both, free Cys groups in proteins as well as free thiols in the DTT or DTE additives). Now, because DTT or DTE are typically used at approximately 40 mM level, this means that 80 to 100 mM iodoacetamide will have to be added. This *per se* does not pose any risk in the subsequent IEF/IPG fractionation, because this molecule is neutral, so large amounts can be tolerated in the focusing step. However, upon prolonged focusing (1–2 days, as often applied in IPG runs), there could be the risk of partial hydrolysis of the amido bond in iodoacetamide, thus provoking undue currents and destabilizing the focusing process. Here is where the use of TBP might be preferred. Because the levels used of TBP, as a reducing agent, are minute (2 mM), the levels of alkylating agent (iodoacetamide) to be added will also have to be small—typically, of the order of 5 to 10 mM (i.e., at least one order of magnitude less than in the case of DTT or DTE reduction). But there is more to it. Due to the fact that TBP does not react with some alkylating agents, such as acrylamide and 4-vinylpyridine [57], it could offer the opportunity of a simplified, one-step reduction and alkylation procedure using TBP and acrylamide (as alkylating agent) simultaneously, as proposed by Molloy [33]. However, even alkylation with acrylamide (or its N-substituted derivatives) is not much more efficient than with iodoacetamide; thus the problem of efficient alkylation still remains an open question. Perhaps the proper solution is the one presented in Fig. 4. Here, a solution of 50 μM of α-lactalbumin (LCA) in 50 mM Tris-acetate, pH 9.0 and 7 M urea, has been reduced with 50 mM dithiothreitol (DTT) for 1 h at 25°C. After reduction, the protein has been alkylated with 100 mM of either DMA or 4-vinylpyridine (4-VP) for 1 h at 25°C. Both reactions were carried out in presence of 2% Triton X-100, one of the several surfactants adopted in proteome analysis. The reaction products were analyzed via mass spectrometry in MALDI-TOF. As shown in Fig. 4, whereas in the case of 4-VP only one reaction product (m/z 15038), corresponding to the adduct of LCA with eight residues of 4VP, is visible, in the case of DMA small amounts of unreacted product (m/z 14213) are still present, together with an impressive series of partial reaction products, of which the most abundant components are the tri and tetra-alkylated species. Because this alkylation reaction can be easily conducted at pH 7 (i.e., at a pH approximately midway between the two pK's of the reactants

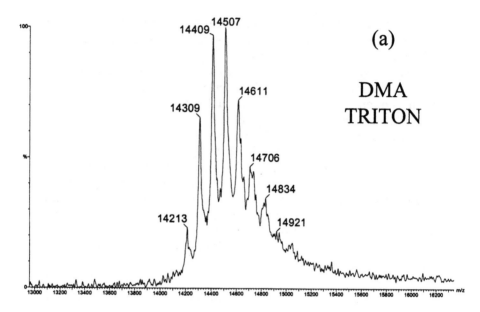

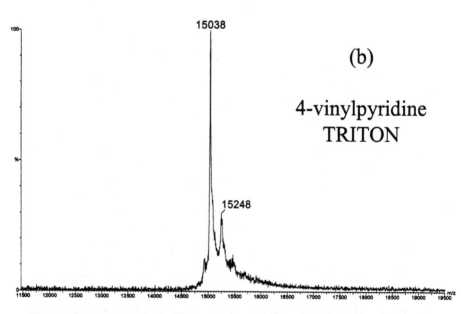

Fig. 4 Reaction of the −SH groups of bovine α-lactalbumin (LCA) with DMA (panel a) or with 4-VP (panel b), in presence of 2% Triton X-100 (unpublished experiments).

[-SH group and nitrogen in the pyridine moiety]), this 100% efficiency of reaction is coupled to 100% specificity, because at this pH no side reaction (e.g., on Lys) can possibly occur.

VI. MASS SPECTROMETRY AND THE PROTEOME: AN INDISSOLUBLE CONNUBIUM

The basic requirements of proteome analysis are a wide dynamic detection range, high-throughput and high-confidence protein identification, protein quantification, the ability to deal with multiple proteins in a single spot, and the possibility of identifying posttranslational modifications. Mass spectrometry (MS) can fulfill all of these requirements and has thus become a fundamental tool in any laboratory dealing with proteome research. Matrix-assisted laser desorption ionization (MALDI) and electrospray ionization (ESI) are the most commonly used methods to generate gas phase ions of both small and large biomolecules. Both techniques can be coupled to a variety of ion analyzers, including time-of-flight (TOF), quadrupoles, ion traps, and magnetic and electrostatic sectors. For structural analysis, such as peptide sequencing, two steps of mass spectrometry are effected in tandem (MS/MS). It is not always necessary to use a multiple-component mass spectrometer to perform such a task: ion traps manipulate electric fields to select a particular ion and monitor its post-collision fragments. In this chapter we discuss a number of mass spectrometers commonly used in proteome research; however, more details and other experimental arrangements can be found in James [5].

A. Triple-Stage Quadrupole and Ion-Trap Mass Spectrometers

Mass spectrometry analysis based on the motion of ions in an oscillating electric field has evolved over a period of more than 30 years [70]. Quadrupole ion traps and triple-stage quadrupole (TSQ) instruments are among the most versatile mass spectrometers yet developed, because of their power in the structural analysis of molecules through tandem mass spectrometry analysis (MS/MS). Both types of instruments are capable of product ion scanning—the isolation of a given precursor ion and the detection of its product species, via collision-induced dissociation (CID) [71]. Ion traps can also perform multiple stages of precursor isolation and fragmentation in a single experiment (MS^n). Much of the value of MS analysis lies in its ability to generate reliable m/z values of peptide derived from proteolytic fragmentation of isolated spots from a 2-D map. Such fragments can be exploited for identifying proteins by comparing them with those expected from each entry in the rapidly expanding primary-sequence databases [72,73]. In addition to peptide mass, sequence information obtained from MS/MS can be used for this purpose; these data can be obtained from both triple-quadrupole and ion-trap instruments [74,75]. An extension of this approach is the de novo sequencing of currently unknown proteins [76].

A key to the versatility of triple-quadrupole and ion-trap instruments is their ability to perform MS/MS. Valuable structural information can be gathered by subjecting a precursor ion (isolated at an initial stage of MS) to CID, followed by a second stage of MS analysis of the product ions. CID requires that sufficient energy is imparted to the ion, so that the vibrational energy in a given bond exceeds its bond strength. Low-energy CID is particularly useful in the analysis of peptides, because fragmentation occurs frequently at amide bonds, leading to product ions characteristic of the peptide's sequence [77]. CID can produce fragments that contain the N-terminal portion of the peptide (designated as a, b, c, or d ions, depending on the bond cleaved) or fragments that contain the C-terminus (termed w, x, y, or z ions) [78]. In both types of instruments, the most commonly used mode of ionization is ESI, which has the virtue of sampling directly from a liquid phase, imparting relatively little internal energy to the analytes (so that intact macromolecules can be detected) and efficiently bringing polar and charged molecules to the gas phase. Proteins with molecular mass values far in excess of the mass range possessed by a mass spectrometer can usually be detected, because they bear enough charges to produce an envelope of peaks with m/z less than 2000 [79]. One reason ESI has found such extensive use is its compatibility with on-line separations by chromatography and capillary electrophoresis. A recent extension has been the spraying from ever-smaller ESI emitters with the lowest achievable stable flow rate; this technique is known as nanospray [80], and it is usually performed using glass (or fused-silica) capillaries with tips pulled to inner diameters of less than 5 μm and flow rates of a few nanoliters per minute.

B. Merits and Limits of TSQ and Ion-Trap Spectrometers

Both types of instruments have been used extensively for protein and peptide analysis; they both share the ability to perform MS/MS for generating peptide-sequence information that can be used to search protein (and DNA, if needed) databases. Both instruments are capable of full mass-range scanning with detection limits for peptides at low femtomole levels when capillary LC-MS is employed [81]. For the quadrupole, however, the highest sensitivity is gained at the cost of resolution; so, if the instruments are compared when operating at equal resolution, the ion trap may appear to be more sensitive. Product-ion spectra of peptides at the attomole level can be achieved by using capillary LC-MS [82], compared with low femtomole levels with triple quadrupoles.

C. Matrix-Assisted Laser Desorption, Ionization Time-of-Flight Mass Spectometry

In MALDI-TOF MS, nanosecond pulses of laser are fired at matrix-protein sample mixtures (protein-doped matrix crystals). The resulting gas-phase ions are accelerated to keV energy into a field-free flight tube. At the end of this journey, the ions are detected and their arrival times are converted into their corresponding m/z values. For improving mass resolution, modern MALDI-TOF instruments have an ion reflector, which turns the ions around in an electric field, sending them toward a second "reflectron" detector. The reflector allows partial peptide sequence analysis by post-source decay [83].

Some important aspects of the solutions in which the protein is dissolved should be borne in mind. First of all, strong ionic detergents (such as SDS) should be thoroughly removed, but nonionic surfactants, such as Triton X or octylglucoside, are tolerated. Sodium azide in buffers also should be avoided because its presence suppresses protein ion formation in the mass spectrometer. Exposure to formic acid solutions should be avoided because this compound reacts with amino groups (especially Lys residues), resulting in a formyl derivative of the protein. Also, concentrated trifluoroacetic acid can react with free amino groups. Finally, it should be remembered that, although the best stain, in view of MS analysis, is Coomassie Blue, even the latter is not immune from artifacts. Some proteins bind tenaciously to Coomassie Blue, resulting in multiple peaks in MALDI-TOF analysis. This, per se, is not dangerous, because such extra peaks can be clearly identified, due to the regular mass increments per each adduct, corresponding to the mass Coomassie. More disturbing, though, is the fact that, during digestion with trypsin, some Coomassie residues might cover the potential hydrolytic sites, resulting in anomalous fragmentation (it should be remembered that because the Coomassie family contains two sulfate groups, they tend to bind to positively charged sites on proteins, such as Lys and Arg residues—that is, to the typical sites of attack by trypsin).

It is here recalled that peptide-mass fingerprinting (PMF) compares an experimental profile of peptide masses against a theoretical profile calculated from the known sequences in a non-redundant database. The effectiveness of a PMF search is strongly influenced by the accuracy with which the peptide masses are measured. Because MALDI is today performed on TOF machines, and such instruments today incorporate both delayed extraction and reflectrons, both resolution and mass accuracy are dramatically improved. Modern electronics has increased the sampling rate, which is now as high as 4 GHz. Such improvements and good calibration routines allow databases to be searched with peptide masses typically measured to an accuracy of 10 ppm. The confidence in such searches is directly correlated with mass accuracy. At least four or five peptides must be matched and 20% sequence coverage is required for proper identifications. As any other analytical technique MALDI-TOF MS has its limitations:

First of all, selective signal suppression. In an equimolar pool of peptides from the digest of a protein, some peptides will not be seen and there will be wide variation in signal intensity from the remainder. For example, peptides with C-terminal arginine generally result in higher signals than peptides with a C-terminal lysine, an effect that hinders relating peptide peak height with the quantity of the sample present unless a suitable internal standard is used.

Second, as the amount of protein in the gel decreases, so does the number of peptides and, at very low levels of protein, only a few peptides may be observed. Thus, identifying a protein with confidence from just a few peptide masses is impossible and yet this situation often arises with very small amounts of protein;

Third, MALDI-TOF MS has only a limited ability to deal with mixtures. In proteomics, mixtures are common to the point at which, even in the case of apparently "clean" spots eluted from 2-D gels, a few principal components and several minor ones might be present.

Given the above caveats, why does MALDI-TOF continue to be so popular? One of the reasons must certainly be its very high throughput. The other reason could be the possibility of adding partial peptide sequence to the mass information. In general, just one mass, accurately measured, plus partial sequence, is sufficient for identifying a protein with confidence. It just so happens that, with a judicious use of MALDI-TOF MS post-source decay (PSD), some sequence information can be obtained, although the fragmentation pattern could be difficult to interpret. MALDI-PSD (as well as ESI) have now started to replace classic analytical approaches such as Edman degradation, in certain cases. MALDI-PSD, for example, offers higher sensitivity over chemical processing (Edman) and has the potential to analyze multicomponent samples. In addition, it has a higher tolerance to sample impurities and thus is ideally suited for modern approaches of 2-D separation and 2-D sample handling and storage. PSD analysis is a method that allows one to mass analyze fragment ions that are formed in the field-free region following the ion source of a TOF mass spectrometer. These fragment ions are produced from the decay of MALDI-generated ions. Mass determination of these PSD ions is based on using an electrostatic ion reflector as an energy analyzer. In fact, during the linear ion path (the first part of the flight), all ions leaving the source have the same nominal kinetic energy. Most of them are still unfragmented precursor molecular ions; however, during the subsequent flight time (tens of microseconds), through the field-free drift region, a certain number of analyte ions undergo post-source decay into product ions. Due to their lower mass, such PSD ions have a considerably lower kinetic energy than their precursor ions. The ion reflector in PSD instruments is thus used as an energy filter and therefore as a mass analyzer for PSD ions. Due to their mass-dependent kinetic energies, PSD ions are reflected at different positions within the reflector, thus impinging at different zones within the area of the reflector detector; this in turn leads to mass-dependent total flight times through the instrument. The majority of applications employing PSD-MALDI MS have concentrated on peptide characterization. The sensitivity of the method is in the range of 30 to 100 fmol of peptide. Mass resolving power is in the range of 6000 to 10000 for precursor ions and in the range of 1500 to 3000 for PSD ions. An example of this partial sequence determination by PSD is given in Fig. 5. This is an interesting case, which deserves further comments. When performing digestion of bovine β-lactoglobulin B, Bordini et al. [84] obtained a peptide of 2113.6 Da, which could not be found in the Swiss-Prot database. On the assumption that this could be a wrong cut obtained by trypsin digestion, an attempt was made at obtaining partial amino acid sequence by the PSD method. PSD analysis in fact resulted in the prod-

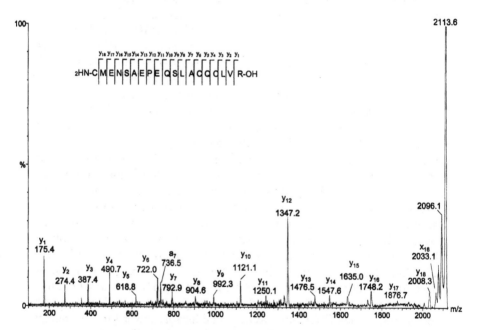

Fig. 5 Example of sequence determination by PSD of a peptide of m/z 2113.6 obtained by tryptic digestion of β-lactoglobulin B. On the basis of the PSD spectrum and the sequence determined, it was possible to assign this fragment to the amino acid region [106–124] of the protein chain. For a detailed explanation, see text. (From Ref. 84.)

uct ion spectrum of Fig. 5. This spectrum was obtained by operating the instrument in the reflectron mode, during which appropriate ion-gating pulses were applied, allowing the selection of the parent ion (i.e., the 2113.6 Da peptide) and the monitoring of its unimolecular and background collision-induced fragments. The spectrum shows the complete series of y_n-fragment ions [78], which allowed the verification of the sequence given in the same figure. By this procedure, which resembles an Edman degradation, except that this degradation occurs in the mass spectrometer rather than being chemically induced via phenylisothiocyanate attack, one could construct this entire sequence: $_2$HN-Cys-Met-Glu-Asn-Ser-Ala-Glu-Pro-Glu-Gln-Ser-Leu-Ala-Cys-Gln-Cys-Leu-Val-Arg-OH—a sequence of 19 amino acids found to correspond to the tryptic fragments [106–124] of β-lactoglobulin B. This is an extraordinary sequence, due to both its accuracy and its length. Although it is rare to obtain such a long sequence (typically restricted, under normal operating conditions to 8–10 amino acids), this figure proves that, when obtaining high quality spectra, such sequencing process can give much improved results and longer readings.

Another way of controlling fragmentation would be to insert the MALDI source onto a tandem mass spectrometer. A recent approach has been to use a tandem TOF-MS, although this method has been shown to work only with standard peptides and not yet with protein digests [85]. A second way is to use MALDI with a hybrid quadrupole TOF mass spectrometer, in order to generate the needed sequence information. Two recent articles [86,87] have described this approach, by using a MALDI rather than an ESI ion source.

D. Quadrupole–Time-of-Flight-Mass Spectrometry

Although first introduced commercially only six years ago, quadrupole–time-of-flight (Q-TOF) has been rapidly embraced by scientists working in proteomic research. These instruments can accommodate both MALDI and ESI techniques; however, the latter ionization technique is more commonly used. The development of Q-TOF followed closely the development of ESI-TOF, and therefore the current configuration can be considered either as the addition of a mass-resolving quadrupole and collision cell to an ESI-TOF or as the replacement of the third quadrupole in a TSQ instrument by a TOF analyzer. In full scan mode these instruments possess high sensitivity and high mass accuracy for both precursor and product ions; one of the main advantages of Q-TOF instruments over triple quadrupoles is the high mass resolution of TOF, which can be as high as 12000 (m/Δm, where Δm is the full peak width at half-maximum (FWHM). As a result of this capability, interfering peaks of ions having the same nominal mass can be resolved, and the charge stage of multiply charged ions can in many cases be determined from their isotopic spacing. These instruments are commonly used in conjunction with capillary LC and nano ESI, which render them ideal for peptide characterization derived from fairly complex protein digests.

E. Fourier-Transform Ion-Cyclotron Resonance Mass Spectrometry

FT-ICR is potentially one of the most sensitive and widely applicable of all MS instruments. The technique is endowed with impressive performance capabilities. A resolving power of over 100,000 has been demonstrated for proteins with a mass of 100 kDa with an accuracy of ±3 Da [88]. The upper limit factor is the distribution of low-abundance isotopes, such as ^{13}C, ^{15}N, ^{18}O. This can be alleviated by the use of maximum-entropy–based deconvolution [89] to narrow the isotopic distribution or by cultivating the organism or cell producing the proteins of interest in a medium depleted of isotopic ^{13}C and ^{15}N [90]. FT-ICR is endowed with a unique resolving power: a resolution of almost one million was claimed with a peptide of m/z 650, allowing the separation of the isotopes ^{12}C and ^{15}N and the complete resolution of the ^{34}S and ^{13}C isotopes, separated by only 0.011 Da [91]. This is coupled to an incredibly high mass range: for example, in the analysis of coliphage T4 DNA, Chen et al. [92] trapped an ionized species whose mass was determined to be 1.1 × 10^8 amu. Sub-attomole sensitivity (<600,000 molecules) of intact peptides and proteins with masses as high as 30 kDa was also claimed [93]. Another unique feature of FT-ICR is that in this technique, ion detection is nondestructive [94]. The ions can be visualized by applying a radio-frequency electric field with the same frequency as the ion cyclotron frequency, thus producing resonance [95]. This has an importance consequence: after mass determination, it is theoretically possible to purify the sample by molecular mass selection and to recover it from the instrument for further uses, a feature of interest if one is dealing with DNA molecules, which can be subsequently amplified via PCR.

The main limitations of the FT-ICR MS technique are the need to operate at very high vacuum (10^{-8} Torr), the low duty cycle, and the necessity for long data-accumulation times. For example, if one needs to couple FT-ICR to external atmospheric ionization techniques, such as ESI, the ion source must be separated from the trap by several stages of differential pumping. Thus, the time required for pumping the instrument to operating pressures, after ion injection, is responsible for the poor duty cycle. In addition, the data sets generated by FT-ICR are extremely large because a huge number of data points must be collected in order to take advantage of the resolving power; hence, a second limiting factor is the time required to store the data, even though the actual measurement times are quite short (a few hundred microseconds).

F. 2-D or Not 2-D?

If one considers the way 2-D maps are run today (IEF in the first dimension, coupled with SDS-PAGE in the second, orthogonal space), one is confronted with a kind of a paradox. SDS-PAGE separates the focused zones according to a mass parameter, but in reality it gives a mass assessment of no real value, considering the poor performance of the technique in terms of mass accuracy and the many exceptions to the canonical binding rules, which would require 1.4 mg SDS per mg of protein [96]. In order to identify the protein, moreover, one would still have to stain it, elute it, digest it, and then characterize all fragments in a mass spectrometer. It would thus seem logical that one should replace the very low mass accuracy of the SDS-PAGE dimension with the high-resolution mass spectrometer. Initial steps in this direction were already taken in 1997 by Ogorzalek Loo et al. [97] by using direct scanning of the first-dimension IPG gel strip with an infrared laser. One could even replace the first dimension gel by capillary IEF, which can then be directly connected to a mass spectrometer with an ESI interface [98]. Via the latter technique, by injection only 300 ng protein from an *E. coli* extract, four hundred to one thousand putative-peptide/proteins with a mass range between 2 kDa and 100 kDa could be characterized.

There is even more to it: it is known that one of the major limiting factors in biological protein analysis is its enzymatic digestion. When protein concentration falls below 500 fmol/µL, trypsin digestion is very inefficient, since proteases exhibit 50% maximal activity in the 5- to 50-pmol/µL range. And this without even accounting for sample losses at such dilute protein concentration, when eluting spots from a 2-D gel. Modern MS technologies allow performing the fragmentation directly in the mass spectrometer. MS/MS of intact proteins electrosprayed into various MS instruments has been achieved with a wide range of techniques, such as CID [99], infrared multi-photon dissociation [100], and UV-laser–induced and surface-induced fragmentation [101]. It will be curious to see how 2-D mapping will evolve in the near future.

VII.   INFORMATICS AND PROTEOME: INTERROGATING DATABASES

The field of informatics in proteome analysis is growing at an impressive rate and it would be impossible here to summarize the present state of the art because this summary would be rapidly outdated. It is important, however, for readers to understand the basics of it, in order to be able to screen properly the databases and derive the relevant information. To this aim, one should consult at least three major chapters in the book of Wilkins et al. [9], which give a unique reconstruction and provide excellent guidelines for database browsing. They are the chapter by Bairoch [102] on proteome databases, the one by Appel [103], on interfacing and integrating databases, and the third one by Peitsch and Guex [104] on large-scale comparative protein modeling. In addition, for understanding how these databases started and evolved over the years, it is of great interest to read some reviews highlighting the evolution of the informatization of this area. These are an article by Sanchez et al. [105] offering a tour inside Swiss 2-D PAGE database and its update by Hoogland et al. [106]; a review by Bairoch and Apweiler [107] on the SWISS-PROT sequence database and its supplement TrEMBL; a paper by Crawford et al. [108] on databases and knowledge resources for proteomics research; and, finally, a report by Beavis and Fenyö [109] on database searching with mass-spectrometric information. While this list is certainly not exhaustive, it will provide the readers with at least the basic tools for a good start.

Identifying protein components from physical evidence, such as mobility on gels or the masses of the intact protein, is often the starting point. After this comes the full protein sequence, usually predicted from the known DNA sequence, and then structural and func-

Table 1 Primary and Secondary Attributes for Protein Recognition

Protein attribute	Origin of analytical data
Primary	
Species of origin	Starting biological material
Isoelectric point (pI)	From first dimension IPG
Apparent mass	From second dimension SDS-PAGE
Real mass	From mass spectrometry, e.g. MALDI-TOF of intact protein
Protein specific sequence at N- and C-termini	From chemical sequencing of proteins immobilized onto membranes
Extended N-terminus sequence	From chemical sequencing of proteins immobilized onto membranes
Secondary	
Fingerprinting of peptide masses	MALDI-TOF or ESI-MS of digested peptides
Peptide fragmentation and de novo sequencing via MS	Fragmentation of peptides via MS-MS or PSD MALDI-TOF
Amino acid composition	Multiple radio-labeling analysis; chromatography of total protein hydrolysates

tional predictions. These initial steps are summarized in Table 1. In model organisms such as yeast, more than half of the proteins have already been functionally analyzed and, of the proteins coded for by the human genome, roughly 10% have already been studied in one or more laboratories. The big revolution in proteomics, though, has come from MS analysis. Highly accurate masses can be obtained for peptide fragments isolated from proteolytic digests of gel-separated proteins, and these are sufficient for protein identification, given effective databases. Many resources today exist for identifying proteins by peptide masses, and these are listed in Table 2. Perhaps the most commonly used databases are SWISS-PROT, TrEMBL and the nonredundant (nr) collection of protein sequences at the US National Centre for Biotechnology Information (NCBI). SWISS-PROT is an annotated collection of protein sequences on the Expasy server; TrEMBL is a large collection of predicted protein sequences given automatic annotation until they can be fully annotated and entered

Table 2   URLs for Proteomic Resources

Name of resource	Web location
Protein identification	
SWISS-2DPAGE	http://www.expasy.ch/ch2d/
PeptIdent	http://www.expasy.ch/tools/ peptident.html
PROWL	http://prowl.rockefeller.edu/
SEQUEST	http://thompson.mbt.washington.edu/ sequest/
MS-FIT/TAG	http://prospector.ucsf.edu/
Sequence databases	
NCBI/BLAST tools	http://www.ncbi.nlm.nih.gov(BLAST/
SWISS-PROT	http://www.expasy.ch/sprot/
3-D structure	
SCOP/PDBL-ISL	http://scop.mrc-Imb.cam.ac.uk/scop/
3Dcrunch	Http://www.expasy.ch/swissmod/ SM_3DCrunch.html
SWISS-MODEL	Http://www.expasy.ch/swissmod/ SWISS-MODEL.html
MODBASE	http://guitar.rockefeller.edu/modbase/
VAST	Http://www.ncbi.nlm.nih.gov(Structure/ VAT/vast.html.
PDB	http://www.rcsb.org.pdb/
Domain structure	
PROSITE	http://www.expasy.ch/tools/ scnpsit1.html
PRINTS	http://www.biochem.ucl.ac.uk/bsm/ dbbrowser/PRINTS/PRINTS.html
BLOCKS	http://www.blocks.fhcrc.org/
Pfam	http://pfam.wustl.edu/
ProDOM	http://protein.toulouse.inra.fr/ prodom.html
SMART	http://smart.embl-heldelberg.de/ index.shtml
InterPro	http://www.ebi.ac.uk/interpro/
Subcellular-location prediction	
PSORT-II	http://psort.nibb.ac.jp/
Model organism resources	
SGD	http://genome-www-stanford.edu/ Sccharomyces/
WormBase/AceDB	http://www.wormbase.org/
FlyBase	http://www.fruitfly.org/
Literature-based databases	
YPD/WormPD/PombePD/	http://www.proteome.com/databases/

into SWISS-PROT; the NCBI nr database contains translated protein sequences from the entire collection of DNA sequences kept at GenBank, and also protein sequences in the PDB, SWISS-PROT, and PIR databases. Such databases also offer additional information, including brief functional descriptions (if known), an annotation of sequence features (e.g., modification signals), secondary and tertiary structure predictions, key references, and links to other databases.

VIII. QUANTITATIVE PROTEOMICS

One of the main problems with current 2-D map protocols, including the most sophisticated ones, is the lack of proper quantitation of the resolved and detected polypeptide spots. For example, specific classes of proteins have long been known to be excluded or underrepresented in 2-D gel patterns. These include very acidic or basic proteins, excessively large or small polypeptides, and membrane proteins. In addition, it is now clear that 2-D maps cannot detect low-abundance proteins without a pre-enrichment step. Thus, quantitative proteomics still has to come of age. Yet it is thoroughly needed because the ability of accurately detecting and quantifying potential protein changes induced in an organism by a specific perturbation is an essential part of the study of dynamic biological processes. It might be argued that, in principle, quantitative proteomics might be assessed by measuring the levels (and its changes) of specific mRNAs, as present in an organism at rest or as induced by given stimuli. Yet this approach might be fallacious, given the following facts:

> In the few studies that measured the copies of mRNA and protein levels in the same system, the correlation of these two levels was very poor and not strong enough to enable prediction of one value from the other [110,111].
>
> Protein activation or inactivation by post-translational mechanisms cannot be detected by looking at the mRNA levels.

One elegant way out of this impasse has been recently proposed by Gygi and Aebersold [112]: quantifying proteins by using stable-isotope dilution. This "venerable" technique involves the addition to the sample of a chemically identical form of the analyte(s) containing stable heavy isotopes (such as ^{2}H, ^{13}C, ^{15}N) as internal standards. The most suitable internal standard for a candidate peptide is the same peptide labeled with stable isotopes. Therefore, proteins can be profiled in a quantitative manner as two protein mixtures are compared, with one serving as the reference sample, by containing the same proteins as the other sample but at different abundance and labeled with heavy stable isotopes. In theory, then, all peptides from the combined samples exist as analyte pairs of identical sequences but with different masses. Thus, when analyzed by MS techniques, the ratio between the intensities of the lower and upper mass components of these pairs of peaks provides an accurate measure of the relative abundance of such peptides (or proteins) in the original cell lysates. There are two strategies for achieving this goal: labeling either before or after extraction.

A. Labeling Before Extraction

This approach has been followed by Oda et al. [113] and Pasa-Tolic et al. [114]. The first authors grew one yeast culture on a medium containing the natural nitrogen isotope distribution (^{14}N 99.6%; ^{15}N 0.4%) and another one on the same medium enriched in ^{15}N (>96%). After a suitable growth period, the cell pools were combined and the proteins of interest extracted and separated by RP-HPLC and then by SDS-PAGE. After extraction of the proteins of interest and digestion, the peptides were subjected to MS analysis. Each incorporated ^{15}N atom shifted the mass of any given peptide upward by one mass unit, leading to a pair of peaks from each peptide. Pasa-Tolic et al. [114], on the other hand, used stable isotope media for giving proteins a specific isotope signature. They compared the cadmium stress response in *E. coli* grown in normal and rare-isotope–depleted media (lacking ^{13}C, ^{15}N and ^{2}H). Intact-protein mass measurements were carried out by Fourier-transform ion-cyclotron-resonance (FT-ICR) MS, a technique of growing importance for ultra-high resolution MS analysis of biopolymers [115]. Although no protein could be positively identified, the expression ratio for 200 different proteins could be compared.

B. Labeling After Extraction

This is the approach taken by Gygi and Aebersold [112]. It is based on so-called isotope-coded affinity tags (ICAT). The method is illustrated in Fig. 6. In this novel procedure, the stable isotopes are incorporated after isolation by the selective alkylation of Cys residues with either a "heavy" or "light" reagent, after the two protein pools to be compared are mixed. The ICAT reagent (whose general structure is depicted in Fig. 6a) is composed of three parts: a biotin portion, used as an affinity tag; a linker, which can incorporate either the heavy or light isotopes; and a third terminal group, which contains a reactive iodine atom able to alkylate specifically thiol groups (Cys residues). The "heavy" ICAT contains eight deuterium atoms, which in the "light" one are replaced by standard hydrogen atoms. Proteins from two different cell states are harvested, denatured, reduced, and labeled at Cys residues with either light or heavy ICAT reagent. The samples are then combined and digested with trypsin. ICAT-labeled peptides can be further isolated by biotin-affinity chromatography and then analyzed by on-line HPLC coupled to tandem MS. The ratio of the ion intensities for any ICAT-labeled pair quantifies the relative abundance of its parent protein in the original cell state. In addition, the tandem MS approach produces the sequence of the peptide, and thus can unambiguously identify the protein of interest. This strategy, ultimately, results in the quantification and identification of all protein components in a mixture and, in principle, could be applied to protein mixtures as complex as the entire genome. The ICAT strategy appears to be endowed with the following advantages:

First, the method is compatible with any amount of protein harvested from body fluids, cells, or tissues under any growth conditions.

Second, the alkylation reaction is highly specific and occurs in the presence of salts, detergents, and a large body of solubilizers.

Third, the complexity of the peptide mixture to be analyzed is reduced by isolating only the Cys-tagged peptides.

As a last bonus, the ICAT strategy permits almost any type of biochemical, immunological, or physical fractionation, which renders it compatible with the analysis of low-abundance proteins.

Some drawbacks appear to be due to the fact that the ICAT label is quite large (approximately 600 Da), a fact that can complicate

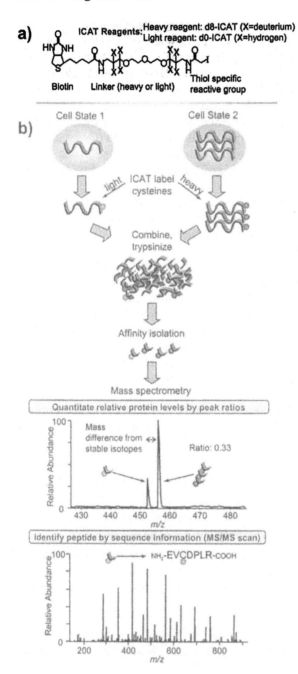

a)

ICAT Reagents: Heavy reagent: d8-ICAT (X=deuterium)
Light reagent: d0-ICAT (X=hydrogen)

Biotin   Linker (heavy or light)   Thiol specific reactive group

b)

Cell State 1   Cell State 2

light ICAT label *heavy*
cysteines

Combine, trypsinize

Affinity isolation

Mass spectrometry

Quantitate relative protein levels by peak ratios

Mass difference from ↔ stable isotopes   Ratio: 0.33

Identify peptide by sequence information (MS/MS scan)

NH₂-EVCDPLR-COOH

database searching for small peptides. Another problem could result from the fact that small proteins, lacking Cys residues, will not be labeled. Finally, it remains to be seen if the reaction is quantitative. Although, in a recent article, these authors [116] claim to have achieved 100% derivatization of all Cys residues in proteins, this appears quite strange, indeed, considering that our group has amply demonstrated that iodoacetamide-type, or acrylamide-based, reactants rarely permit >80% blocking of −SH groups in a protein [68,69]. Moreover, when insisting on the reaction for longer periods of time (e.g., overnight), there is no progression in the −SH alkylation, whereas parasitic reactions take over (such as blocking of Lys residues). If this occurs with such a relatively small molecule as iodoacetamide, one wonders why much larger iodoacetamide-based reactants should exhibit better reaction kinetics! To our knowledge, the only −SH labeling reagent able to derive the reaction to 100% (even under the most adverse conditions, such as in the presence of surfactants) is vinyl pyridine (see Fig. 5), just because, as the reaction is conducted at a pH half way between the two pK values (8.3 for the −SH group and 5.6 for the nitrogen in 4-VP), its progression and completion is strongly driven by the coulombic attraction between the residual negative and positive charges on the two species.

On a similar line of thinking, Peters et al. [117] have proposed a novel, multifunctional label, 2-methoxy-4,5-dihydro-1H-imidazole, for use in quantitative proteomics studies. This compound is a lysine-specific labeling reagent that also affects a number of properties desirable in MS proteomic studies. The increased basicity of the imi-

---

Fig. 6 Scheme of the isotope-coded affinity tag (ICAT) strategy for quantification of protein expression. (A). Structure of the ICAT reagent. It consists of three segments: an affinity tag (biotin); a linker, which can incorporate either deuterium or hydrogen; and a reactive tail specific for thiol groups. (B). ICAT strategy. Protein from two different cell states is harvested, denatured, reduced and labeled at Cys with the light or heavy ICAT reagents. The samples are then combined and digested with trypsin. After affinity isolation of the ICAT-labeled peptides, they are analyzed in a tandem mass spectrometer. The ratio of the ion intensities for each ICAT-labeled pair quantifies the relative abundance of its parent proteins in the original state. Finally, the MS spectrum enables sequencing and protein identification. (From Ref. 112.)

dazol-2-yl, as compared with lysine, greatly increases sequence coverage obtained in peptide-mapping experiments. In addition, this moiety greatly simplifies the tandem mass spectrum of the resulting derivatized peptide, yielding primarily an easily interpretable series of y-ions. As an extra bonus, quantitative differentiation studies can also be performed, as this reagent possesses four sites for deuterium substitution.

C. Quantitative Analysis via Differential Fluorescence Labeling

The methods described in Sections B and C rely on MS analysis for a quantitative differential display of protein expression. A way to circumvent that is to label the control and experimental samples with fluorescent labels, able to be imaged separately due to different emission wavelengths. Tonge et al. [118] have recently described such a protocol, which exploits derivatization of the ε-amino groups of proteins with a mass and charge-matched set of fluorophores (called Cy2, Cy3, and Cy5). These are a family of cyanine dyes, containing a quaternary nitrogen, which form an amide group upon reaction on a Lys residue. The dyes have very similar molecular masses and are positively charged so as to match the charge on the Lys group, since the latter will be abolished upon reaction. The dye to protein labeling ratio is deliberately kept low, so that only proteins containing a single dye molecule are visualized on the gel. Charge matching ensures that there is little shift over the unlabeled protein in the pI dimension. The procedure is then analogous to that of the ICAT protocol [112]: the two pools of proteins are separately labeled with either fluor A or B, mixed, and run in a 2-D gel. At the end of the separation, the gel is imaged by excitation at two different wavelengths, so one is able to obtain, from the same gel, two distinct images. These digitized maps will be overlaid, the spots will be quantified, the signals normalized, and the potential differential expression of proteins in the two samples obtained by direct comparisons of the normalized spot intensities. Although this approach could also be of interest, it should be appreciated that, in addition to the costs of the fluorescent labels, one would have to buy specialized (and costly) scanners for collecting the fluorescent signal. Most scanners today used in the vast majority of proteome labs can only digitize images taken by transmission under visible light.

IX. PRE-FRACTIONATION IN PROTEOME ANALYSIS

Although both the first (IEF/IPG) and second (SDS-PAGE) separation dimensions appear to be performing at their best, there remains
the fact that, because of the vast body of polypeptide chains present
in a tissue lysate, especially at acidic pH values (pH 4–6), resolution
is still not good enough. Some recent articles suggest that this problem could be overcome by running a series of narrow-range IPG
strips (covering no more than 1 pH unit) on large size gels, which
would dramatically increase the resolution [119]. The entire wide-
range 2-D map would then be electronically reconstructed by
stitching together the narrow-range maps. This might turn out to
be a will-o'-the-wisp though: there remains the fact that even when
very narrow IPG strips are used, they have to be loaded with the
entire cell lysate, containing proteins focusing all along the pH scale.
Massive precipitation will then ensue, due to aggregation among un-
like proteins, with the additional drawback that the proteins which
should focus in the chosen narrow-range IPG interval will be
strongly underrepresented, because they will be only a small fraction
of the entire sample loaded. That this is a serious problem has been
debated in a recent work by Gygi et al. [120]. These authors analyzed
a yeast lysate by loading 0.5 mg total protein on a narrow-range IPG
pH 4.9–5.7. Although the authors could visualize by
silvering approximately 1500 spots, they lamented that a large
number of polypeptides simply did not appear in such a 2-D
map. In particular, proteins from genes with codon bias values of
<0.1 (low abundance proteins) were not found, even though
fully one half of all yeast genes fall into that range. Thus, they
concluded that, in reality, when analyzing protein spots from 2-D
maps by mass spectrometry, only generally abundant proteins (codon bias >0.2) could be properly identified. Therefore, the number
of spots on a 2-D gel is not representative of the overall number or
classes of expressed genes that can be analyzed. This is a severe
limitation, because it is quite likely that the portion of proteome we
are currently missing is the most interesting one from the point of
view of understanding cellular and regulatory proteins, since such
low-abundance polypeptide chains will typically be regulatory proteins.

Due to such major drawbacks, we believe that the only way out
of this impasse will be prefractionation. At present, two approaches

have been described: chromatographic and electrophoretic. They will be illustrated below.

A. Sample Pre-Fractionation via Different Chromatographic Approaches

This is mostly the work of Fountoulakis' group. In a first procedure, Fountoulakis et al. [121] and Fountoulakis and Takàcs [122] adopted affinity chromatography on heparin gels as a pre-fractionation step for enriching certain protein fractions in the bacterium *Haemophilus influenzae*. Because of its sulfate groups, heparin also functions as a high-capacity cation exchanger. In fact, about 160 cytosolic proteins bound with different affinities to the heparin matrix and were thus highly enriched prior to 2-D PAGE separation. As a result, more than 110 new protein spots, detected in the heparin fraction, were identified, thus increasing the total identified proteins of *H. influenzae* to more than 230. In a second approach [123], the same lysate of *H. influenzae* was pre-fractionated by chromatofocusing on Poly-buffer Exchanger. In the eluate, two proteins, major ferric iron–binding protein (HI0097) and 5′-nucleotidase (HI0206) were obtained in a pure form with another hypothetical protein (HI0052) purified to near homogeneity. Four other proteins, aspartate ammonia lyase (HI0534), peptidase D (HI0675), elongation factor Ts (HI0914), and 5-methyltetrahydropteroyltriglutamate methyltrasferase (HI1702), were strongly enriched by the chromatofocusing process. Approximately 125 proteins were identified in the eluate collected from the column. Seventy of these were for the first time identified after chromatography on the Polybuffer Exchanger, the majority of them being low-abundance enzymes with various functions. Thus, with this additional step, a total of 300 protein could be recognized in *H. influenzae* by 2-D map analysis, out of a total of approximately 600 spots visualized on such maps from the soluble fraction of this microorganism. In yet another approach, the cytosolic soluble proteins of *H. influenzae* were pre-fractionated [124] by hydrophobic interaction chromatography (HIC) on a TSK Phenyl column. The eluate was subsequently analyzed by 2-D mapping, followed by spot characterization by MALDI-TOF MS. Approximately 150 proteins, bound to the column, were identified, but only 30 for the first time. In total, with all the various chromatographic steps adopted, the number of total proteins identified could be increased to 350.

The same heparin chromatography procedure was subsequently applied by Karlsson et al. [125] to the pre-fractionation of human fetal brain soluble proteins. Approximately 300 proteins were analyzed, representing 70 different polypeptides, 50 of which were bound to the heparin matrix. Eighteen brain proteins were identified for the first time. The polypeptides enriched by heparin chromatography included both minor and major components of the brain extract. The enriched proteins belonged to several classes, including proteasome components, dihydropyrimidinase-related proteins, T-complex protein 1 components, and enzymes with various catalytic activities.

In yet another variant, Fountoulakis et al. [126] reported enrichment of low abundance proteins of *E. coli* by hydroxyapatite chromatography. The complete genome of *E. coli* has now been sequenced and its proteome analyzed by 2-D mapping. To date, 223 unique loci have been identified and 201 protein entries were found in the SWISS-PROT 2-D PAGE on ExPASy server using the Sequence retrieval System query tool (http://www.expasy.ch/www/sitemap. html). Of the 4289 possible gene products of *E. coli*, about 1200 spots could be counted on a typical 2-D map when approximately 2-mg total protein was applied. Possibly, most of the remaining proteins were not expressed in sufficient amounts to be visualized following staining with Coomassie Blue. Thus, it was felt necessary to perform a pre-fractionation step, this time on hydroxyapatite beads. By this procedure, approximately 800 spots, corresponding to 296 different proteins, were identified in the hydroxyapatite eluate. About 130 new proteins that had not been detected in 2-D gels of the total extract were identified for the first time. This chromatographic step, though, enriched both low-abundance as well as major components of the *E. coli* extract. In particular, it enriched many low-Mr proteins, such as cold-shock proteins. On a similar line of thinking, Harrington et al. [127] reported the use of cation-exchange chromatography for enriching DNA-binding proteins on tissue extracts, prior to 2-D analysis. Here, too, the basic domains do not seem to influence the overall pI of such proteins, so one sees a general spread of the pre-fractionated sample across the complete pI range of a 2-D map.

Although these approaches are impressive and truly innovative in proteome analysis, they are not free from drawbacks, such as:

The need for high amounts of salts (up to 2.5 M) for elution

The potential loss of groups of proteins during the various manipulation steps

The fact that the eluted fractions do not represent narrow pI cuts, but are in general constituted by proteins with pIs in the pH 3–10 range.

As a possible alternative, fractionation steps fully compatible with the subsequent 2-D mapping, could be adopted, such as a preparative focusing step based on multicompartment electrolyzers with isoelectric membranes.

B. Sample Pre-Fractionation via Electrophoretic Approaches

The multicompartment electrolyzer (MCE), developed at the end of the 1980s [128–130], is based on the principle of "isoelectric traps"; isoelectric membranes, arranged in order of increasing pI values between anode and cathode, produced with the same Immobiline technology adopted for casting IPG strips, are used as a bait for capturing sets of proteins having pI values in between the pIs of two neighbouring membranes. Such "narrow pI-cuts" can be directly interfaced with 2-D map analysis, because these fractions are already equilibrated in the same solubilizing cocktail and emerge from the MCE salt-free. Although in principle any set of traps could be made, the narrowest workable pI fractionation is limited (due to time limitations in the process) to approximately 1 pH unit; the pI precision in making the membranes is typically of the order of 0.01 pH unit.

Herbert and Righetti [131] have recently applied precisely this technology to sample pre-fractionation in preparation for 2-D map analysis. Figure 7 displays the analysis of the pI 4.0–5.0 fraction collected from the pI 4.0–5.0 chamber of the MCE apparatus in the analysis of an *E. coli* lysate (right side). As a comparison, the left panel shows the proteins that could be detected in the same pH interval from an unfractionated sample, as available on the Internet at the SWISS-2D-PAGE site. Although the gel on the right is stained with Coomassie Brilliant Blue, it is apparent that more spots are

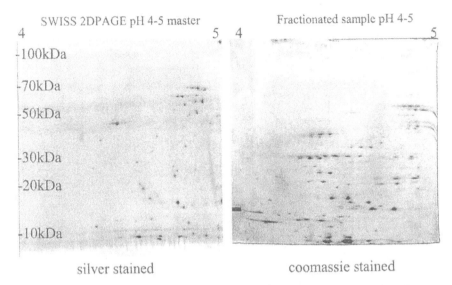

Fig. 7 Silver-stained 2-D map of an *E. coli* entire cell lysate (left panel) vs. the colloidal Coomassie G250 stained gel of the sample pre-fractionated with the MCE instrument (right panel), run in a 18 cm pH 4–5 gradient in the first dimension. (From Ref. 131.)

visible, as compared with the silver-stained SWISS-2DPAGE *E. coli* map. Moreover, because the gel in the right panel is stained with Coomassie Brilliant Blue and not with silvering, almost all the visible spots will be present in sufficient quantities for mass spectrometry. Although this is a clear indication that the pre-fractionation procedure should not lead to loss of components via trapping into the isoelectric membranes, Herbert and Righetti [131] performed a protein assay on the starting and ending products and could confirm a 95% protein recovery. Additionally, a similar assay performed on the ground and extracted isoelectric membranes failed to reveal any appreciable amount of proteinaceous material bound or adsorbed onto those surfaces. The advantages of such a pre-fractionation procedure can be appreciated from Fig. 8: a new 2-D map was made, with five times more loaded protein as compared with Fig. 7 (right panel). Yet even under such overloading conditions, no sample streaks or smears appeared, but simply more in-

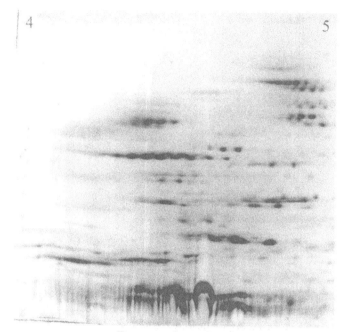

Coomassie stained

Fig. 8 2-D map of an *E. coli* sample as in Fig. 7, but run with a five-fold increased protein concentration (gel stained with colloidal Coomassie Blue (G250). Approximately 50 spots, chosen at random, were eluted and identified by MALDI-TOF MS of the recovered peptides, after trypsin digestion. Approximately 2–3 times as many proteins could be identified as compared to the official SWISS-2DPAGE *E. coli* map. (From Ref. 131.)

tense spots and more low-abundance proteins could be detected [132] (note that the streaking at the gel bottom, in the low Mr region, is due to mixed ASB14-SDS micelles migrating with the salt front).

In conclusion, the present pre-fractionation method could prove a formidable tool in proteome analysis, because it will provide not only the much needed improvement in resolution but also the highly desirable increment in sensitivity, due to the possibility of loading a much higher sample amount in any desired narrow pH interval.

X. CONCLUSIONS: *QUO VADIS,* PROTEOME?

When poor Petrus, hastily retreating from Rome along the via Appia, met an angel at a sharp hairpin turn, bearing a cross and a sign reading "quo vadis," he knew it was hopeless: he meekly retraced his steps back to the metropolis to meet his fate. He was crucified, just like his Master, and thus entered legend and immortality, as Saint Peter. Where is the proteome leading to? Up to the present, it has largely been the domain of electrophoreticists, because 2-D maps are run by the sheer force of a voltage gradient. But now the holy citadel is under siege from foreign troops, all eager to conquer it—for the sake of science, of course, and for the sake of the green dollar, the polychromatic euro, the yen, you name it (although the ruble and the kopeck are defunct by now). So chromatographers today claim that present-day 2-D maps are nonsense and that chromatographic separations will do a much better job in proteome analysis, notwithstanding the superb separation power of 2-D electrophoresis. The ICAT protocol [112], in fact, is based on an affinity chromatographic step, able to capture only the biotinilated peptides in the vastly heterogeneous concoction of the tryptic digest of an entire cell lysate. This is a monodimensional chromatographic step, but, of course, a number of two-dimensional approaches have been already devised [133–139]. It might be of interest to see how the last-mentioned techniques would score in proteome analysis, as compared to the well-ingrained 2-D PAGE mapping methodology. This has been assessed by Washburn and Yates III [140]. According to these authors, none of the above alternative methods has yet attained the proficiency of 2-D PAGE, but they believe that they will in the near future. In their opinion, additionally, the most promising system appears to be the one developed by Link et al. [138,139]—the one hyphenating a binary strong cation exchanger-reversed phase column to MS-MS for on-line protein identification. Were this not enough, affinity protein chips, able to capture single protein species (or families) in the vast sea of a total cell lysate, are now coming of age [141,142]: will the *Elective Affinities* of Johann Wolfgang Goethe take over, or will *Les Liasons Dangereuses* of Choderlos de Laclos prevail? It is hard to give an answer now, at the onset of the games. Meanwhile, as a precaution, we poor electrophoreticists are hiding a cross in our closet: one never knows!

ACKNOWLEDGMENTS

Supported in part by a grant from MURST (Coordinated Project 40%, Proteome Analysis), by ASI (Agenzia Spaziale Italiana, grant No. I/R/28/00) and by a contract from the European Community (Microproteomics, grant No. QLG2-CT-2001-01903).

REFERENCES

1. B. Bjellqvist, K. Ek, P. G. Righetti, E. Gianazza, A. Görg, W. Postel, and R. Westermeier, *J. Biochem. Biophys. Methods, 6*: 317 (1982).
2. R. Aebersold and J. Leavitt, *Electrophoresis, 11*: 517 (1990).
3. S. D. Patterson and R. Aebersold, *Electrophoresis, 16*: 1791 (1995).
4. H. W. Lahm and H. Langen, *Electrophoresis, 21*: 2105 (2000).
5. P. James, *Proteome Research: Mass Spectrometry*, Springer, Berlin, 2001, pp. 1–274.
6. F. Lottspeich, T. Houthaeve, and R. Kellner, in R. Kellner, F. Lottspeich and H. E. Meyer, (Eds.), *Microcharacterization of Proteins*, Wiley-VCH, Weinheim, 1999, p. 141–158.
7. M. R. Wilkins, C. Pasquali, R. D. Appel, K. Ou, O. Golaz, J. C. Sanchez, J. X. Yan, A. A. Gooley, G. Hughes, I. Humphrey-Smith, K. L. Williams, and D. F. Hochstrasser, *BioTechnology, 14*: 61 (1996).
8. S. R. Pennington, M. R. Wilkins, D. F. Hochstrasser, and M. J. Dunn, *Trends Cell Biol., 7*: 168 (1997).
9. M. R. Wilkins, K. L. Williams, R. D. Appel, and D. F. Hochstrasser (Eds.), *Proteome Research: New Frontiers in Functional Genomics*, Springer, Berlin, 1997, pp. 1–243.
10. R. Kellner, F. Lottspeich, and H. E. Meyer (Eds.), *Microcharacterization of Proteins*, Wiley-VCH, Weinheim, 1999, pp. 1–399.
11. T. Rabilloud (Ed.), *Proteome Research: Two-Dimensional Gel Electrophoresis and Identification Methods*, Springer, Heidelberg, 2000, pp. 1–248.
12. P. G. Righetti, A. V. Stoyanov, M. Y. Zhukov, *The Proteome Revisited: Theory and Practice of All Relevant Electrophoretic Steps*, Elsevier, Amsterdam, 2001, pp. 1–395.

13. R. Westermeier, *Electrophoresis in Practice*, VCH, Weinheim, 1997, pp. 213–228.
14. S. Hanash, in B. D. Hames (Ed.), *Gel Electrophoresis of Proteins*, Oxford University Press, Oxford, 1998, pp. 189–212.
15. E. Gianazza, P. Giacon, B. Sahlin, and P. G. Righetti, *Electrophoresis*, *6*: 53 (1985).
16. E. Gianazza, S. Astrua-Testori, and P. G. Righetti, *Electrophoresis*, *6*: 113 (1985).
17. E. Gianazza, P. Giacon, S. Astrua-Testori, and P. G. Righetti, *Electrophoresis*, *6*: 326 (1985).
18. E. Gianazza and P. G. Righetti, *J. Chromatogr.*, *193*: 1 (1980).
19. P. G. Righetti, M. Chiari, P. K. Sinha, and E. Santaniello, *J. Biochem. Biophys. Methods*, *16*: 185 (1988).
20. A. Bossi, P. G. Righetti, G. Vecchio, and S. Severinsen, *Electrophoresis 15*: 1535 (1994).
21. A. Bossi, C. Gelfi, A. Orsi, and P. G. Righetti, *J. Chromatogr. A*, *686*: 121 (1994).
22. P. K. Sinha, M. Praus, E. Köttgen, E. Gianazza, and P. G. Righetti, *J. Biochem. Biophys. Methods*, *21*: 173 (1990).
23. P. G. Righetti and C. Tonani, *Electrophoresis 12*: 1021 (1991).
24. C. Tonani and P. G. Righetti, *Electrophoresis 12*: 1011 (1991).
25. J. A. Lenstra and H. Bloemendal, *Eur. J. Biochem.*, *135*: 413 (1983).
26. M. P. Molloy, B. R. Herbert, B. J. Walsh, M. I. Tyler, M. Traini, J. C. Sanchez, D. F. Hochstrasser, K. L. Williams, and A. A. Gooley, *Electrophoresis*, *19*: 837 (1998).
27. M. P. Deutscher (Ed.), *Guide to Protein Purification. Methods Enzymol.*, 1990, pp. 1–894.
28. N. L. Anderson and B. J. Hickman, *Anal. Biochem.*, *93*: 312 (1979).
29. B. J. Hickman, N. L. Anderson, K. E. Willard, and N. G. Anderson, in B. J. Radola (Ed.), *Electrophoresis '79*, de Gruyter, Berlin, 1980, pp. 341–360.
30. S. L. Tollaksen, J. J. Edwards, and N. G. Anderson, *Electrophoresis*, *2*: 155 (1981).
31. P. H. O'Farrell, *J. Biol. Chem.*, *250*: 4007 (1975).
32. M. R. Wilkins, E. Gasteiger, J. C. Sanchez, A. Bairoch, and D. F. Hochstrasser, *Electrophoresis*, *19*: 1501 (1998).
33. M. P. Molloy, *Anal. Biochem.*, *280*: 1 (2000).

34. T. Rabilloud and M. Chevallet, in T. Rabilloud (Ed.), *Proteome Research: Two-Dimensional Gel Electrophoresis and Identification Methods*, Springer, Heidelberg, 2000, pp. 9–29.

35. T. Rabilloud, C. Adessi, A. Giraudel, and J. Lunardi, *Electrophoresis, 18*: 307 (1997).

36. C. Pasquali, I. Fialka, and L. A. Huber, *Electrophoresis, 18*: 2574 (1997).

37. L. Musante, G. Candiano, and G. M. Ghiggeri, *J. Chromatogr. A, 705*: 351 (1997).

38. T. Rabilloud, *Electrophoresis, 19*: 758 (1998).

39. J. A. Gordon and W. P. Jencks, *Biochemistry, 2*: 47 (1963).

40. P. Hagel, J. J. T. Gerding, W. Fieggen, and H. Bloemendal, *Biochim. Biophys. Acta, 243*: 366 (1971).

41. M. N. Jones, *Int. J. Pharm., 177* 137 (1999).

42. D. F. Hochstrasser, M. G. Harrington, A. C. Hochstrasser, M. J. Miller, and C. R. Merill, *Anal. Biochem., 173*: 424 (1988).

43. P. Holloway and P. Arundel, *Anal. Biochem., 172*: 8 (1988).

44. M. A. Rimpilainen and P. G. Righetti, *Electrophoresis, 6*: 419 (1985).

45. D. Satta, G. Schapira, P. Chafey, P. G. Righetti, and J. P. Wahrmann, *J. Chromatogr., 299*: 57 (1984).

46. M. J. Dunn and A. H. M. Burghes, *Electrophoresis, 4*: 97 (1983).

47. K. E. Willard, C. Giometti, N. L. Anderson, T. E. O'Connor, and N. G. Anderson, *Anal. Biochem., 100*: 289 (1979).

48. E. Gianazza, T. Rabilloud, L. Quaglia, P. Caccia, S. Astrua-Testori, L. Osio, G. Grazioli, and P. G. Righetti, *Anal. Biochem., 165*: 247 (1987).

49. T. Rabilloud, E. Gianazza, N. Cattò, and P. G. Righetti, *Anal. Biochem., 185*: 94 (1990).

50. M. Chevallet, V. Santoni, A. Poinas, D. Rouquie, A. Fuchs, S. Kieffer, M. Rossignol, J. Lunardi, J. Garin, and T. Rabilloud, *Electrophoresis, 19*: 1901 (1998).

51. M. P. Molloy, B. R. Herbert, M. B. Slade, T. Rabilloud, A. S. Nouwens, K. L. Williams, and A. A. Gooley, *Eur. J. Biochem., 267*: 1 (2000).

52. V. Santoni, T. Rabilloud, P. Doumas, D. Rouquie, A. Fuchs, S. Kieffer, J. Garin, and M. Rossignol, *Electrophoresis, 20*: 705 (1999).

53. V. Santoni, S. Kieffer, D. Desclaux, F. Masson, and T. Rabilloud, *Electrophoresis*, *21*: 3329 (2000).
54. P. G. Righetti, B. Barzaghi, E. Sarubbi, A. Soffientini, and G. Cassani, *J. Chromatogr.*, *470*: 337 (1989).
55. E. Bordini, M. Hamdan, and P. G. Righetti, *Rapid Commun. Mass Spectrom.*, *13*: 1818 (1999).
56. P. G. Righetti, G. Tudor, and E. Gianazza, *J. Biochem. Biophys. Methods*, *6*: 219 (1982).
57. U. T. Ruegg and J. Rüdinger, *Methods Enzymol.*, *47*: 111 (1977).
58. B. R. Herbert, M. P. Molloy, A. A. Gooley, J. Walsh, W. G. Bryson, and K. L. Williams, *Electrophoresis*, *19*: 845 (1998).
59. P. G. Righetti, in P. G. Righetti, A. V. Stoyanov, M. Y. Zhukov, *The Proteome Revisited: Theory and Practice of All Relevant Electrophoretic Steps*, Elsevier, Amsterdam, 2001, p. 245.
60. A. Goerg, W. Postel, J. Weser, S. Gunther, J. R. Strahler, S. M. Hanash, and L. Somerlot, *Electrophoresis*, *8*: 122 (1987).
61. A. Goerg, W. Postel and S. Gunther, *Electrophoresis*, *9*: 531 (1988).
62. J. X. Yan, J. C. Sanchez, V. Rouge, K. Williams, and D. F. Hochstrasser, *Electrophoresis*, *20*: 723 (1999).
63. U. K. Laemmli, *Nature*, *227*: 680 (1970).
64. M. J. Dunn, *Gel Electrophoresis: Proteins*, Bio Sci. Publ., Oxford, 1993, pp. 1–154.
65. J. X. Yan, W. C. Kett, B. R. Herbert, A. A. Gooley, N. H. Packer, and K. L. Williams, *J. Chromatogr. A*, *813*: 187 (1998).
66. S. Caglio, M. Chiari, and P. G. Righetti, *Electrophoresis*, *15*: 209 (1994).
67. B. Herbert, M. Galvani, M. Hamdan, E. Olivieri, J. McCarthy, S. Pedersen, and P. G. Righetti, *Electrophoresis*, *22*: 2046 (2001).
68. M. Galvani, M. Hamdan, B. Herbert, and P. G. Righetti, *Electrophoresis*, *22*: 2058 (2001).
69. M. Galvani, L. Rovatti, M. Hamdan, B. Herbert, and P. G. Righetti, *Electrophoresis*, *22*: 2066 (2001).
70. R. E. Finnigan, *Anal. Chem.*, *66*: 969A (1994).
71. D. Arnott, in P. James (Ed.), *Proteome Research: Mass Spectrometry*, Springer, Berlin, 2001, pp. 11–31.
72. W. J. Henzel, T. M. Billeci, J. T. Stultz, S. C. Wong, C. Grim-

ley, and C. Watanabe, *Proc. Natl. Acad. Sci. USA*, *90*: 5011 (1993).

73. J. R. Yates III, S. Speicher, P. R. Griffin, and T. Hunkapiller, *Anal. Biochem.*, *214*: 397 (1993).
74. J. R. Yates III, J. K. Eng, and A. L. McCormack, *Anal. Chem.*, *67*: 3202 (1995).
75. D. Arnott, W. J. Henzel, and J. T. Stults, *Electrophoresis*, *19*: 968 (1998).
76. M. Wilm, G. Neubauer, and M. Mann, *Anal. Chem.*, *68*: 527 (1996).
77. X. J. Tang, P. Thibault, and R. K. Boyd, *Anal. Chem.*, *65*: 2824 (1993).
78. P. Roepstorff and J. Fohlman, *Biomed. Mass Spec.*, *1*: 601 (1984).
79. R. D. Smith, J. A. Loo, C. G. Edmonds, C. J. Barinaga, and H. R. Udseth, *Anal. Chem.*, *62*: 882 (1990).
80. M. Wilm and M. Mann, *Int. J. Mass Spec. Ion Proc.*, *136*: 167 (1994).
81. A. L. McCormack, D. M. Schieltz, B. Goode, S. Yang, G. Barnes, D. Drubin, and J. R. Yates III, *Anal. Chem.*, *69*: 767 (1997).
82. D. Arnott, K. L. O'Connell, K. L. King, and J. T. Stults, *Anal. Biochem.*, *258*: 1 (1998).
83. B. Spengler, D. Kirsch, R. Kaufman, and E. Jaeger, *Rapid Commun. Mass Spectrom.*, *6*: 105 (1992).
84. E. Bordini, M. Hamdan, and P. G. Righetti, *Rapid Commun. Mass Spectrom.*, *13*: 2209 (1999).
85. K. F. Medzihradszky, J. M. Campbell, M. A. Baldwin, A. M. Falick, P. Juhasz, M. L. Vestal, and A. L. Burlingame, *Anal. Chem.*, *72*: 552 (2000).
86. A. Shevchenko, A. Loboda, A. Shevchenko, W. Ens, and K. G. Standing, *Anal. Chem.*, *72*: 2132 (2000).
87. A. N. Krutchinsky, W. Zhang, and B. T. Chait, *J. Am. Soc. Mass Spect.*, *11*: 493 (2000).
88. N. L. Kelleher, M. W. Senko, M. M. Sieggel, and F. W. Lafferty, *J. Am. Mass Spectr. Soc.*, *8*: 380 (1997).
89. Z. Zhang, S. Guan, and A. G. Marshall, *J. Am. Soc. Mass Spectr.*, *8*: 659 (1997).
90. A. G. Marshall, M. W. Senko, W. Li, S. Dillon, S. Guan, and T. M. Logan, *J. Am. Chem. Soc.*, *119*: 433 (1997).

91. T. Solouki, M. R. Emmet, S. Guan, and A. G. Marshall, *Anal. Chem.*, *69*: 1163 (1997).
92. R. Chen, X. Cheng, D. W. Mitchell, S. S. Hofstadler, Q. Wu, A. L. Rockwood, M. G. Sherman, and R. D. Smith, *Anal. Chem.*, *1159 (1995)*.
93. G. A. Valaskovic, N. L. Kelleher, and F. W. McLafferty, *Science*, *273*: 1199 (1996).
94. S. Guan and A. G. Marshall, *Anal. Chem.*, *69*: 1 (1997).
95. A. G. Marshall, C. L. Hendrickson, and G. S. Jackson, *Mass Spectr. Rev.*, *17*: 1 (1998).
96. R. Pitt-Rivers, and F. S. A. Impiombato, *Biochem. J.*, *109*: 825 (1968).
97. R. R. Ogorzalek Loo, C. Mitchell, T. I. Stevenson, S. A. Martin, W. M. Hines, P. Juhasz, D. H. Patterson, J. M. Peltier, J. A. Loo, P. C. Andrews, *Electrophoresis*, *18*: 382 (1997).
98. P. K. Jensen, L. Pasa-Tolic, G. A. Anderson, J. A. Horner, M. S. Lipton, J. E. Bruce, and R. D. Smith, *Anal. Chem.*, *71*: 2076 (1999).
99. M. W. Senko, J. P. Speir, and F. W. McLafferty, *Anal. Chem.*, *66*: 2801 (1994).
100. D. P. Little, J. P. Speir, M. W. Senko, P. B. O'Connor, and F. W. McLafferty, *Anal. Chem.*, *66*: 2809 (1994).
101. E. R. Williams, K. D. Henry, F. W. McLafferty, J. Shabano-wittz, and D. F. Hunt, *J. Am. Soc. Mass Spectr.*, *1*: 413 (1990).
102. A. Bairoch, in M. R. Wilkins, K. L. Williams, R. D. Appel, D. F. Hochstrasser (Eds.), *Proteome Research: New Frontiers in Functional Genomics*, Springer, Berlin, 1997, pp. 93–148.
103. R. D. Appel, in M. R. Wilkins, K. L. Williams, R. D. Appel, D. F. Hochstrasser (Eds.), *Proteome Research: New Frontiers in Functional Genomics*, Springer, Berlin, 1997, pp. 149–175.
104. M. C. Peitsch, N. Guex, in M. R. Wilkins, K. L. Williams, R. D. Appel, D. F. Hochstrasser, *Proteome Research: New Frontiers in Functional Genomics*, Springer, Berlin, 1997, pp. 177–186.
105. J. C. Sanchez, R. D. Appel, O. Golaz, C. Pasquali, F. Ravier, A. Bairoch, and D. F. Hochstrasser, *Electrophoresis*, *16*: 1131 (1995).
106. C. Hoogland, J. C. Sanchez, L. Tonella, P. A. Binz, A. Bairoch, D. F. Hochstrasser, and R. D. Appel, *Nucleic Acids Res.*, *28*: 286 (2000).

107. A. Bairoch and R. Apweilir, *Nucleic Acids Res.*, *28*: 45 (1999).

108. M. E. Crawford, M. E. Cusick, J. I. Garrels, in W. Blackstock, M. Mann (Eds.), *Proteomics: a Trend Guide*, Elsevier, London, 2000, pp. 17–21.

109. R. C. Beavis, D. Fenyö, in W. Blackstock, M. Mann (Eds.), *Proteomics: a Trend Guide*, Elsevier, London, 2000, pp. 22–27.

110. S. P. Gygi, Y. Rochon, B. R. Franza, and R. Aebersold, *Mol. Cell Biol.*, *19*: 1720 (1999).

111. B. Futcher, G. I. Latter, P. Monsardo, C. S. McLaughlin, and J. I. Garrels, *Mol. Cell Biol.*, *19*: 7357 (1999).

112. S. P. Gygi, R. Aebersold, in W. Blackstock, M. Mann (Eds.), *Proteomics: a Trend Guide*, Elsevier, London, 2000, pp. 31–36.

113. Y. Oda, K. Huang, F. R. Cross, D. Cowburn, and B. T. Chait, *Proc. Natl. Acad. Sci. USA*, *96*: 6591 (1999).

114. L. Pasa-Tolic, P. K. Jensen, G. A. Anderson, M. S. Lipton, K. K. Peden, S. Martinovic, N. Tolic, J. E. Bruce, and R. D. Smith, *J. Amer. Chem. Soc.*, *121*: 7949 (1999).

115. R. D. Smith, L. Pasa-Tolic, M. S. Lipton, P. K. Jensen, G. A. Anderson, Y. Shen, T. P. Conrads, H. R. Hudset, R. Harkewicz, M. E. Belov, C. Masselon, and T. D. Veenstra, *Electrophoresis*, *22*: 1652 (2001).

116. M. B. Smolka, H. Zhou, S. Purkayastha, and R. Aebersold, *Anal. Biochem.*, *297*: 25 (2001).

117. E. C. Peters, D. M. Horn, D. C. Tully, and A. Brock, *Rapid Commun. Mass Spectrom.*, *15*: 2387 (2001).

118. R. Tonge, J. Shaw, B. Middleton, R. Rowlinson, S. Rayner, J. Young, F. Pognan, E. Hawkins, I. Currie, and M. Davison, *Proteomics*, *1*: 377 (2001).

119. G. L. Corthals, C. V. Wasinger, D. F. Hochstrasser, and J. C. Sanchez, *Electrophoresis*, *21*: 1104 (2000).

120. S. P. Gygi, G. L. Corthals, Y. Zhang, Y. Rochon, and R. Aebersold, *Proc. Natl. Acad. Sci. USA*, *97*: 9390 (2000).

121. M. Fountoulakis, H. Langen, S. Evers, C. Gray, and B. Takacs, *Electrophoresis*, *18*: 1193 (1997).

122. M. Fountoulaki and B. Takacs, *Protein Expr. Purif.*, *14*: 113 (1998).

123. M. Fountoulakis, H. Langen, C. Gray, and B. Takacs, *J. Chromatogr. A*, *806*: 279 (1998).

124. M. Fountoulakis, M. F. Takacs, and B. Takacs, *J. Chromatogr. A, 833*: 157 (1999).

125. K. Karlsson, N. Cairns, G. Lubec, and M. Fountoulakis, *Electrophoresis, 20*: 2970 (1999).

126. M. Fountoulakis, M. F. Takacs, P. Berndt, H. Langen, and B. Takacs, *Electrophoresis, 20*: 2181 (1999).

127. M. G. Harrington, J. A. Coffman, F. J. Calzone, L. E. Hood, R. J. Britten, and E. H. Davidson, *Proc. Natl. Acad. Sci. USA, 89*: 6252 (1992).

128. P. G. Righetti, E. Wenisch, and M. Faupel, *J. Chromatogr., 475*: 293 (1989).

129. P. G. Righetti, E. Wenisch, A. Jungbauer, H. Katinger, and M. Faupel, *J. Chromatogr., 500*: 1 (1990).

130. P. G. Righetti, E. Wenisch, M. Faupel, in A. Chrambach, M. J. Dunn and B. J. Radola (Eds.), *Advances in Electrophoresis* Vol. 5, VCH, Weinheim, 1992, pp. 159–200.

131. B. Herbert and P. G. Righetti, *Electrophoresis, 21*: 3639 (2000).

132. P. G. Righetti, A. Castagna, and B. Herbert, *Anal. Chem., 73*: 320A (2001).

133. P. Davidsson, A. Westman, M. Purchades, C. L. Nilsson, and K. Blennow, *Anal. Chem., 71*: 642 (1999).

134. C. L. Nilsson, M. Puchades, A. Westman, K. Blennow, and P. Davidsson, *Electrophoresis, 20*: 860 (1999).

135. M. Raida, P. Schulz-Knappe, G. Heine, and W. G. Forssmann, *J. Am. Soc. Mass Spectrom., 10*: 45 (1999).

136. G. J. Opiteck, K. C. Lewis, J. W. Jorgenson, and R. J. Anderegg, *Anal. Chem., 69*: 1518 (1997).

137. G. J. Opiteck, S. M. Ramirez, J. W. Jorgenson, and M. A. Moseley III, *Anal. Biochem., 258*: 349 (1998).

138. A. J. Link, E. Carmack, and J. R. Yates III, *Int. J. Mass Spectrom. Ion Proc., 160*: 303 (1997).

139. A. J. Link, J. Eng, D. M. Schieltz, E. Marmack, G. J. Mize, D. R. Moris, B. M. Garvik, and J. R. Yates III, *Nature Biotech., 17*: 676 (1999).

140. M. P. Washburn and J. R. Yates III, in W. Blackstock and M. Mann (Eds.), *Proteomics: a Trend Guide*, Elsevier, London, 2000, pp. 27–30.

141. C. A. K. Borrebaeck, S. Ekstrom, A. C. Malmborg Hager, J. Nilsson, T. Laurell and G. Marko-Varga, *BioTechniques, 30*: 1126 (2001).

142. D. Dedelkov and R. W. Nelson, *Int. Lab., 31*: 8 (2001).

7

Improving Our Understanding of Reversed-Phase Separations for the 21st Century

Patrick D. McDonald   *Waters Corporation, Milford, Massachusetts, U.S.A.*

I. INTRODUCTION

> *There is nothing more difficult to take in hand, more perilous to conduct, or more uncertain in its success, than to take the lead in the introduction of a new order of things.*
> —*The Prince, Niccolò Machiavelli (1469–1527), published 1532*
> *One of the most insidious and nefarious properties of scientific models is their tendency to take over, and sometimes supplant, reality.*
> —*Erwin Chargaff, 1978*

Experiments designed to explore the actual physical reality of the interaction of analyte, mobile phase, and stationary phase are fraught with practical limitations that impede drawing significant conclusions from the results. It has been easier for most practitioners to foster the folklore forged through the citation generations of HPLC literature than to refine the "cartoon-model" view of the chromatographic process to a higher "art." In the spirit with which, Phyllis Brown reminds us, Calvin Giddings envisioned this *Advances* series, perhaps it is time now to introduce a new order into our chromatographic thinking and encourage future investigation.

Presented in this personal account are results of some renewed experiments, allied with recent ideas from related disciplines, which challenge traditional concepts of *accessible* surface, ligand density, hydrophobic *collapse*, and silanol interaction. Particular attention should be paid to the role of the structure and physiochemical properties of both particle substrate and mobile phase elements as they, together, determine the constitution and function of the "stationary phase."

II. BACKGROUND

A. Conventional Wisdom

> *Nothing is more dangerous than an idea, when it's the only one we have.*

—Émile Auguste Chartier (1868–1951), Libres-Propos
We has met the enemy, and it is us.
—Walt Kelly, Pogo (1913–1973)

Notions of the nature of stationary bonded phases and how analytes interact with them during the chromatographic process have been illustrated in various chemical "cartoons" that have remained virtually the same over the past two decades [see some recent ones in Fig. 1]. These usually depict a simple, almost flat, silica surface,

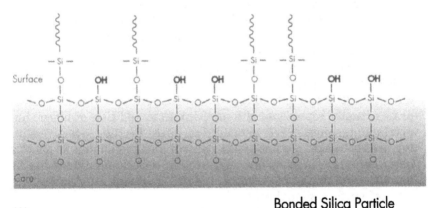

Bonded Silica Particle

(A)

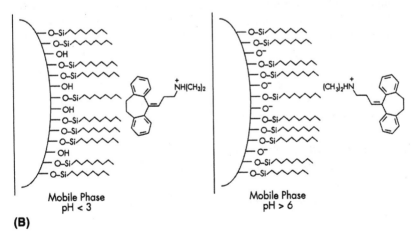

(B)

Fig.1 Traditional cartoons are misleading, implying incomplete coverage of *accessible* silica surface (A) and *direct* analyte–silanol interaction (B). (© 2002 Waters Corporation; used with permission.)

where some, but not all, of the silanol moieties have been bonded with a larger, primary silyl group (e.g., C_{18}).* Sometimes, in a second step, a few more, but not all, of these silanols are "end-capped" with a smaller (usually trimethylsilyl-) alkylsilyl function.

However, though maximally bonded, reversed-phase packings may still exhibit acidity and/or retention characteristic of free, unbonded silanol groups. So the cartoons illustrate these free silanols as interspersed among, and presumably hindered from derivatization by, the bonded silanols. Yet the pictures also imply that analytes interact directly, "face-to-face," not only in a hydrophobic manner with the bonded alkyl chains but also in a hydrogen-bonding, ionic, or electrostatic manner with the free silanols remaining on the incompletely bonded silica surface.

Whatever its nature, secondary silanol interaction is particularly aggravating when trying to separate basic analytes at intermediate pH. It is thought that in a chromatogram, a visible, "tailed" peak actually represents the mathematical addition of offset, but overlapping, more nearly symmetrical peaks for the respective eluting concentration profiles of the neutral and protonated forms of a basic analyte. The cartoons suggest that a basic analyte molecule becomes protonated by forming a "salt" via direct interaction with a "free" surface silanol acid.

In early 2000, my colleagues Dr. Charles Phoebe and Joseph Arsenault, while preparing new seminar programs, asked me if I could explain some anomalies in certain experimental results they were examining. To do so, I resurrected some radical notions that grew out of some work I directed more than twenty years ago. Revisiting these ideas has led to the repetition of seminal experiments as well as a review of recent corroborating ideas from other disciplines. Preliminary reports of my notions were presented in part at ISC 2000 in London [1] and at HPLC 2001 in Maastricht [2].

B. Origin of Ideas: Preparing the Mind

In the fields of observation, chance favors only the mind that is prepared.
    —Louis Pasteur (1822–1895) [Quoted by R. Vallery-Radot, 1927]

---

* "C_{18}" is used throughout this chapter to mean "octadecyldimethylsilyl-".

Happenstance usually plays a part, to be sure, but there is much more to invention than the popular notion of a bolt out of the blue. Knowledge in depth and in breadth are virtual prerequisites. Unless the mind is thoroughly charged beforehand, the proverbial spark of genius, if it should manifest itself, probably will find nothing to ignite.
—Paul Flory, Perkin Medal award address, 1981

I believe that it is instructive and encouraging to students and young researchers to understand the context in which ideas originate and develop. For this reason, I will relate a bit of my background and demonstrate how "breaking the rules" may sometimes lead to discovery.

My great-great-grandfather Elliot Robley owned a sand mine in the small central Pennsylvania town of Mapleton. I was born and raised only 5 miles from the expansive sandy Atlantic Ocean beaches on the south shore of Long Island. Yet, as a young boy, I remember how exciting it was to visit Mapleton, stop alongside that plant, and fill a bag or pail with the whitest, cleanest sand I had ever seen. In the 1930s, this silica was fused by Corning Glass Works to form the mirror blank for the 200" reflecting telescope atop Mt. Palomar in California. The mine is owned today by the U.S. Silica Corporation, and its sand, some of the purest silica found anywhere in the world, is highly prized as a source of silicon for the semiconductor industry. So I believe I was destined to do separations on silica.

No wonder then that in 1963, only five years after Prof. Egon Stahl at Saarbrücken had standardized the technique [3], and a couple of years after thin-layer chromatography [TLC] had been introduced into their colleagues' U.S. research laboratories by Swiss pharmaceutical chemists, I was making my own analytical and preparative TLC plates. While only a junior at Brown University, I learned under the tutelage of graduate student Richard Bartsch* who had purchased his equipment directly from Serva GmbH. in Heidelberg. Prof. John Neumer had sent me to Dick to learn TLC, feeling guilty because, among the box of tubes containing simple unknown compounds from which my Qualitative Organic Chemistry classmates drew their weekly assignments, I was "lucky" enough to

---

* Ph.D. Brown University 1967; currently Paul Whitfield Horn Professor and Chairman, Department of Chemistry and Biochemistry, Texas Tech University.

choose *first* the tube he had "planted" therein on a dare from his colleagues.

I spent most of that semester trying to identify an unknown mixture of 3α- and 3β-hydroxy-Δ⁵-cholestenol. After much trial and error, perfecting the concentration of gypsum binder in the silica slurry and the drying conditions for preparing crack-free, thick-layer, preparative plates, I managed to isolate and purify the two components. I confirmed the identity of the 3β-hydroxy isomer (cholesterol) accidentally, when, after measuring its melting point, during the cool-down cycle, I felt inspired, out of curiosity, to continue to watch the capillary. At the moment of resolidification, I observed the flash of blue color characteristic of this liquid crystal. As a reward for my success, I was assigned to learn the practice of gas chromatography in another graduate lab and deliver two lectures to our class on GC. Little did I suspect then that my "misfortune" would foreshadow my studies in steroid biochemistry and my career in chromatography.

As Prof. Gordon Hamilton guided my doctoral research at Penn State University, I had to isolate and purify phenolic substances for my investigations of bioorganic reaction mechanisms [4]. I did large-scale open column liquid chromatography on silica, learning much from the latest tomes on adsorption LC by Lloyd Snyder [5], and on LC dynamics by Cal Giddings [6]. I also made TLC plates using a mixture of silica and microcrystalline cellulose to achieve liquid–liquid partition separations [7]. Gordon and Prof. Stephen Benkovic also taught me the wonders of enzyme catalysis and proton transfer reactions in protein chemistry.

In the fall of 1970 I moved on to the College of Physicians and Surgeons at Columbia University, as a postdoctoral trainee in endocrinology. My mentor there, Prof. Seymour Lieberman, had first practiced chromatography with Prof. Louis Fieser at Harvard, and then with master chromatographer and Nobel laureate Prof. Thaddeus Reichstein at the ETH in Zürich. Liquid–liquid partition open column chromatography [LLC] had been developed to a high art in Seymour's laboratories, after being introduced there by a graduate student, Pentti "Finn" Siiteri,† who had learned the craft working

---

† Finn, now retired, spent many years doing research in steroid biochemistry as a professor on the medical faculty of the University of California, San Francisco.

with Charles Pidacks at Lederle Laboratories. Variations of LLC such as reversed-phase and paired-ion [8] had been refined or developed in our lab at Columbia to resolve all manor of steroid structures [9–11].

Then, one day in 1971, James L. Waters telephoned me and sold us one of the first ALC-100 HPLC systems. I did pioneering steroid analyses on that system (which I reengineered myself for preparative scale separations) [12,13]. Even though I installed all the latest instrumentation and column developments as they became available, such as the first dual-reciprocating piston HPLC pump (M6000), the first septumless injector (U6K), and the first μBondapak™ C_{18} column,[‡] I could never quite achieve the selectivity that I routinely obtained by judicious choice of an open column liquid–liquid partition system on a low efficiency Celite™ diatomaceous earth support.[§] At this point, I was a highly skilled practical, rather than theoretical, chromatographer, having developed instincts about how to achieve separations based on my background in organic synthesis and mechanistic bioorganic chemistry. I could refer to a documented host of successful separations in our laboratory "bible" to draw significant correlations between a molecule's structure and its possible interactions with potential mobile and stationary phases, by which I could most often predict the best separation system to choose for a new mixture.

Rather than pursuing an academic career in steroid biochemistry, I decided instead to join my friends at Waters Associates in 1974 and continue to develop the art and practice of chromatographic separations [14–16]. I was delighted to find Charles Pidacks working there and to reminisce with him about our experiences doing liquid–liquid chromotography, while watching Charlie revolutionize pharmaceutical analysis with his separations on reversed-phase packings [17,18]. Early on, I profited from reading the latest work of Joe Huber, Istvan Halász, Csaba Horváth, John Knox, Cal Giddings, George Guiochon, and so many other pioneers. Yet, I enjoyed so much more having the opportunity to discuss ideas with them informally at international symposia or during their visits to Milford. I sat silently on the sidelines of the seemingly eternal "adsorption versus partition" debate over the mechanism of retention in reversed-

---

[‡] μBondapak is a trademark of Waters Corporation.
[§] Celite is a trademark of Johns-Manville Corporation.

phase chromatography [19–31], knowing, with my experience, there was no question: partition was paramount.

C. Breaking the Rules: The Key Experiment

Do the experiment first!
    —James L. Waters (1925–)
    Even when the experts all agree, they may well be mistaken.
    —Bertrand Russell (1872–1970)

Back in 1979–80, we were working to develop a reliable chromatographic test, using a mixture of simple, stable, organic compounds, encompassing acidic, basic, and neutral molecules, for measuring the performance characteristics of various bonded reversed-phase HPLC stationary phases. Barbara Kogut and I were pursuing an approach using a simple methanol–water mobile phase, wishing to minimize any chemical alteration of the stationary phase surface, beyond what the mobile phase might effect, so that differences of any sort between phases might become apparent. Colleagues were using an acetonitrile–water mobile phase containing a dissolved salt, buffered to a selected pH, thinking that by masking secondary interactions in this way, it would be easier to distinguish various phases on the basis of hydrophobic interaction alone.

Barbara's results with μBondapak™ C_{18} columns were quite promising. All the analytes in the proposed test mixture eluted in just a few minutes with reasonable, reproducible retention times, and the peaks were nicely resolved. As would be expected from a phase made from silica with its porosity optimized for chromatography by a series of physiocochemical process treatments, and its surface bonded and end-capped, the peaks showed good symmetry.

We turned our attention next to another phase, based on a spherical silica, made by a sol–gel process, with a smaller average pore size, maximally bonded with octadecyldimethylchlorosilane, but not end-capped. Our test mixture did not behave as well on this packing. The key difference, as expected, was the pronounced tailing and longer retention time for the basic analyte, quinoline.

So far, everything seemed to agree with the cartoons. However, a unique situation presented itself. I decided to have Barbara break the rules of accepted HPLC practice and try a novel experiment. We had found on the lab bench an unused cartridge column containing this same spherical, C_{18}-bonded packing. The column had been packed several weeks earlier, using a highly volatile, nonaqueous packing medium, but someone had forgotten to cap and seal the ends of the tube. As a result, the column had thoroughly dried out.

Normally, when beginning to use a new reversed-phase HPLC column, it is recommended to pump 20–30 column volumes, or more, of mobile phase through the bed to "equilibrate" the separation system so that column performance will be stable and reproducible [32]. I wondered what would happen, if, instead, we took this dry column and tried our test separation on it before "equilibrium" was reached. So I instructed Barbara to connect the HPLC pump to the column inlet, begin the flow of mobile phase, and, as soon as the first full drop of effluent emerged from the column outlet, connect the tubing leading to the detector, immediately make an injection, and repeat injections as close together as practical (every few minutes) during the equilibration period.

What we observed was totally unexpected. At first, all the peaks, *even that for quinoline*, were tall, sharp, and symmetrical. With each subsequent injection, though, only the quinoline peak got shorter, with a broadening tail, until after an hour, the peak had gradually flattened and all but disappeared into the baseline. Peaks for the other analytes remained well-behaved.

Barbara recorded these seemingly inexplicable observations in her notebook while I filed the information in my memory bank. As often happens, we had to move on to more pressing matters. Time passed; Barbara married another Waters employee, Bob Grover, and left to begin a new career in real estate. And 20 years later, I told Chuck and Joe that what I thought had happened in our old experiment might well explain some of their observations. Taking a cue from that master of aptly acronymic phrases, Prof. Csaba Horváth, I wrote on Chuck's blackboard my explanation:

Hydro-Linked Proton Conduit™ [**HLPC**™].*

III. CURIOUS INCIDENTS

Just do it!™
—*Nike Corporation, Advertising Slogan, 1989*

---

* Hydro-Linked Proton Conduit and HLPC are trademarks of Waters Corporation.

A. Air-Dried Column: Curiosity Redux

While Chuck and Joe were receptive to my ideas, some other colleagues were not so sure. Barbara's old notebook could not be found, and my vivid recollections were not enough to sway the skeptics. Such a result certainly could not be repeated, they opined. But my colleague Pamela Iraneta graciously permitted me to use the resources of her HPLC evaluation group to reinvestigate what had happened. Bonnie Alden, whom I had hired, fresh from college, in 1984, had developed into an excellent chromatographer. With the clues I provided her, she was able to reconstruct the critical experiment.

Rather than use the entire text mixture, we injected only the analyte of interest, quinoline. Bonnie managed to find a column containing the same type of spherical, C_{18}-bonded, but not end-capped, silica Barbara had used 20 years earlier, which, fortuitously, had loose end fittings and had been drying, unintentionally, for at least one year. She chose a mobile phase composition [methanol:water 70:30] that would fully wet the packing and also give a reasonable retention time for quinoline. As soon as the initial bubbling stopped and a continuous flow of eluate emerged from the column outlet, she connected the detector and began making injections every 1.5 minutes. The result is shown in Fig. 2.

The outcome was just as it had been 20 years earlier. The peak from the first injection was tall, sharp, and symmetrical. With each successive injection, the peak height diminished and the tail broadened until after an hour, the peak had nearly disappeared into the baseline. A key observation here is that the retention time, as measured at the peak maximum, remained *constant* throughout the series of quinoline injections.

B. Oven-Dried Column: Additional Neutral Analyte

Out of the stockroom, Bonnie took a second column, identical to the first, except that it was properly capped and filled with the shipping solvent. She dried it carefully with a continuously flowing argon stream in an oven at 80–90°C. for 18–20 hours. A mixture of naphthalene and quinoline was injected into this super dry column every 3 min. Naphthalene was chosen as the neutral marker because it is nominally the same size and shape as quinoline and has a reasonable, but distinctly longer, retention time in the same mobile phase. The result is shown in Fig. 3.

Unexpected Peak Shape

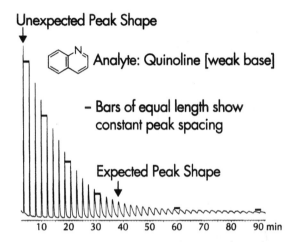

Analyte: Quinoline [weak base]

- Bars of equal length show constant peak spacing

Expected Peak Shape

10   20   30   40   50   60   70   80   90 min

Fig. 2 Experiment A. An experiment first done in 1980 was successfully repeated. Breaking the rules by not waiting for the dry column to equilibrate before making injections led to this unexpected result. Notice the sharp, symmetric shape for peaks from the first several injections which steadily degrades after about 30–40 min to the broad, tailed peak shape typical for a basic analyte passing through a non-end-capped, C_{18}-bonded phase. Column: 3.9 × 150 mm, packed with fully octadecyldimethylsilyl-bonded, but *not* end-capped, spherical silica (made by a sol–gel process with few pores larger than 100 Å in diameter; see Fig. 11), which had air-dried for more than one year. Mobile phase: 70:30 methanol:water (a good wetting solvent). Flow rate: 0.5 mL/min. Protocol: Column inlet was connected and flow was initiated. As soon as the first full drops (not bubbles) of fluid began to emerge from the column outlet (about 1 minute), the detector was connected, and the first injection was made. Identical injections were repeated at 1.5 minute intervals. Sample analyte: Quinoline (weak base). Instrument: Waters® HPLC System consisting of 600 Multiple Solvent Delivery System; 2487D UV detector; 717Plus Autosampler; Millennium³²™ Chromatography Manager Software v.3.05.01. (© 2002 Waters Corporation; used with permission.)

As in the earlier experiment, the quinoline peak from the first injection was sharp and symmetrical, but with each successive injective, the quinoline peak displayed further degradation, though constant retention time, just as before. However, the naphthalene peak, eluting later, remained consistent in sharpness, symmetry, and retention time. Column efficiency and performance was constant throughout the experiment.

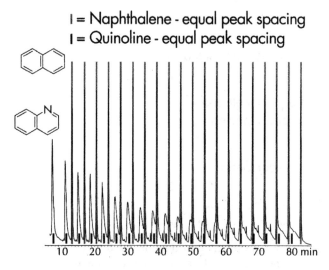

I = Naphthalene - equal peak spacing
I = Quinoline - equal peak spacing

Fig. 3 Experiment B. A pair of analytes was used to investigate the integrity of a super dry column. Since naphthalane peaks remain consistently sharp and tall, quinoline peak shape degradation is *not* due to a loss of column efficiency. A constant retention time for naphthalene indicates *no* change in phase ratio. This result suggests that analyte-accessible pores wet quickly. Column: same packing and dimensions as in Fig. 2 except a fully wetted column was dried at 80–90°C for 18–20 h with a continuous flow of argon through the bed to prevent oxidation. Mobile phase, Flow rate and Instrument: same as in Fig 2. Sample analytes: mixture of quinoline and naphthalene (approximately same molecular size and shape). Protocol: same as in Fig. 2 except identical injections were repeated at 3-min intervals. (© 2002 Waters Corporation; used with permission.)

C. Oven-Dried Column: Different Silica

Another type of silica and bonding protocol was tested next. A column packed with a high purity spherical silica, synthesized from tetraethoxysilane, which had been fully C_{18}-bonded and end-capped, was argon-dried as before. The same fully wetting mobile phase, sample (quinoline and naphthalene) injection, and column flow protocols were used. The result is shown in Fig. 4A.

To the eye, as drawn on this time scale, it appears that peak sharpness and symmetry were maintained throughout for both analytes. However, computer-generated peak asymmetry measurements on each peak were plotted. As shown in Fig. 4B, asymmetry for naphthalene was constant, while peak tailing steadily increased

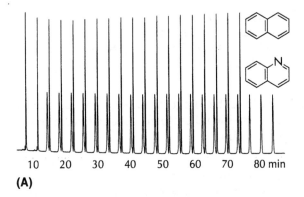

(A)

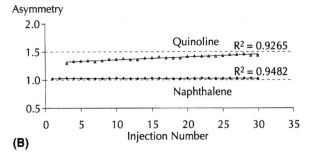

(B)

Fig. 4 Experiment C. (A) Using a C_{18}-bonded, end-capped, high-purity synthetic silica (which has more surface area in larger pores; see Fig. 12) shows a subtle but similar result. Retention times for naphthalene and quinoline remain constant. (B) While visual inspection of quinoline peaks shows little change, a plot of peak asymmetry versus injection number shows a slow, but steady, degradation of peak shape. Meanwhile, the naphthalene peak retains excellent symmetry throughout the experiment. Column: 4.6 × 150 mm, treated as in Fig. 3. Mobile phase, Flow rate, Protocol, Sample, Instrument, and Software same as in Fig. 3. (© 2002 Waters Corporation; used with permission.)

for quinoline. This effect appears to be similar, though diminished in magnitude, to that in the experiments with the previous silica substrate.

D. Questions

> *Watchman, what of the night?*
> *—Isaiah, 21:11*

". . . the curious incident of the dog in the nighttime."
"The dog did nothing in the nighttime."
"That was the curious incident," remarked Sherlock Holmes.
—Sir Arthur Conan Doyle, Silver Blaze, 1894

Trying to explain the results of the foregoing experiments by means of the ideas illustrated in traditional cartoons does not seem possible. Consider the following points:

1.   Suppose that the wetting of a dry column occurs in stages, with the largest pores filling first, followed in turn by successively smaller pores and passages in a time sequence that takes minutes, not fractions of a second. And assume that the quinoline or naphthalene analytes, upon each successive injection, access more and more of the surface area within these pores, as their wetted state enables interaction. This means that the effective phase ratio for the separation system, the ratio of available stationary phase to mobile phase, would be larger for each successive injection. This would result in a retention time increase for quinoline or naphthalene with each injection, especially during the early course of the experiment. This did NOT happen.

2.   Suppose the column was somehow being degraded or altered by repeated injection of quinoline or by some other means. This would also affect the column's performance with other analytes. But the simultaneous injection of naphthalene, with consistent good performance, showed this did NOT happen.

3.   If only 50–60% of the free silanols, at best, are bonded, and unbonded surface silanols in between bonded alkyl chains interact with basic analytes to cause peak tailing, then even the initial injection of quinoline on a newly wetted column should show this effect. This did NOT happen.

4.   If exhaustively end-capping a fully bonded silica minimizes the availability of surface silanols for interaction with basic analytes, then quinoline peak symmetry on such a phase should not change during the wetting process. This did NOT happen.

Such observations lead to several questions whose traditional answers may not be entirely satisfactory. For example, we may ask:

1. How much of the surface in a porous particle can be accessed by bonding reagents? by solvent molecules? by analyte molecules?
2. What is the ligand density in a fully bonded phase?
3. Where are the surface silanols in a bonded phase?
4. What is the effect of wetted pores in the stationary phase?
5. What is the role of surface silanols in reversed-phase separations?
6. Can designed modifications to the surface or substrate structure of the stationary phase particle teach us anything about how reversed-phase separations take place?

Let's review what we know and don't know and see what evidence we have that might help answer some of our questions.

IV. INTERDISCIPLINARY IDEAS AND INFERENCES

An idea is a feat of association.
—Robert Frost (1874–1963)
Things should be as simple as possible, but not simpler.
—Albert Einstein (1879–1955)
When you have eliminated the impossible, whatever remains, however improbable, must be the truth.
—Sir Arthur Conan Doyle, The Sign of Four, 1890

A. Pore Accessibility

A glance through the chromatographic literature finds pores variously modeled as ink bottles, craters, cylinders, bubbles, channels, or plates. These assumed geometries lend themselves to simplified mathematical description, but they may not necessarily represent reality. The building block of silica is a simple tetrahedron with a silicon atom at its center and oxygen atoms at each apex. Tetrahedra may interlink via siloxane bonds, progressing from the micro to the macro level, to create, in aggregate, the "unfinished" irregular structure of silica gel. Regions with all possible siloxane bridges in place are highly crystalline. The ultimate form of crystalline, rigid silicon dioxide is quartz. But in silica gel, many potential siloxane bridges haven't formed, leaving silanol moieties on the exposed surfaces. This results in a structural skeleton with spaces and passages, some interconnected, some dead-ended, of varying size having an active, hygroscopic, acidic, interior/exterior surface.

Anyone who has seen a Buckminster Fuller geodesic dome or played with models of buckyballs may imagine what an unfinished, open structure formed from tetrahedral building blocks must look like. I turned to Dr. Yuehong Xu in our analytical laboratory to see if we might have in our extensive archive an electron photomicrograph that might illustrate the structure of silica I envisioned. Her colleague Pamela Richards had taken the photo shown in Fig. 5. The scale of this view of an experimental silica is not important for our discussion. What is relevant is the realization that porous silica gel used for chromatography, whether at the nano-, micro-, or mesoporous level, likely appears as such a labyrinth of open spaces and inter-

Fig. 5 Models for silica pores such as ink bottles, craters, cylinders, bubbles, channels, or plates may lend themselves to simplified mathematical description. However, this photo of an experimental silica, made in our laboratories, demonstrates that reality is a labyrinth formed, like geodesic domes, with tetrahedral building blocks. All surfaces have terminal silanol (—Si—OH) groups. Electron photomicrograph: Taken with JEOL JSM-5600 Scanning Electron Microscope, Acceleration voltage: 5 kV; Working distance: 8 mm. (© 2002 Waters Corporation; used with permission.)

connected passages, shaped by a substructure of tiny atomic tetrahedra.

With this image of silica in mind, it doesn't take a mathematician to convince us that most of the surface in a porous particle is on the inside, not the outside. In chromatographic applications of such materials, the important aspect as far as solvent, reagent, and analyte molecules are concerned is the *accessibility* of this surface [33]. In a simple view, I envision three arbitrary levels of accessibility:

Surface Access Level One: This region consists of the outer particle surface and the surface in pore spaces greater than about 100 Å in diameter that are *nearest* to the outside of the particle. Level One surface is most accessible to large silyl bonding reagents under favorable reaction conditions. It is thought that this easiest-to-access surface reacts first, and completely (see Section IV.B.). I have attempted to illustrate Level One bonded surface in Fig. 6.

Surface Access Level Two: Accessible to smaller silyl end-capping reagents and solvent molecules (see Fig. 7), the surface in pore spaces greater than about 50Å is directly connected to that in Level One pores. Diffusional distance from the outer surface is still minimal.

Surface Access Level Three: As indicated in Fig. 8, only small solvent molecules may access the surface deep within the labyrinth. Hydrogen-bonding molecules, especially water, are most favorably attracted. Infusion or diffusion of water or other small polar solvent molecules into these pores may take minutes. Once there, they may be difficult to remove.

B. Surface Coverage

Chromatographers seek simple measures by which comparisons between column packings and column performance may be made. One such parameter for bonded-silica phases is surface coverage—expressed either in units of percent carbon or of micromoles of ligand per square meter. But a review of a generation of literature reveals that too much "clean" thinking may have been based on "dirty" parameters.

Surface Access Level One

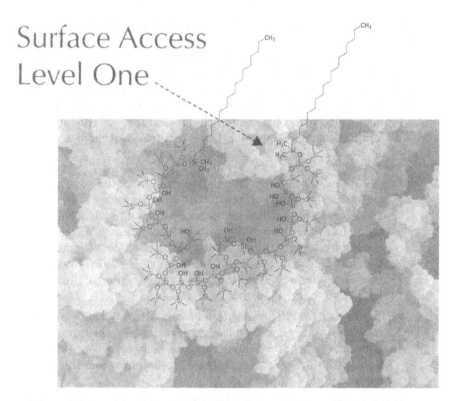

Fig. 6 Surface Access Level One includes outer surface and surface in pore spaces larger than about 100 Å in diameter nearest to the outside of the particle. This surface is easily accessible to large silane bonding reagents under favorable reaction conditions. It reacts rapidly and completely. (© 2002 Waters Corporation; used with permission.)

First, consider what "percent carbon" means. Carbon–hydrogen [C–H] analysis typically takes a careful dried and weighed sample of bonded-silica, thoroughly burns off the organosilane moieties, thereby converting all the carbon to carbon dioxide, which is then trapped and weighed. Results are reported as a weight percent of carbon relative to the original total sample weight.

In order to compare properly two bonded phases on a percent carbon basis, they need to have identical silica substrates (same pore structure and surface area) and identical bondings. Separation properties or performance of two packings prepared on disparate silicas

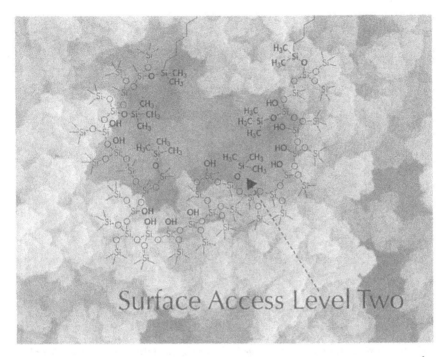

Fig. 7 Surface Access Level Two is in pore spaces larger than about 50 Å in diameter, which are directly connected to Level One pores. Diffusion distance from the outer surface is still reasonable. Under proper conditions, this surface level is accessible to smaller silane end-capping reagents and solvent molecules. (© 2002 Waters Corporation; used with permission.)

of different densities with silanes of different molecular weights using different bonding protocols cannot logically be compared solely on the basis of their respective weight percent carbon. Yet this is often done. So even as a relative measure of column behavior, percent carbon is not very useful [34].

More important, for the purposes of our discussion here, is the difficulty in comparing surface coverage on the basis of micromoles of bonded silane per square meter. A formula typically used for this calculation is shown in Fig. 9 [35]. Key variables include the molecular weight of, and mole percent of carbon in, the silyl function, the percent carbon in the packing, and the silica's surface area. Determination of percent carbon from C-H analysis is straightforward, as is

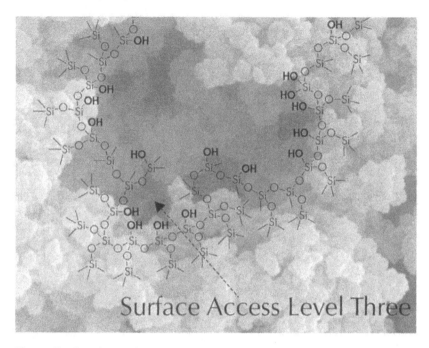

Surface Access Level Three

Fig. 8 Surface Access Level Three is deeper in the labyrinth, accessible only to small solvent molecules. Hydrogen-bonding molecules, especially water, are most favorably attracted. Infusion of water into these pores can take minutes. Once there, it may be difficult to remove. (© 2002 Waters Corporation; used with permission.)

a calculation of the silane's molar composition. The problem is surface area measurement.

Surface area is typically measured by methods that rely on intrusion of nitrogen molecules [BET nitrogen adsorption] or elemental mercury [mercury porosimetry] into pores [36]. Nitrogen is a small molecule that is favorably attracted into, and rapidly penetrates, the tiniest of silica or bonded-silica pore spaces (down to a few angstroms in diameter). Mercury also wets these surfaces and under controlled pressure will penetrate pores as small as about 30Å in diameter. Level Three pore space is inaccessible to larger solvent molecules and silane bonding reagents. So the value for total surface area determined by either of the above methods may be much higher than that for reagent-accessible surface (Level One and/or Two). Us-

$$M_s/m^2 = \frac{C \times 10^6}{(S_C - C) \times MW_s \times SA}$$

in µmoles/square meter

where
  C = % Carbon [from CH analysis]
  MW_s = Mol. Wt. of silyl function
  S_C = mole % Carbon in silyl function
  M_s = Micromoles of Silyl Function
  SA = Surface Area Accessible to Chlorosilane
    [in sq. meters/gram]

For octadecyldimethylsilyl:

$$M_s/m^2 = \frac{C \times 10^6}{(77.4 - C) \times 310 \times SA}$$

For trimethylsilyl:

$$M_s/m^2 = \frac{C \times 10^6}{(50 - C) \times 72 \times SA}$$

Fig. 9 Formula used to calculate surface coverage, with specific values for the most commonly used octadecyldimethylsilyl- and trimethylsilyl-ligands. While measuring the percent carbon value is a straightforward analytical technique, determination of accessible surface area is not so easy. Traditional use of a value for total surface area measured by BET nitrogen adsorption can give a value for surface coverage that is at least 2× too low, depending on the morphology and pore size distribution of the silica particle being tested. (© 2002 Waters Corporation; used with permission.)

ing a BET surface area value in the formula in Fig. 9 may result in a surface coverage number that is at least 2–3× smaller than it should be. This was especially true 25 years ago when surface areas of silicas used for chromatography were on the order of 500–600 square meters per gram or more [37,38].

Let's take another approach to solving this problem. If one silanol group in each square nanometer of silica surface were bonded, then Avogadro's number translates that into 1.66 micromoles per square meter. Leo de Galan's group has estimated from crystal models that a typical chromatographic silica gel has 4.8 silanol groups within each square nanometer of surface [39,40, 35], a value in general agreement with that estimated for maximally hydrated silica as well as those from other crystal studies [41]. This means that, if all the silanols reacted, the maximum surface coverage for any bonded silica would be about 8 micromoles per square meter.

Recent studies in the semiconductor field challenge the myth that high ligand densities on silica surfaces cannot be achieved. In self-assembled monolayers of octadecyltrichlorosilane on silica wafers, each molecule occupies about 0.2 square nanometers [42]. Ulman states: "Very strong molecule-substrate interactions . . . result in . . . formation of chemical bonds. . . . at the interface, molecules try to occupy every available binding site on the substrate. . . . in this process they push together molecules that have already adsorbed, thus eliminating free volume. . . . in all self-assembled monolayers the spontaneous adsorption at the organic material–substrate interface, together with the strong van der Waals attraction amongst the alkyl chains, are the driving force for the formation of highly ordered, and closely packed systems [43]."

Does this imply that about five ligand molecules could bond to all accessible silanol sites in each square nanometer of a porous silica gel substrate? The key word in this question is *accessible*. It is possible that on Level One accessible surface, as defined earlier, this limit may be approached with typical large ligands such as octadecylsilyl-chains. Likewise, on Level Two accessible surface, small end-capping ligands may approach maximal bonding density. This would be consistent with the points raised earlier. But to answer our question, we also need to know more about the pore substructure of the silica substrate and how the total surface area may be apportioned between the three levels of accessibility.

C. Importance of Silica Substrate Structure

Dr. Xu and her colleagues in our analytical laboratory routinely run surface area and pore size distribution measurements on a host of porous materials. One technique they use is BJH (Barrett, Joyner, Halenda) mesopore area and volume distribution calculations using data from nitrogen desorption. Computer algorithms convert equilibrium gas pressure isotherms to surface area and pore dimension distributions, and the results are reported as the cumulative percent of surface area versus the average pore diameter in angstroms.

I asked them to test under identical conditions four different types of Waters commercial silicas, spanning the history of modern

HPLC-grade, small-partile silica manufacturing technology. What we discovered from these data was that the distribution of surface area vs. pore diameter varies significantly with silica type and method of preparation. Results are shown in Figs. 10 through 14.

To compare these various silicas on the basis of accessible surface, I have indicated on each curve, with perpendicular lines, three arbitrary points:

First, what percent of surface area is in pores larger than 50 Å in diameter. Presumably, surface in *accessible* pores of this size can be bonded with small end-capping reagents. This number also gives an indication, by subtraction from 100%, of the percent of surface area in a particle's smaller meso- and micropores [Level Three].

Second, what percent of surface is in pores above 100 Å [Level One]. *Accessible* surface in these pores may be bonded by larger silanes. Subtracting from this number the cumulative percent value from the first point above provides an *estimate* of the percent of Level Two surface in a particle.

Third, at what pore diameter is the mid-point of cumulative surface area. The higher this number is above 100 Å, the more surface a particle may have that might be accessible to analytes and participate in the partitioning of analyte molecules between mobile and stationary phases.

Keep in mind that these data are not an absolute indicator of chromatographically or chemically accessible surface. It is not only a pore's size but also its location within the labyrinth and the diffusion distance required for an analyte or bonding reagent to reach it that determine accessibility. For example, it is possible to have a large 100 Å diameter pore within the particle, perhaps not even very far from the outer surface, which is inaccessible to larger molecules. Either molecules may be unable to fit through the much smaller diameter intervening openings or passageways necessary to reach such a pore or to pass over the entire distance to this destination which is too large relative to their diffusional speed and/or the chromatographic process kinetics.

Irregular silica, still made now as it was almost 30 years ago, is ground from large filter cakes formed by acid precipitation of silicate, followed by proprietary treatments to provide a more consistent pore structure and surface for bonding. Note in Fig. 10 that it has only 7% of its surface in micropores, 19% between 50 Å and 100 Å, and

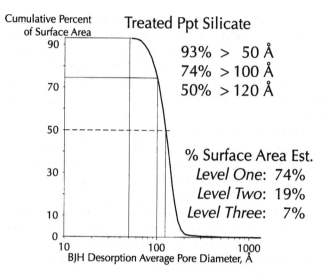

Fig. 10 Distribution of surface area versus pore diameter varies with silica type and method of preparation. Keep in mind that in addition to pore size, diffusion distance may also limit surface accessibility. In this, and the following four figures, cumulative percent of surface area is plotted against the average pore diameter in Å determined by BJH nitrogen desorption. Follow each curve from right to left, noting three arbitrary points marked on each respective graph. The values at 100Å and 50 Å can be used to estimate the percent of surface area in each of the three levels illustrated in Figs. 6 through 8. The silica shown here represents an irregular silica, first used for HPLC packings in 1974. Note the high percentage of Level One Surface Area. Instrument: Micromeritics ASAP 2405 Accelerated Surface Area and Porosimetry System. (© 2002 Waters Corporation; used with permission.)

nearly three-quarters of its surface areas in pores above 100 Å. Half of the surface area lies in pores whose diameter is larger than the median point of 120 Å.

One of the earliest successful spherical silicas, developed in the late 1970s, is represented in Fig. 11. Surprisingly, while only 6% of the surface is in micropores, the median point is only 83 Å, and only 3% of the cumulative surface area lies in pores above 100 Å in diameter. By difference, 91% of the surface lies in pores between 50 Å and 100 Å in diameter. This is the type of silica used to make the C_{18} phase used in Experiments A (see Fig. 2) and B (see Fig. 3) above.

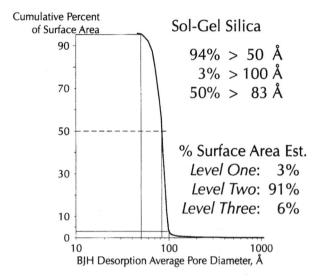

Sol-Gel Silica

94% > 50 Å
3% > 100 Å
50% > 83 Å

% Surface Area Est.
Level One: 3%
Level Two: 91%
Level Three: 6%

Fig. 11 One of the first spherical silicas, developed in the late 1970s, has very little Level One Surface. Method and Instrument same as in Fig. 10. (© 2002 Waters Corporation; used with permission.)

Analysis of one of the first modern, high purity, highly reproducible synthetic spherical silicas, introduced in 1994, is shown in Fig. 12. Its median point is right at 100 Å while 13% of its surface is in pores below 50 Å in diameter, leaving 37% of the surface between 50 Å and 100 Å. This is the silica substrate for the C_{18} phase used in Experiment C (see Fig. 4) above.

A new type of hybrid silica, introduced in 1999, made by controlled copolymerization of tetraethoxysilane and methyltriethoxysilane was also tested. Results in Fig. 13 show how successful were the attempts to create a more desirable pore structure. Less than 1% of the surface areas rests in micropores, while 70% is in pores above 100 Å and the median point, 120 Å, matches that of the 30-year-old irregular silica. Only 29% of the surface lies in pores between 50 Å and 100 Å in diameter.

To test further our notions of accessible surface area levels, we examined the cumulative percent of surface area versus pore diameter for the silica shown in Fig. 12 *after* it had been bonded with a long chain silane. As seen in Fig. 14, the median point moved from

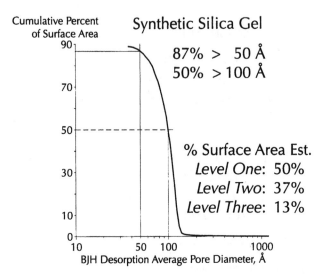

Fig. 12 A modern, high-purity, synthetic silica, developed in 1994, has only half its surface area above 100 Å. Method and Instrument same as in Fig. 10. (© 2002 Waters Corporation; used with permission.)

100 Å down to 70Å. Consistent with our ideas, bonding apparently occurs in pores above 100 Å in diameter, since the total surface area has been cut in half (from 340 to 170 m²/g)* and the percent of surface in pores above 100 Å has decreased drastically, from 50% to 3%. Note also that the average pore diameter has changed from 91 to 65 Å, which, like the 30 Å shift downward in the median point, interestingly, is about the same distance as the length of all-*trans* form of the bonded silyl function.

Thus, it appears that in modern bonded phases, the accessible surface coverage, expressed as micromoles per square meter, may indeed be higher than that calculated by using BET surface areas, by a factor of 2× or more.

---

* A similar change [−54%] in BET surface area after bonding on a synthetic silica prepared from poly(ethoxysiloxane) was reported by Prof. Klaus Unger and colleagues in 1975 [22].

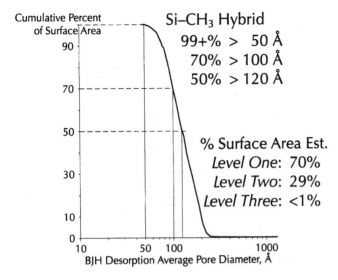

Fig. 13 A recently developed organic–inorganic hybrid silica, developed in 1999, has 70% of its surface above 100 Å, very similar to the corresponding value for the irregular silica from an earlier generation (see Fig. 10). Incredibly, though, it has less than 1% of its surface in micropores. Method and Instrument same as in Fig. 10. (© 2002 Waters Corporation; used with permission.)

D. Wetting Pores

Another old idea that has been depicted in models and cartoons by Lochmüller [26] and others has been termed *hydrophobic collapse* [44]. It has been argued that wetted alkyl chains of bonded phase ligands extend fully outward and provide an opportunity for maximum interaction with analyte molecules. Conversely, it was thought that in predominantly aqueous mobile phases, C_{18} ligands collapse into a hydrophobic tangle of alkyl chains and provide minimal opportunity for analyte adsorption.

However, if ligand density on accessible surface is maximal, long alkyl chains might be so densely packed as to be self-supporting and would not be likely to collapse easily, no matter what type of mobile phase surrounded them. So there must be another explanation.

My colleague (and fellow Brown alumnus) Mark Capparella

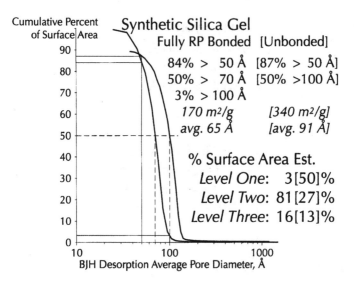

Fig. 14 These results illustrate the change in surface area distribution after fully bonding synthetic silica particles (characterized in Fig. 12) with an RP 18 ligand (see Fig. 18). Total surface area is cut in half, corresponding to the change in the percentage at 100 Å. This is consistent with the notion that dense bonding of large ligands occurs on Level One surface. Some micropores are blocked. The average pore diameter is reduced by about the same distance as the length of the fully extended (all-*trans*) form of the silyl ligand (26 Å). Method and Instrument same as in Fig. 10. (© 2002 Waters Corporation; used with permission.)

made an interesting observation in 1995. He was running repetitively a test separation on a C_{18}-bonded silica column in a 100% aqueous mobile phase with good reproducibility. After 20 hours and 77 injections, he stopped the pump for about 10 minutes to do some system maintenance. Then, upon restarting the pump and making the next injection, he noticed that all retention for the test analyte had been lost, as shown in Fig. 15.

Dr. Thomas Walter, Pamela Iraneta, and Mark decided to investigate this phenomenon further and presented their findings in a poster at HPLC 1997 in Birmingham, England [45]. According to the Young-Laplace equation, the pressure required to force a liquid into a capillary is proportional to the surface tension, the cosine of the contact angle of the fluid with the surface, and, inversely, to the cap-

Observation:

100% aqueous mobile phase

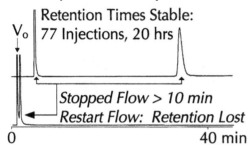

Fig. 15 Observation of hydrophobic collapse. A separation run on a reversed-phase column was stable and reproducible for 77 injections made over a period of 20 h. After stopping flow by turning off the pump for about 10 min, and then restarting the pump, retention for both analytes was lost. For experimental details, see Walter *et al.* [45]. (© 2002 Waters Corporation; used with permission.)

illary diameter [see Fig. 16]. Contact angles for alcohol solutions on a hydrophobic surface have been studied by Janczuk and coworkers [46]. In methanol, with its contact angle of 39.9°, cosine theta is *positive*, so the equation predicts a *negative* value for the pressure required for wetting. Thus, methanol wets pores easily and rapidly with no force required. In contrast, water has a higher surface tension and a contact angle of 110°; cosine theta is *negative*, so the pressure required to force water into the pores is very high.

   Tom, Pam, and Mark reasoned that when Mark had stopped the flow in his earlier experiment, water was expelled from the pores of the C_{18}-bonded phase, because there was no longer sufficient pressure to keep it there. Then, when flow was resumed, the back pressure was insufficient to rewet the pores. This prevented interaction of the analyte with the majority of the accessible surface area, thereby causing a loss of retention. When the pores were rewet with 100% methanol, and then the mobile phase composition was changed to 100% aqueous without flow interruption, maintaining pressure on the column, the orginal retention was restored. They were able to devise a protocol with which they could induce various states of partial dewetting. They tested this protocol on C_{18}-bonded

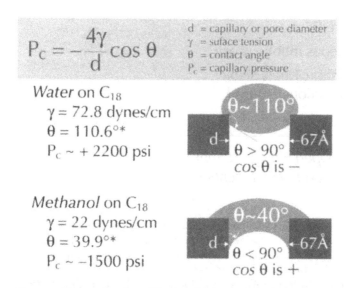

$$P_c = -\frac{4\gamma}{d}\cos\theta$$

d = capillary or pore diameter
γ = suface tension
θ = contact angle
P_c = capillary pressure

Water on C_{18}
  γ = 72.8 dynes/cm
  θ = 110.6°*
  P_c ~ + 2200 psi

θ~110°
d→
67Å
θ > 90°
cos θ is −

Methanol on C_{18}
  γ = 22 dynes/cm
  θ = 39.9°*
  P_c ~ −1500 psi

θ~40°
d→
67Å
θ < 90°
cos θ is +

Fig. 16 Young-Laplace Theory. If the value of the contact angle, θ, for a given fluid on a particular surface (*e.g.*, water on C_{18}) is more than 90° [46], then cosine θ is negative, and a high pressure is required to force fluid into the pores. Conversely, if the contact angle is less than 90°, then pores wet readily without requiring additional pressure. (© 2002 Waters Corporation; used with permission.)

phases with different degrees of ligand coverage. Some of their data is shown in Fig. 17. A reduction in the void volume of the column correlates with the percent of dewetting. These data suggest that as the stationary phase dewets, mobile phase is extruded from the bonded-surface–containing pores. Further, it is probably the surface in these pores that directly participates *via* some mechanism in the chromatographic separation.

Last year in Maastricht, a team from Nomura Chemical Co. in Seto, Japan, reported similar experimental results [47]. They even went so far as to weigh a column before and after loss of retention and found a weight decrease that corresponded to the loss of mobile phase from the pores. Clearly, now, the phenomenon called *hydrophobic collapse* is, in fact, a symptom of *pore dewetting*.

There are other questions that come to mind in light of these reports. It would be interesting to know if, and with how much difficulty, water molecules find their way into the hygroscopic micro-

Confirmation:

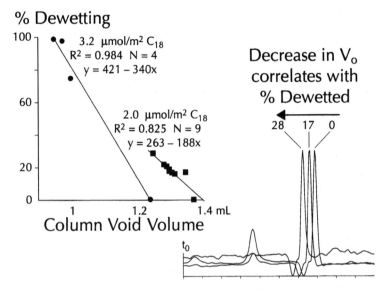

Fig. 17 These data suggest that as stationary phase pores dewet, mobile phase is extruded from the pores. Thus, the phenomenon termed *hydrophobic collapse* is really caused by *pore dewetting*. (© 2002 Waters Corporation; used with permission.)

pores of a C_{18}-bonded phase packing while it is fully wetted. Large pores with a hydrophobic surface may expel water when pressure is relieved, but what happens to water in hygroscopic (hydrophilic-surfaced) micropores? Does it diffuse out into larger, empty pore spaces in nature's attempt to reestablish an equilibrium distribution?

A more important question, as yet largely unconsidered by practitioners, may be: What ramifications does pore dewetting have in the modern world of high speed HPLC done on shorter, wider-diameter columns whose packings are prone to hydrophobic collapse? System back pressure, as measured during operation, usually represents the resistance to flow offered by the column (including frits and end fittings) and any other parts of the fluid pathway (tubing, valves, detector cells, etc.) downstream of the point of pressure measurement (usually near the pump). Consider for the moment, just the column as a flow-resistive element. The back pressure at the outlet is nearly equal to atmospheric pressure, unless the detector cell and connecting tubing downstream of the column offer significant resistance. So, in longer, small-particle–containing columns, there may be a significant pressure gradient inside the bed from inlet to outlet. But making a column shorter and/or wider in diameter reduces the magnitude of this pressure gradient.

Although with smaller particle sizes and longer bed lengths, under certain mobile phase composition and flow conditions, there may be sufficient pressure near the column inlet to keep pores wetted, at some point not too far down the bed, the pressure may drop below this critical value. Thus, it may be quite possible that most of such a reversed-phase column's accessible surface area *does not actively participate* in the chromatographic separation process in highly aqueous mobile phases. Under these conditions, most of the bed may behave as if it were packed with nonporous particles. In shorter columns used for ultra high speed analyses, pores throughout the bed of a fully bonded reversed-phase column may never remain wetted in aqueous mobile phases. For such columns, optimal gradient peak capacity is usually achieved at high flow rates and smaller particle sizes (<3 μm) [48]; both these conditions increase back pressure, which may serve to sustain pore wetting with aqueous mobile phases.

Another way to avoid the problem of pore dewetting is to use a stationary bonded-phase, which, while hydrophobic, retains its wetted state in water. Ligands with polar groups embedded near their point of surface attachment offer such a possibility. In Waters research laboratories, Drs. Uwe Neue, Carsten Niederländer, and John Petersen invented a series of phases using an alkyl ligand chain containing a carbamate group [49,50]. As shown in Fig. 18, Tom, Pam, and Mark demonstrated that such phases [designated RP], even with a high ligand density and a small average pore diameter [65 Å], exhibited minimal loss of retention in 100% aqueous mobile phases under conditions where C_{18} and C_8 packings dewetted virtually completely [45].

E. Proton Transfer via Water Wires

If one accepts our notions of the three levels of surface area accessibility, and maximal ligand density in the most accessible portions of bonded porous particles, how then are analyte molecules affected

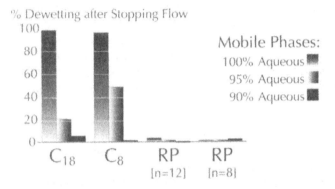

Fig. 18 Effect of embedded polar group. Even with a high ligand density and a small average pore diameter [65 Å], a phase bonded with a long-chain alkyl ligand that has a polar carbamate group [49,50] embedded near its point of surface attachment exhibits dramatically little dewetting in aqueous mobile phases (see Walter et al. [45] for experimental details). (© 2002 Waters Corporation; used with permission.)

by subterranean surface silanols with which they cannot possibly interact directly? The answer may lie in enzyme chemistry and recent studies in quantum dynamics.

Protein transfer processes are ubiquitous in natural systems. Protons are, by several orders of magnitude, the fastest moving species in solution. While I was in graduate school, exciting studies were elucidating enzyme-catalyzed proton transfer. Today, how protons move from place to place is still an active subject of investigation [51–61].

A series of dramatic, three-dimensional textbook illustrations are available on the Internet which show how light drives a classic proton pump, bacteriorhodopsin, a protein that captures retinol at its active site and is integral to the mechanism of vision in the eye [62]. Protons enter the protein from the cytoplasm, move rapidly through the hydrophobic regions to the active site, and, ultimately, exit into the extracellular space. The *proton path* or *proton conduit* is formed by a continuous strand of hydrogen-bonded moieties within strategically located amino acids, bridged, where necessary, by *water* molecules.

Recently, Sadeghi and Cheng have modeled the quantum dynamics of proton movement [63]. A single proton does not simply move from one place to another. Rather, when a string of water molecules have their oxygen–hydrogen bonds in the proper relative orientation, protons hop back and forth within these *water wires* by means of a series of rapid hydrogen bond length fluctuations [see Fig. 19]. Intermediate species such as H_3O^+ and $H_2O_5^+$ are formed during this process, and the time scale is amazingly fast—*sub-picosecond*!

For protons to move in a given direction, rather than bounce to and fro, momentum is crucial. As shown in Fig. 20, a basic molecule may provide such momentum; Sadeghi and Cheng used ammonia as the base species in their calculations. When the base comes into contact with the end of the water wire, it immediately brings order to the sequence of bond breaking/formation, effectively pulling a proton toward itself from the opposite end of the wire in less than a picosecond! An analogy might be touching the end of a live wire, thereby closing the open circuit, and receiving a shock.

In addition to its speed, this process of proton movement is temperature sensitive. Furthermore, Sadeghi and Cheng point out, two physical conditions are required for water wire formation:

1. Interaction between water molecules and their surroundings must be weak.
2. A single strand of water molecules is responsible for conduction of protons.

A hydrophobic region in a protein or in the hydrophobic pore of a bonded-silica packing might well be just the sort of environment conducive to proper bond orientation within a single strand of water molecules. Recently, Richmond and colleagues have shown with vibrational studies that dipolar interactions between weakly hydrogen-bonded water molecules and hydrophobic surfaces cause strong orientation effects in the interfacial region [64,65]. They state that this has "important implications" for a molecular-level understanding of "some of the most important technological and biological processes" [65].

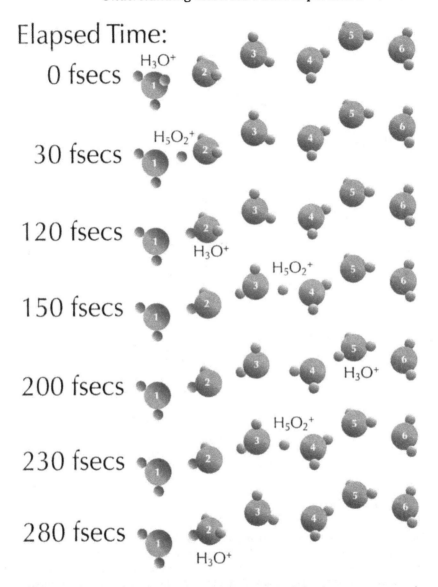

Fig. 19 Proton conduits. Proton transfer is one of the fastest processes in solution. Protons hop back and forth, almost aimlessly, through *water wires*, by means of rapid hydrogen-bond length fluctuations. Intermediate species formed include H_3O^+ and $H_5O_2^+$. (Data and illustration adapted from Sadeghi and Cheng [63].) (© 2002 Waters Corporation; used with permission.)

Elapsed Time:

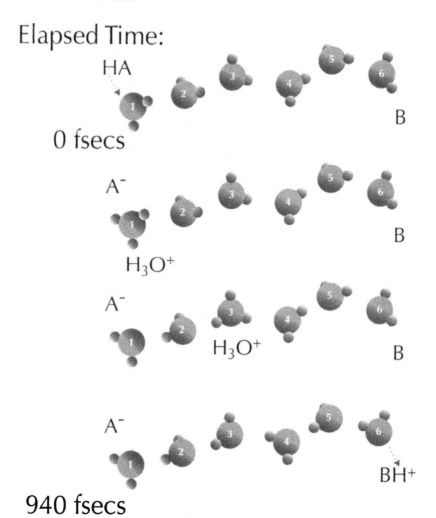

0 fsecs

940 fsecs

Fig. 20 Base creates momemtum. Relative O—H bond orientation is important to create a water wire. The process is sensitive to temperature, and the time scale is rapid—sub-picosecond! For proton movement in a given direction, momentum is crucial. Sadeghi and Cheng calculated from model studies in quantum dynamics that a basic molecule may readily associate with the end of a water wire and create such momentum. (Data and illustration adapted from Sadeghi and Cheng [63].) (© 2002 Waters Corporation; used with permission.)

It is imagined that the organic component of a chromatographic mobile phase facilitates the wetting of the stationary phase pores, bringing water molecules into the hydrophobic regions of the Level One surface, where energies may favor their slow, but steady, attraction and diffusion deeper into the labyrinth to less accessible, unbonded, hygroscopic regions. From a dry state, in a flowing system, it may take minutes to form these *water wires*. The hydrocarbon chains provide the environment to support water wires that extend from Level Three surface all the way out to the partitioning area where a roaming basic analyte molecule might occasionally touch a "live wire," instantly drawing a proton to itself. Protic solvent molecules, like those of methanol, might even form links in the proton conduit, as shown in Fig. 21. This explanation for the *silanol effect* in reversed-phase HPLC had occurred to me back in 1980 as a possible explanation for the observations we made in Barbara's original experiment. But it was my conversation with Chuck and Joe, on March 31, 2000, as well as reading the recent work just cited, which prompted me to proffer on Chuck's blackboard a proposed brand name for this concept:

***H**ydro-**L**inked **P**roton **C**onduit*™ [HLPC™].

F.   Surface by Design: Hybrid Technology

> *I look for what needs to be done. . . . After all, that's how the universe designs itself.*
> —*R. Buckminster Fuller (1895–1983)*

As I stated at the outset, it is difficult to control all the variables in, and thereby draw proper conclusions from, experiments whose aim is to improve our understanding of reversed-phase chromatography. So many of the measurement techniques employed in years past only give us an indirect view of reality, requiring significant assumptions for reasonable interpretation. When a new tool or technology emerges, we must examine its potential to provide meaningful answers to our longstanding questions. One such development is *hybrid particle technology*.

I took the opportunity in several 1996 hallway conversations to relate my notions of how reversed-phase chromatography worked to a young, newly hired materials scientist, Dr. Zhiping Jiang. My experience in doing LLC and my chemical intuition, I told him, were confirmed by John Knox's proposal in 1975 that reversed-phase chro-

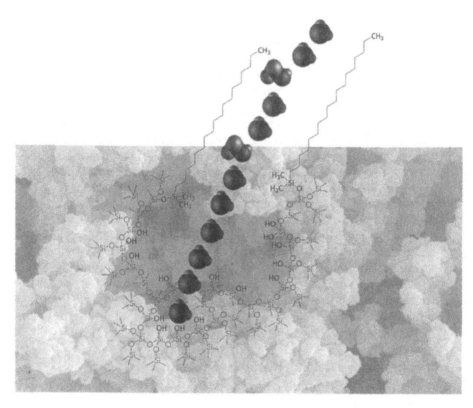

Fig. 21 **Hydro-Linked Proton Conduit**™ Concept. Sadeghi and Cheng point out that two physical conditions are required for water wire formation: (1) interaction between water molecules and their surroundings is weak and (2) a single strand of water molecules is responsible for protons [63]. Fully bonded Level One pores in a reversed-phase packing may provide both of these conditions. As shown here, in an analogy to the proton pathways through hydrophobic regions within protein structures, the organic mobile phase component not only facilitates pore wetting; protic solvent molecules such as methanol can form links in the proton conduit which begins at the hydrophilic Level Two (non-end-capped) or Level Three surface. The alkyl ligands may form narrow channels that satisfy both conditions required for supporting the wire structure (© 2002 Waters Corporation; used with permission.)

matography was a partitioning process, wherein a layer (which he roughly estimated to be 10 to 50 Å thick) of mobile phase solvent molecules was bound to the surface of a porous bonded-phase particle [19]. Because of the carbonaceous nature of the surface ligands, John believed, as I did, that the composition of the *stationary* solvent layer would be enriched in the less polar organic component(s) of the mobile phase. Analyte molecules would then partition between the flowing mobile phase and the stationary phase liquid layer, just as they might if the two phases had been immiscible layers in a separatory funnel.

It must be further realized, I told Zhiping, that analyte molecules in solution are not single molecular entities, but rather are solvated selectively by mobile phase molecules that may have a particular affinity for certain functional groups. Such solvent cages may dramatically change the effective size and shape of analyte molecules in solution and determine their diffusion and partition characteristics. For example, olefins may favor other pi-electron-system–containing solvent molecules like acetonitrile; alcohols and amines (1° or 2°) may be attracted to aprotic ethers like tetrahydrofuran; ketones or aldehydes may favorably attract protic methanol or water molecules. Ionizable molecules, like steroid sulfates, could be partitioned as ion pairs [8] (John had speculated about ion pairs of tetracyclines and perchloric acid [19]). I had often used this chemically intuitive approach to develop successful LLC and reversed-phase HPLC separation systems in the early 1970s.

If only we could somehow change, at will, the structure of the surface in a porous particle, especially in the micropores as well as the mesopores, I told Zhiping, we might have a way of investigating systematically how reversed-phase separations really work. We might be able, by design, to vary the affinity of pore surfaces for specific mobile phase components and thereby modify the nature of the stationary bound-liquid phase layer. Then we could effect, simply, significant changes in selectivity. Zhiping politely endured the ravings of this old chromatographer and quietly escaped back to his laboratory.

I was pleased, however, when, some months later, he told me he had invented a new hybrid silica porous particle that not only had all the desirable mechanical properties (rigidity, pore volume, surface area) but also embedded methyl groups that impart resistance to siloxane bond hydrolysis or dissolution at pH extremes. This particle could have its accessible surface modified, just as we bonded that of normal silica, with all the usual chlorosilanes. But one third of the silanol groups *throughout the backbone** were replaced with —Si—CH$_3$ (see Fig. 22) [66–68].

Increasing the hydrophobicity, or conversely, reducing the hygroscopicity, of the surface in micropores as well as mesopores should certainly change in some way the nature of this hybrid's behavior in reversed-phase separation systems. Uwe devised, Chuck designed, and KimVan Tran executed experiments to compare this hybrid with a synthetic silica. In Fig. 23, we see the variation with pH of the retention factor for a basic analyte when injected into an *unbonded* silica column using an acetonitrile–water mobile phase. Over the pH range of these experiments, note that amitriptyline is fully ionized at one extreme and neutral at the other, whereas the inverse is true for surface silanols. The retention of the strongly basic analyte reaches a maximum at a different pH value for each kind of silica, one an organic–inorganic hybrid (see Fig. 13) and one synthesized from tetraethoxysilane alone (see Fig. 12). Hybrid silica is clearly different. Presumably, retention of amitriptyline on hybrid silica does not fall off as rapidly at higher pH due to the surface hydrophobicity within the accessible pores. One explanation for the hybrid's shift in pH value for maximum retention, as compared with that for silica, is the lower acidity, on balance, of the surface silanol population. Methyl groups interspersed among the 33% smaller population of silanols may decrease the percentage of adjacent silanols that would ordinarily hydrogen bond to one another (and thereby become more acidic).

A similar experiment with naproxen, an acidic compound, is shown in Fig. 24. Here both the analyte and the silanols ionize at

---

* Klaus Unger and his colleagues had reported the synthesis of inorganic-organic silica hybrid phases in 1976 using, for example, benzyltriethoxysilane in an elegant attempt to build the desired surface ligand into the copolymer structure in a single step [22]. The bulky ligand, however, probably impeded the formation of a particle backbone sufficiently strong to withstand the rigors of HPLC. Zhiping's approach embeds a small ligand in the backbone, appending the larger ligand to the accessible surface in a subsequent bonding step. The resultant hybrid particle is very sturdy. (A U.S. patent application on this process has been approved.)

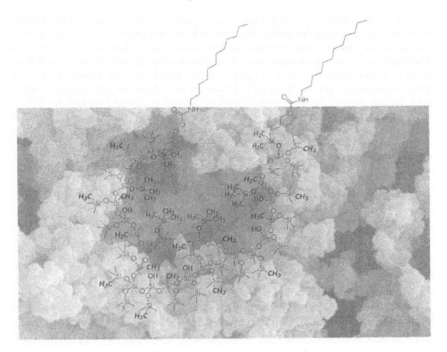

Fig. 22 Organic-inorganic hybrid silica structure. A 2 to 1 mixture of tetra-ethoxysilane and methyltriethoxysilane is used to create a particle in which, *throughout* the backbone, one-third of — Si — OH groups are replaced with — Si — CH₃. Some surface silanols still remain available for bonding and end-capping (RP- and trimethylsilyl- ligands shown here, respectively). Un-like traditional bonded-silicas, surface hydrophobicity exists in pores of all sizes. (© 2002 Waters Corporation; used with permission.)

the same pH extreme. The key difference between silica and hybrid silica is at low pH. Once again, the innate hydrophobicity of the hy-brid surface increases the retention of the neutral analyte dramati-cally when compared to that of silica.

What effects do a hybrid organic–inorganic silica substructure and accessible surface modification by ligand bonding have, respec-tively, on the silanol interaction mediated by proton conduits? In Fig. 25, I have tried to illustrate what might happen. First, a water-rich region or *shield* may encircle the embedded polar groups near the silica surface such as the one shown here, fully bonded with an RP-type ligand. If proton conduits began to form within the laby-

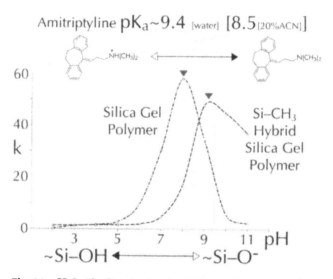

Fig. 23 Hybrid silica is clearly different. Retention of a strongly basic analyte reaches a maximum value at different pH values for hybrid and high-purity silicas. A smaller population of —Si—OH groups on the hybrid surface may reduce the opportunity for hydrogen bonding between adjacent silanols, thereby changing the net acidity of the surface. Note that the retention of amitriptyline does not fall off rapidly at higher pH for the hybrid silica, presumably because the innate surface hydrophobicity induces retention of the neutral form of the analyte molecule, despite the presence of ionized silanols. Conditions: Column, 3.9 × 20 mm; 5 µm particle size. Mobile phase: acetonitrile: aqueous buffer [phosphate (for pH 2, 3, 6, 7, 8, 11), acetate (pH 4, 5), bicarbonate (pH 9, 10)] 20:80. Instrument: Waters 2690 Separations Module, 996 Photodiode Array Detector, Millennium[32™] software. (© 2002 Waters Corporation; used with permission.)

rinth, this water shield would have the effect of *short-circuiting* the wires. Protons would bounce back and forth among water molecules within this layer, greatly reducing the frequency with which they could zip up in wires through the hydrophobic region to reach basic analytes. One would predict that such RP-type phases would exhibit reduced peak tailing and improved peak symmetry for basic analytes. This, in fact, is the case. [50]

Hydrophobic methyl groups throughout hybrid silica particles should diminish their attraction for protic molecules like water or

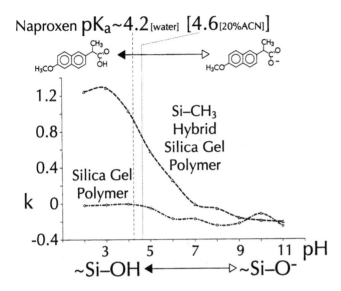

Fig. 24 Similar results using an acidic analyte. Naproxen is retained at low pH on hybrid silica whereas it is virtually unretained at all pH values on high-purity silica. Presumably, naproxen, in its neutral form at low pH, partitions significantly into the neutral stationary phase due to the innate hydrophobicity of the hybrid surface. Conditions: Column, 3.9 × 20 mm; 5 µm particle size. Mobile phase: acetonitrile: aqueous buffer [phosphate (for pH 2, 3, 6, 7, 8, 11), acetate (pH 4, 5), bicarbonate (pH 9, 10)] 20:80. Instrument: Waters HPLC System consisting of an Alliance™ 2690 Separations Module, a 996 Photodiode Array Detector, and Millennium³²™ software. (© 2002 Waters Corporation; used with permission.)

methanol, even in micropores and smaller mesopores where ligand bonding is not possible. Aprotic solvent molecules such as acetonitrile would most likely be preferentially attracted to the smaller pores in such particles. Thus, water wire formation would be inhibited by such a particle surface. Bonding an RP-type ligand onto the surface of a hybrid particle might be expected to be the best combination for inhibiting the formation, and/or disrupting the action, of proton conduits. In fact, such phases exhibit the best peak symmetry and performance for basic analytes, over a wide pH range, of any HPLC packings tested to date [69].

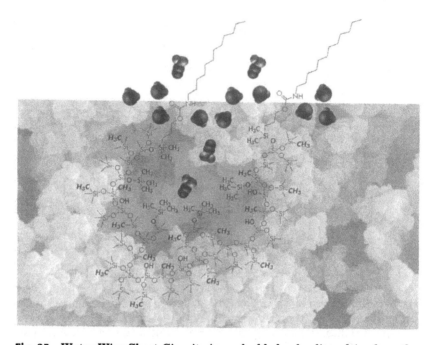

Fig. 25 Water Wire Short Circuit: An embedded polar ligand (such as the carbamate group in the RP phase shown here) may attract and retain a *shield* layer of water molecules near the silica surface. If a water wire were to form a bridge from this shield to silanol groups in the labryinth, any protons that reached this shield would have too many opportunities to hop back and forth within the layer, instead of moving beyond it out toward a basic analyte. Also, the hydrophobicity built into the pores of a *hybrid* particle would lower the attraction for water molecules. This would inhibit formation of water wires in the first place. As shown here, molecules of acetonitrile might be more favorably attracted to the surface than those of methanol. A combination of an RP-type ligand on the surface of a hybrid particle does indeed, as predicted, provide the best peak shape for basic analytes [69]. (© 2002 Waters Corporation; used with permission.)

G. What of the Curious Incidents?

Experiment A [see Fig. 2]

The packing material used was built upon the silica characterized in Fig. 11 with only 3% of its surface area in pores above 100 Å in diameter. So even though the particles were bonded to completion with C_{18} silane, most of the total surface area remains unbound. When mobile phase flow was initiated, the accessible surface in Level One pores, where chromatographic separation takes place, was immediately and completely wetted. Minimal pressure was required to wet these high-ligand-density, hydrophobic pores because the percentage of methanol in the mobile phase was 70%. This explains why the peak shape was excellent for the first injection.

Slowly, methanol mediated the transport of water molecules through the hydrophobic pore regions to the bare Level Two surface. Water wires began to form. With each subsequent injection, more conduits became available to protonate quinoline molecules, causing an ever increasing degree of degradation of peak shape. Because the analyte molecules could not access Level Two pores, the stationary phase volume available for analyte partition was not increased by these proton conduits, and the retention time for quinoline remained constant. If a criterion of constant peak shape for quinoline is used as a measure of achieving equilibrium, then, such a state was not reached, even after one hour of mobile phase flow.

Experiment B [see Fig 3]
Adding naphthalene to the sample mixture did not alter the process just outlined as an explanation for Experiment A. It seems that naphthalene partitions into the same Level One surface, and since it has more affinity for the hydrophobic stationary phase layer than does quinoline, it has a predictably longer retention time. No degradation of column performance is apparent, as evidenced by the constant retention and excellent peak shape for naphthalene throughout the series of injections. Being a neutral, nonpolar molecule, naphthalene is unaffected by proton conduits.

Experiment C [see Fig 4]
Here the high purity base silica used was that whose surface distribution analysis is shown in Fig. 12. After bonding, a much higher proportion of the total surface is available for chromatographic partition than in the case of the packing in the previous two experiments. Additionally, this packing has been fully end-capped, so ligand density is high even in Level Two pores. Thus, for proton conduits to form, water must migrate much further into the labyrinth. And, with a lower number of water-accessible silanols available, many fewer

proton conduits will form. Thus, peak tailing for quinoline increases slowly over the time frame for the experiment and is measurable only by careful computation of asymmetry (see Fig. 4B). Good performance for naphthalene confirms that no problems of a chemical or mechanical nature developed with the column. A chromatographer looking at such performance would assume that phase equilibrium was rapidly established in this column.

So it appears that the foregoing discussion in Section IV is consistent with the four points and may provide reasonable answers to the six questions raised in Section III.D above. The quest now is to daringly design and effectively execute experiments that further substantiate, ultimately refine, or radically subvert, these ideas and inferences. It is my hope that students of chromatographic science reading this chapter will be prepared and inspired to meet this challenge.

H. Summary of New Notions

The way to get good ideas is to get lots of ideas and throw the bad ones away.
—Linus Pauling (1901–1994)

Traditional cartoons do not depict the reality of reversed-phase packing structure and performance in HPLC.

A bold experiment, successfully repeated, forced a fresh look at old views and catalyzed new ideas.

Pore surface area *accessibility* is key to effective surface modification and reversed-phase HPLC performance.

Maximal ligand density in properly bonded phases may be achieved in *accessible* pores. Traditional total surface area measurements can lead to erroneously low surface coverage values.

Pore *dewetting*, not hydrophobic collapse, causes retention loss when using aqueous mobile phases in reversed-phase HPLC.

Silanols, even in micropores, may affect analytes *via water wires* or *hydro-linked proton conduits* [HLPC].

Water wires may be disrupted by embedded polar groups in bonded-phase ligands. Their formation may be inhibited by hybrid organic–inorganic particles with carbon atoms throughout their molecular backbone.

V. CONCLUSIONS

Man's mind stretched to a new idea never goes back to its original dimensions.
—Oliver Wendell Holmes (1809–1894)

The mutual confidence on which all else depends can be maintained only by an open mind and a brave reliance upon free discussion.
—Learned Hand, Speech to the Board of Regents. University of the State of New York, Oct 24, 1952

The joy of insight is a sense of involvement and awe, the elated state of mind that you achieve when you have grasped some essential point; it is akin to what you feel on top of a mountain after a hard climb or when you hear a great work of music.
—Victor Weisskopf, The Joy of Insight, 1991

With the invention of Hybrid Particle Technology, we stand at the threshold of creating chromatographic substrates with interesting, specifically designed chemical modifications that reside not simply at the accessible surface, but, rather, throughout the entire backbone of a particle's molecular makeup. This then will permit rational exploration of structure–activity relationships with a view toward understanding and explaining observations that previously seemed incongruous. Accessible-surface modifications in the traditional manner may add further variables to the study of the ways in which analyte molecules interact with the mobile and stationary phases on their "random walk" through the chromatographic bed. It is hoped that these new capabilities enable, and the ideas set forth herein inspire, fruitful investigations and fervent discussions that will succeed in drawing a realistic view of the chromatographic process. Perhaps, also, those studying other scientific disciplines, if they chance to read this story, will heed the quoted advice, lead in introducing a new order of things, and succeed in creating a future filled with insightful ideas that stretch all our minds and hearts.

ACKNOWLEDGMENTS

I am blessed with a wonderful wife, Kathryn, whose love and inspiration make my life worthwhile and who patiently allows me to infringe on our personal time to invent things and write such stories.

I am proud to have been the son of Dorothy Emma Doyle McDonald (1907–1996), a great contralto and one of the finest high school teachers of English literature ever in the state of New York, who studied with Robert Frost at Breadloaf, who found Sir Arthur Conan Doyle in our family tree, who insisted I learn how to type, who rejoiced in fine musical performance, whether hers, mine, or someone else's, and who taught me about the science of grammar, the lyricism of words, and the power of language.

I am privileged to know, to have known those no longer with us, and to work with wonderful colleagues and friends who listen, who teach, who discover, who discuss, who share, and who encourage me to achieve bigger and better things. Among those I've already acknowledged by the roles they play in my story, I must single out Dr. Charles "Chuck" Phoebe and Joseph Arsenault for enthusiastically encouraging me to pursue and publish my ideas in the first place and for continuing to uncover interesting experimental observations that will lead to more ideas for our next story. Special thanks also go to Prof. Phyllis R. Brown, my former Brown University classmate, who has graciously furnished this forum for my story; Dr. John O'Gara, who served as a sane sounding board for my ideas and led me to several key publications that corroborated them; Carla J. Clayton, Manager, Grace M. Lavallee, Assistant Librarian, and, most recently, Maureen Allegrezza, Database Administrator, all of Waters Information Center, who, respectively, took the time (or gave me the tools) to locate, procured, and maintained a database of the raft of references upon which I have relied for ideas and inspiration for nearly a quarter of a century; Burleigh Hutchins who taught me how to be a successful inventor; and Jim Waters who, like Linus Pauling, demonstrated to me the value in having lots of ideas, many of which were good ones.

REFERENCES

1. P. D. McDonald, B. A. Alden, K. Tran, C. H. Phoebe, Jr., P. C. Iraneta, M. Capparella, T. H. Walter, U. D. Neue, B. K. Grover, J. E. O'Gara, J. C. Arsenault, Y. Xu, and P. A. Richards, Lecture 154, ISC 2000, London, 3 Oct 2000). [URL for free online access to full text version (file size: 16.5 MB): <*http://www.waters.com/pdfs/pmISC-MMns.pdf*>].
2. P. D. McDonald, B. A. Alden, K. Tran, C. H. Phoebe, Jr., P. C.

Iraneta, M. Capparella, T. H. Walter, U. D. Neue, B. K. Grover, J. E. O'Gara, J. C. Arsenault, Y. Xu, and P. A. Richards, Poster P0236, *HPLC 2001*, Maastricht (18–19 June 2001). [URL for free online access to full text version (file size: 9.4 MB): <*http://www.waters.com/pdfs/PM_LC01.pdf*>]. For a review of this poster, see pp. 52–54 in: "The World of Separation Science," R. Stevenson, Am. Lab., *33*(16), 50, 52–56 (2001).

3. Egon Stahl (Ed.), *Thin-Layer Chromatography, A Laboratory Handbook*, Springer-Verlag, 1962; English translation: Academic Press, 1965.
4. P. D. McDonald, *Ph.D. Thesis*, The Pennsylvania State University (1970).
5. L. R. Snyder, *Principles of Adsorption Chromatography*, Wiley, 1965.
6. J. C. Giddings, *Dynamics of Chromatography, Part 1*, Marcel Dekker, 1965.
7. P. D. McDonald and G. A. Hamilton, *J. Am. Chem. Soc.*, *95*(23): 7752 (1973).
8. H. Mickan, R. Dixon, and R. B. Hochberg, *Steroids*, *13*(4): 477 (1969).
9. B. Luttrell, R. B. Hochberg, W. R. Dixon, P. D. McDonald, and S. Lieberman, *J. Biol. Chem.*, *247*(5): 1462 (1972).
10. R. B. Hochberg, P. D. McDonald, M. Feldman, and S. Lieberman, *J. Biol. Chem.*, *249*(4): 1274 (1974).
11. R. B. Hochberg, P. D. McDonald, S. Ladany, and S. Lieberman, *J. Steroid Biochem.*, *6*(3-4): 323 (1975).
12. R. B. Hochberg, P. D. McDonald, M. Feldman, and S. Lieberman, *J. Biol. Chem.*, *251*(7): 2087 (1976).
13. R. B. Hochberg, P. D. McDonald, L. Ponticorvo, and S. Lieberman, *J. Biol. Chem.*, *251*(23): 7336 (1976).
14. P. D. McDonald and C. Rausch, U.S. Patent 4,250,035 (1981).
15. P. D. McDonald, R. V. Vivilecchia, and D. R. Lorenz, U.S. Patent 4,211,658 (1980).
16. E. S. P. Bouvier, R. E. Meirowitz, and P. D. McDonald, U.S. Patents 5,882,521 (1996); 5,976,367 (1998); 6,106,721 (1999); 6,254,780 (2001).
17. D. J. Phillips, P. D. McDonald, and U. D. Neue, American Chemical Society 222nd National Meeting, Chicago, 26 August 2001. [URL for free online access to full text version (file size: 2.6 MB): <*http://www.waters.com/pdfs/WA10785.pdf*>]

18. U. D. Neue, Chap. 12 in H. J. Issaq (Ed.) *A Century of Separation Science*, Marcel Dekker, 2002, p. 203.
19. J. H. Knox and A. Pryde, *J. Chromatogr.*, *112*: 171 (1975).
20. C. Horváth, W. Melander, I. Molnár, *J. Chromatogr.*, *125*: 129 (1976).
21. K. Karch, I. Sebestian, and I. Halász, *J. Chromatogr.*, *122*: 3 (1976).
22. K. K. Unger, N. Becker, and P. Roumeliotis, *J. Chromatogr.*, *125*: 115 (1976).
23. H. Hemetsberger, M. Kellerman, and H. Ricken, *Chromatographia*, *10*(12): 726 (1977).
24. H. Colin and G. Guiochon, *J. Chromatogr.*, *141*: 289 (1977).
25. C. Horvath and W. Melander, *J. Chromatogr. Sci.*, *15*: 393 (1977).
26. C. H. Lochmüller, A. S. Colborn, M. L. Hunnicutt, and J. M. Harris, *Anal. Chem.*, *55*(8): 1344 (1983).
27. J. G. Dorsey, K. A. Dill, *Chem. Rev.*, *89*: 331 (1989).
28. P. W. Carr, J. Li, A. J. Dallas, D. I. Eikens, L. C. Tan, *J. Chromatogr. A*, *656*: 113 (1993).
29. A. Vailaya, C. Horváth, *J. Chromatogr. A*, *829*: (1998).
30. I. Rustamov, T. Farcas, F. Ahmed, F. Chan, R. LoBrutto, H. M. McNair, Y. V. Kazakevich, *J. Chromatogr. A*, *913(1+2)*: 49 (2001).
31. Y. V. Kazakevich, R. LoBrutto, F. Chan, T. Patel, *J. Chromatogr. A*, *913(1+2)*: 75 (2001).
32. U. D. Neue, *HPLC Columns: Theory, Technology, and Practice*, Wiley-VCH, 1997, pp. 345–346.
33. P. D. McDonald and B. Bidlingmeyer, in B. Bidlingmeyer (Ed.) *Preparative Liquid Chromatography*, Elsevier, 1987, pp. 64–66.
34. H. Colin and G. Guiochon, *J. Chromatogr.*, *141*: 295 (1977).
35. G. E. Berendsen, L. de Galan, *J. Liq. Chromatogr.*, *1(5)*: 561 (1978).
36. U.D. Neue, *HPLC Columns: Theory, Technology, and Practice*, Wiley-VCH, 1997, pp. 86–89.
37. C. H. Lochmüller and D. R. Wilder, *J. Chromatogr. Sci*, *17*: 574 (1979).
38. H. Colin and G. Guiochon, *J. Chromatogr.*, *141*: 296 (1977), see Table I.

39. G. E. Berendsen, L. de Galan, *J. Liq. Chromatogr.*, *1(4)*: 403 (1978).
40. G. E. Berendsen, K. A. Pikaart, L. de Galan, *J. Liq. Chromatogr.*, *3(10)*: 1437 (1980).
41. L. R. Snyder, *Principles of Adsorption Chromatography*, Wiley, 1965, p. 159.
42. A. Ulman, *Adv. Mater.*, *2(12)*: 573 (1990).
43. *Ibid.*, p. 573.
44. U.D. Neue, *HPLC Columns: Theory, Technology, and Practice*, Wiley-VCH, 1997, pp. 373–374.
45. T. Walter, P. Iraneta, and M. Capparella, Poster P-202/A, *HPLC'97*, Birmingham (1997) [URL for free online access to full text version (file size: 72K): <*http://www.waters.com/pdfs/TWHPLC97.pdf*>].
46. B. Janczuk, T. Bialopiotrowicz, W. Wojcik, *Colloids Surfaces*, *36*: 391 (1989).
47. N. Nagae and T. Enami, Poster #P 0347, *HPLC 2001*, Maastricht (20–21 June 2001).
48. U. D. Neue, J. L. Carmody, Y-F. Cheng, Z. Lu, C. H. Phoebe, and T. E. Wheat, *Adv. Chromatogr.*, *41*: 93 (2001).
49. U. D. Neue, C. L. Niederländer, and J. S. Petersen, U.S. Patent 5,374,755 (1994).
50. J. E. O'Gara, D. P. Walsh, B. A. Alden, P. Casellini, and T. H. Walter, *Anal. Chem.*, *71*(15): 2992 (1999).
51. D. Eisenberg and W. Kauzmann, *The Structure and Properties of Water*, Oxford University Press, New York, 1969, pp. 225–227.
52. G. A. Jeffrey, *An Introduction to Hydrogen Bonding*, Oxford University Press, New York, 1997, pp. 119–123.
53. M. Gutman and E. Nachliel, *Ann. Rev. Phys. Chem.*, 48: 329 (1997).
54. I. Hofacker and K. Schulten, *PROTEINS: Structure, Function, & Genetics, 30*: 100 (1998).
55. H. Koller, G. Engelhardt, and R. A. van Santen, *Topics in Catalysis, 9*: 163 (1999).
56. J. Fitter, R. E. Lechner, and N. A. Dencher, *J. Phys. Chem. B, 103*: 8036 (1999).
57. P. L. Geissler, C. Dellago, D. Chandler, J. Hutter, and M. Parrinello, *Chem. Phys. Lett., 321*: 225 (2000).

58. S. Brandsburg-Zabary, O. Fried, Y. Marantz, E. Nachliel, and M. Gutman, *Biochim. Biophys. Acta, 1458*: 120 (2000).
59. Z. Smedarchina, W. Siebrand, A. Fernández-Ramos, L. Gorb, and J. Leszczynski, *J. Chem. Phys., 112(2)*: 566 (2000).
60. S.-I. Cho and S. Shin, *J. Molecular Structure, 499*: 1 (2000).
61. P. L. Geissler, C. Dellago, D. Chandler, J. Hutter and M. Parrinello, *Science, 291*: 2121 (2001).
62. Biochemistry in 3D—Lehninger Principles of Biochemistry, 3rd ed., web site, Worth Publishers, Inc. (2000) [URL: <*http://www.worthpublishers.com/lehninger3d/index_title.html*>]
63. R. R. Sadeghi, and H.-P. Cheng, *J. Chem. Phys., 111(5)*: 2086 (1999).
64. G. L. Richmond, *Chem. Eng. News, Sept 11*: 29 (2000).
65. L. F. Scatena, M. G. Brown, and G. L. Richmond, *Science, 292*: 908 (2001).
66. T. Walter, *Waters Whitepaper, WD164*: 4 pp (1999). [URL for free online access to full text version (file size: 188K): <*http://www.waters.com/pdfs/XT_WP2.pdf*>]. See also: <*http://www.waters.com/pdfs/Lxterra.pdf*>.
67. T. Walter, B, Alden. P. Iraneta, E. Bouvier, J. Carmody, R. Crowley, R. Fisk, J. Grassi, C. Gendreau, Z. Jiang, J. O'Gara, and D. Walsh, Lecture L073, *HPLC 1999,* Granada, Spain (2 June 1999). [URL for free online access to full text version (file size: 72K): <*http://www.waters.com/pdfs/TW_lc99L.pdf*>].
68. J. O'Gara, D. P. Walsh, A. R. Pelissey, B. A. Alden, C. A. Gendreau, P. C. Iraneta, and T. Walter, Poster #PA2/10, *HPLC 1999,* Granada (31 May 1999). [URL for free online access to full text version (file size: 148K): <*http://www.waters.com/pdfs/JO_LC99p.pdf*>].
69. B. Alden, C. Gendreau, P. Iraneta, and T. Walter, Poster #PA2/8, *HPLC 1999,* Granada (31 May 1999). [URL for free online access to full text version (file size: 192K): <*http://www.waters.com/pdfs/PI-lc99P.pdf*.].

ADDITIONAL READING

About Monolayers, Surfaces, Pores:

M. J. Wirth, R. W. P. Fairbank, and H. O. Fatunmbi, *Science, 275*: 44 (1997).

S. Heid, F. Effenberger, *Langmuir*, *12*: 2118 (1996).

J. H. Moon, J. W. Shin, S. Y. Kim, and J. W. Park, *Langmuir*, *12*: 4621 (1996).

R. S. Drago, D. C. Ferris and D. S. Burns, *J. Am. Chem. Soc.*, *117*: 6914 (1995).

About Silica & Bonded Phases:

K. K. Unger, *Porous Silica*, *J Chrom Library*, *16*, Elsevier, 1979.

K. K. Unger, in K. M. Gooding and F. E. Regnier (eds), *HPLC of Biological Macromolecules*, Second ed. Marcel Dekker, 2002.

K. K. Unger, D. Kumar, M. Grün, G. Büchel, S. Lüdtke, Th. Adam, K. Schumacher, and S. Renker, *J. Chromatogr. A*, 892: 47 (2000).

S. D. Rogers, J. G. Dorsey, *J. Chromatogr. A, 892*: 57 (2000).

J. E. O'Gara, B. A. Alden, C. A. Gendreau, P. C. Iraneta, T. H. Walter, *J. Chromatogr.*, *893*(2): 245 (2000).

Z. Li, S. C. Rutan, and S. Dong, *Anal. Chem.*, *68(1)*: 124 (1996).

I. Chuang and G. E. Maciel, *J. Am. Chem. Soc.*, *118(2)*: 401 (1996).

L. C. Sander, C. J. Glinka, and S. A. Wise, *Anal. Chem.*, *62(10)*: 1099 (1990).

C. J. Glinka, L. C. Sander, S. A. Wise, M. L. Hunnicutt, and C. H. Lochmüller, *Anal. Chem.*, *57*(11): 2079 (1985).

8

Clinical Applications of High-Performance Affinity Chromatography

David S. Hage *University of Nebraska, Lincoln, Nebraska, U.S.A.*

I.   INTRODUCTION

A variety of analytical methods are used in today's clinical laboratory, but one of the most common is that of high-performance liquid chromatography (HPLC). Although the majority of clinical work in HPLC is performed with traditional methods such as reversed-phase, size-exclusion, and ion-exchange chromatography, another approach that is gaining in importance is the technique of *affinity chromatography*. In this chapter the general principles of this method will be examined and the basic components of an affinity chromatographic system will be described. A variety of clinical and pharmaceutical applications of this technique will also be presented, with an emphasis being given to HPLC-based methods for chemical analysis and quantitation. These applications will include traditional affinity methods as well as systems in which affinity columns are used in automated immunoassays or in combination with other techniques to perform on-line extraction or post-column detection.

A.   Definitions and Common Terms

Before examining the various uses of affinity columns in clinical HPLC labs, it is necessary to first consider what is meant by "affinity chromatography" and to define some common terms that are used to describe this method. A good working definition for affinity chromatography is "a liquid chromatographic technique that makes use of a biological interaction for the separation and analysis of specific analytes within a sample" [1,2]. Examples of such binding processes include the interaction of an antibody with an antigen, the binding of a substrate or inhibitor to an enzyme, and the association of a hormone or drug with a target receptor. Affinity chromatography makes use of these highly selective and reversible processes by placing one of two interacting agents within a column for the isolation or quantitation of the other agent in samples (e.g., see the scheme shown in Fig. 1). The immobilized agent that is used within the column is known as the *affinity ligand* and it is this ligand that acts as the stationary phase for the chromatographic system.

Although the affinity ligand is often of biological origin, syn-

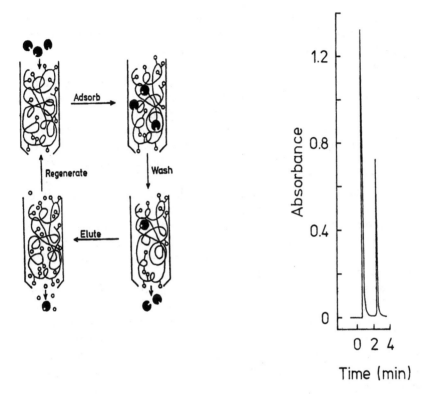

Fig. 1 Typical step gradient operating scheme for affinity chromatography (left), and an illustration of this format through the use of an immobilized protein A HPAC column for the analysis of immunoglobulin G (right). The first peak in the chromatogram represents the non-retained serum components; the second peak is the immunoglobulin G that elutes during a step change to a lower pH buffer. (Reproduced with permission from Refs. 2 and 8.)

thetic compounds that have selective binding or that mimic natural binding agents can also be used as stationary phases in affinity chromatography. Some examples of these synthetic ligands include boronates, immobilized metal ion complexes, and various dyes. The terms *bioaffinity chromatography* or *biospecific adsorption* are sometimes used to specify whether the affinity ligand is really a biological compound. Alternatively, more specific names can be employed to identify the type of ligand within the column. Examples that will be discussed later in this chapter include the methods of *immunoaffinity chromatography*, *lectin affinity chromatography*, and *boronate affinity chromatography*, to name a few [2–6].

Another item that can be used to distinguish between one affinity method and another is the type of support used within the column. In *low-performance* (or *column*) *affinity chromatography*, the support is usually a large diameter, non-rigid gel, like agarose, dextran, or cellulose. In *high-performance affinity chromatography* (*HPAC*), the support consists of small, rigid particles based on silica or synthetic polymers that are capable of withstanding the flow rates and/or pressures that are characteristic of HPLC systems [2–4]. Both low- and high-performance methods have been used in clinical methods. Low-performance affinity chromatography is commonly employed for sample extraction and pretreatment because it is relatively easy to set up and inexpensive to use. However, the better flow and pressure stability of high-performance supports makes HPAC easier to incorporate into instrumental systems, which in turn gives it better speed and precision for the automated quantitation of analytes. The better mass transfer properties of HPAC supports also contribute to the greater speed at which these materials can be used in affinity columns versus those supports found in low-performance affinity columns. It is for these reasons that high-performance methods of affinity chromatography will be emphasized in the remainder of this chapter.

B. Basic Components of HPAC

In order to properly use HPAC, it is essential to understand the role played by each component of a typical affinity system. For example, the most important factor in determining the success of any affinity method is the ligand within the column. As stated earlier, affinity ligands can either be of biological origin (e.g., antibodies, carbohydrates, and lectins) or man-made (e.g., boronates and synthetic dyes) [2–6]. A few general examples of common affinity ligands can be found in Table 1.

Regardless of their origin, all affinity ligands can be placed into one of two categories: high-specificity ligands or general ligands. The term *high-specificity ligand* refers to compounds that bind to only one or a few closely related molecules. This type of ligand is used in chromatographic systems in which the goal is to analyze or purify

Table 1 Examples of Some Common Affinity Ligands [2]

Ligand	Retained substance
Biological ligands	
Antibodies	Antigens (drugs, hormones, peptides, proteins, viruses & cell components)
Inhibitors, substrates, cofactors & coenzymes	Enzymes
Lectins	Sugars, glycoproteins & glycolipids
Nucleic acids	Complementary nucleic acid sequences & DNA/RNA-binding proteins
Protein A & protein G	Antibodies
Synthetic ligands	
Boronates	Sugars, glycoproteins, & *cis*-diol containing compounds
Triazine dyes	Nucleotide-binding proteins & enzymes
Metal chelates	Metal-binding amino acids, peptides & proteins

a specific solute. Typical high-specificity ligands include antibodies (for binding antigens), substrates or inhibitors (for separating enzymes), and single-stranded nucleic acids (for the retention of a complementary sequence). These ligands tend to be of biological origin and have relatively strong binding to sample solutes. *General, or group-specific, ligands* are compounds that retain a family or class of related molecules. These ligands are used in methods in which the goal is to isolate a class of structurally similar solutes. Such ligands can be of either biological or nonbiological origin and tend to have weak to moderate binding to solutes [2,3].

Another factor to consider in affinity chromatography is the material being used to hold the ligand within the column. Two general properties needed for any type of affinity support are (1) it should have low nonspecific binding for sample components and (2) it should be easy to modify for ligand attachment. In HPAC, it is also necessary that the support be stable under the flow-rate, pressure, and solvent conditions to be used in the final application. These materials include various types of modified silica or glass, azalactone beads, and hydroxylated polystyrene media. Besides these commercial supports, suitable materials for HPAC can also be made in-house using underivatized HPLC-grade silica, glass, or a variety of other materials [2–4].

A third item to consider in using affinity chromatography is the way in which the ligand is attached to the solid support, or the *immobilization method*. Several techniques are available for this. For a protein or peptide, this generally involves coupling the molecule through free amine, carboxylic acid, or sulfhydryl residues present in its structure (see Fig. 2). Immobilization of a ligand through other functional sites (e.g., aldehyde groups produced by carbohydrate oxidation) is also possible. Each immobilization method involves at least two steps: an *activation step*, in which the support is converted to a form that can be chemically attached to the ligand, and a *coupling step*, in which the affinity ligand is actually attached to the

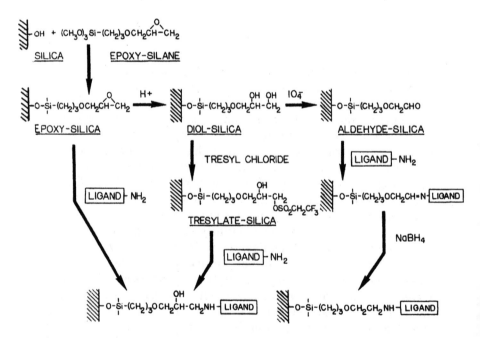

Fig. 2 Examples of immobilization schemes for the coupling of proteins to a silica support. (Reproduced with permission from Ref. 36.)

activated support. Some techniques require an additional third step, in which remaining activated groups on the support are eliminated. Activated supports for HPAC can be prepared in-house according to procedures provided in the literature (see Refs. 4 and 7) or can be obtained in preactivated forms through various commercial suppliers [2].

The application buffer is the fourth factor to consider in the use of an affinity column. Most application buffers in affinity chromatography are solvents that mimic the pH, ionic strength, and polarity experienced by the test solute and ligand in their natural environment. Any cofactors or metal ions required for solute-ligand binding should be present in this solvent. Under these conditions, the solute will probably have its strongest binding to the ligand and its highest degree of retention on the column. The proper choice of an application buffer can also help minimize nonspecific binding. For example, ionic interactions between undesired solutes and the column can often be decreased by altering the concentration and pH of the application buffer. In addition, surfactants or blocking agents may be added to the buffer to prevent nonspecific adsorption to the support or affinity ligand.

Finally, the conditions to be used for analyte elution have to be considered. Such elution can be achieved either by using a solvent that produces weak solute binding to the affinity ligand (i.e., a method known as *nonspecific elution*) or by using a competing agent that displaces solute from the column (i.e., a technique known as *biospecific elution*). Biospecific elution is the gentler of these two methods because it is carried out under essentially the same solvent conditions as used for sample application. This makes this approach attractive for purification work, in which a high recovery of active solute is desired. But this approach tends to produce slow elution times and broad solute peaks. As a result, most analytical applications of HPAC instead use nonspecific elution. In this approach, the solute–ligand interactions are directly altered by changing the pH, ionic strength, or content of the solvent passing through the column. This results in an alteration in the structure of the solute or ligand, leading to solute dissociation and elution. Nonspecific elution tends to be much faster than biospecific elution in removing analytes from affinity columns, but care must be employed to avoid using conditions that are too harsh for the column. If this is not considered, it may result in long column regeneration times or an irreversible loss of ligand activity [2,3].

C. Separation Format

Most ligands in affinity chromatography have relatively strong interactions with their target substances. This results in prohibitively long elution times when such columns are used under isocratic conditions or even with many types of linear gradient elution methods. As a result, it is common in affinity chromatography to instead use a *step gradient*, or *"on/off"* application and elution, as is illustrated in Fig. 1.

In the step gradient method, the sample of interest is first injected onto the affinity column under conditions in which the analyte has strong binding to the immobilized ligand. This is usually done at a pH and ionic strength that mimic the natural environment of the ligand and analyte. Due to the specificity of the analyte-ligand interaction, other solutes in the sample tend to have little or no binding to the ligand and wash off quickly from the column. After these nonretained solutes have been removed, an elution buffer is applied to dissociate the retained analyte; this commonly involves changing the pH or buffer composition of the mobile phase (to decrease the strength of the analyte-ligand interaction) or adding a competing agent to the mobile phase (to displace the analyte from the ligand). As the analyte elutes, it is then detected or collected for further use. Later, the initial application buffer is reapplied to the system, and the column is allowed to regenerate prior to the next sample injection. The overall result is a separation that is selective and easy to perform. It is this feature that makes this format appealing for the quantitation of compounds.

Although step gradients are used in most affinity methods, there are examples in which affinity columns can be run under isocratic conditions or by using a more traditional type of gradient scheme. Examples of this include the use of affinity columns in isoenzyme separations, in chiral separations, and in boronate affinity chromatography. An example is shown in Fig. 3, where a boronic acid column and pH 7.5 phosphate buffer were used to separate various nucleosides under isocratic conditions. This approach, known as *weak affinity chromatography*, requires that the binding of the affinity ligand be mild-to-weak in strength for the analyte to elute from the column in a reasonable amount of time [10]. This approach also re-

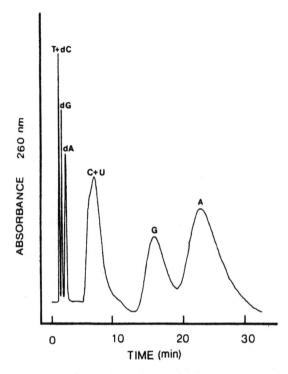

Fig. 3 Separation of various nucleosides by a boronate affinity column under isocratic conditions. These results were obtained with a 5 mm ID × 10 cm boronic acid column using pH 7.4, 0.10 M sodium phosphate buffer as the mobile phase. The injected sample contained thymidine (T), deoxycytidine (dC), deoxyguanosine (dG), deoxyadenosine (dA), cytidine (C), uridine (U), guanosine (G), and adenosine (A). (Reproduced with permission from Ref. 9.)

quires that relatively fast association and dissociation kinetics be present so that sharp analyte peaks and reasonable peak resolution are produced. The advantage of this approach is that it is more applicable than step elution to the separation of multiple substances by a single affinity column. If performed under isocratic conditions, there is the added advantage of eliminating the need for any reequilibration between the application of different solvents to the chromatographic system.

II. TRADITIONAL METHODS

Many analytes can be detected directly as they elute from affinity columns. In HPAC this can be done by using an on-line detector. In low-performance affinity columns this is accomplished by using fraction collection and analysis by a second method. Examples of both approaches will be considered in this section for methods that involve the use of boronate ligands, immobilized antibodies, and other affinity media.

In HPAC direct detection for moderate-to-high concentration analytes is often accomplished by on-line monitoring of UV/V absorbance or fluorescence. In addition, other methods like electrochemical detection or even LC/MS can be used. For lower concentration analytes, it is possible to employ precolumn sample derivatization, post-column reactors, or fraction collection followed by off-line detection by another method (e.g., an immunoassay). Later sections in this chapter will consider other approaches for the detection of trace compounds, including on-line methods based on chromatographic immunoassays, multidimensional systems, and post-column affinity detection.

A. Boronate Affinity Chromatography

One of the first affinity methods to become popular in clinical testing was *boronate affinity chromatography*. This has been used for two decades to determine glycated hemoglobin in serum as a means of assessing the long-term management of blood sugar levels in diabetic patients. In general, at a pH greater than 8.0, boronate ligands have the ability to form covalent bonds with compounds that contain *cis*-diol groups in their structure. Because sugars like glucose possess such diol groups, this makes boronates useful for resolving glycoproteins (e.g., glycated hemoglobin) from nonglycoproteins (e.g., normal hemoglobin).

Boronate affinity columns were first used to examine glycated hemoglobin in 1981 by Mallia and coworkers [11]. This was done by using a low-performance agarose support and absorbance measurements at 414 nm to detect and measure the retained and nonretained hemoglobin fractions in human hemosylate samples. Elution in this early work was performed by passing sorbitol (a diol-containing sugar) through the column to displace any retained glycated hemoglobin from the column; however, a decrease in mobile-phase pH can also be used for elution [12]. Since the work by Mallia et al., similar methods have been reported [13–24] and modified for use in HPLC systems [12,14,17,21,24]. One example is shown in Fig. 4, in which nonglycated and glycated hemoglobin were separated in less than 10 min by a phenylboronic acid HPAC column [12].

In addition to their use with glycated hemoglobin, boronate columns have been used to detect and study other types of glycoproteins. For example, by using absorbance detection at 280 nm, a boronate affinity column can be used to determine the relative amount

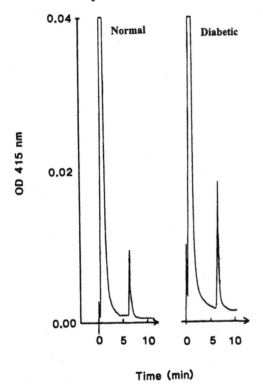

Fig. 4 Separation of normal hemoglobin (first peak) from glycosylated hemoglobin (second peak) by a phenylboronic acid column. The left chromatogram was obtained for a hemolysate sample from a normal individual; the chromatogram on the right represents a sample from a diabetic patient. These results were obtained with a 6 mm ID × 2.5 cm phenylboronic acid column using pH 7.1, 0.05 M sodium phosphate buffer with 0.1 M sorbitol as the mobile phase. (Reproduced with permission from Ref. 12.)

of all glycated proteins in a sample [17]. By employing a more specific detection method, this approach can be made more selective for a particular type of protein (e.g., as is done by using an absorbance of 410–415 nm to detect glycated hemoglobin). For instance, boronic acid columns followed by immunoassays have been used to measure glycated albumin in serum or urine [25] and glycated apolipoprotein B in serum [26,27]. Boronate columns have similarly been employed with mass spectrometry to detect glycated proteins in lens proteins [28].

Boronate affinity columns can also be used to examine other diol-containing substances. One popular example is the catechols, which show strong binding to boronate columns. This has led to the development of numerous methods in which HPAC boronate columns have been combined with other HPLC columns for the clinical analysis of epinephrine, norepinephrine, and dopamine [29–31], dihydroxyphenylalanine [32], dihydroxyphenylacetic acid [32,33], 5-S-cysteinyldopa [34], and vanilmandelic acid [35]. In addition, the presence of cis-diol groups on the sugar ribose gives boronic acid columns the ability to separate ribonucleotides (see Fig. 3). This approach has been reported for the profiling [36] or quantitation of ribonucleotides in human urine [37,38] and serum [38]. An off-line boronic acid column has similarly been used for the reversed-phase analysis of modified nucleosides in patients with gastrointestinal cancer [39] and the LC/MS analysis of diadenosine polyphosphates in human platelets [40].

B. Immunoaffinity Chromatography

Another diverse group of affinity methods are those that make use of antibodies or antibody-related agents as ligands. This approach is referred to as *immunoaffinity chromatography* or, when performed as part of an HPLC system, as *high-performance immunoaffinity chromatography (HPIAC)*. Advantages of this approach include the high selectivity of antibodies in their interactions with other molecules and the ability to produce antibodies against a wide range of solutes [41,42].

Direct detection of analytes in HPIAC is usually performed with application and elution by means of a step gradient, but with weak binding antibodies it is also possible to use isocratic elution [10]. Fig. 1 shows a typical chromatogram generated by HPIAC with a step elution scheme for the determination of immunoglobulin G in serum samples. Other examples of direct analyte detection by HPIAC are listed in Table 2. As can be seen from Table 2, applications for this method range from small carbohydrates and peptides to antibodies and other proteins. In many cases these separations can be obtained in less than 5–6 min. In addition, if the application and elution conditions are selected properly, there have been several reports in which such columns have been used for several hundred to a thousand injections with no serious signs of degradation [41].

The most common clinical application for low-performance immunoaffinity columns is their use for the isolation of a particular sample component prior to the off-line analysis of these components by another method. This technique is referred to as *immunoextraction* or *immunoaffinity extraction*. Sample preparation by off-line immunoextraction has been the subject of several recent reviews [41,58–61], most of which have emphasized its applications in the area of drug analysis. A few examples involving human samples include the use of immunoextraction followed by reversed-phase liquid chromatography for the determination of albuterol [62] and vitamin

Table 2 Direct Detection of Clinical Analytes by HPIAC[a]

Analyte	Label/Monitoring Method	Sample [Refs.]
Anti-idiotypic antibodies	UV absorbance	Serum [43,44]
Fibrinogen	UV absorbance	Plasma [45]
Glucose-containing tetrasaccharide	PAD or radiolabels	Serum, urine [46,47]
Granulocyte colony stimulating factor	Fluorescence derivatization	Plasma, CSF, BMAF [48]
Human serum albumin	UV absorbance	Serum [49], urine [50]
Immunoglobulin G antibodies	UV absorbance	Serum [51], CSF [52]
Immunoglobulin E antibodies	UV absorbance/ELISA	Plasma, serum [53]
Interferon	UV absorbance	Plasma/serum, urine, saliva, CSF [54]
Interleukin-2	Fluorescence derivatization, receptor assay	Tissue samples [55]
β_2-Microglobulin	UV absorbance/EIA	Plasma [56]
Transferrin	UV absorbance	Serum [57]

[a] Abbreviations: BMAF, bone marrow aspirate fluid; CSF, cerebrospinal fluid; EIA, enzyme immunoassay; ELISA, enzyme-lined immunosorbent assay; PAD, pulsed amperometic detection.

D-3 metabolites [63] in serum or plasma; the measurement of human chorionic gonadotropin [64], polycyclic aromatic hydrocarbon metabolites [65], LSD [66], β2-agonists [67] and tetrodotoxin [68] in urine; and the analysis of ochratoxin A in serum, plasma, or milk [69]. Immunoextraction has also been combined with off-line gas chromatography or GC/MS to measure prostaglandins, thromboxanes [60–73] and alkylated DNA adducts in urine [74,75], along with furans and polychlorinated dibenzo-*p*-dioxins in serum [76].

In some reports several different types of antibodies have been used together to detect or isolate more than one analyte from the same sample. For instance, ten different immunoaffinity columns were recently connected in series for the analysis of cytokines in clinical samples (see Fig. 5) [77]. This approach has further been used within a microanalytical system as a means for examining the immune response under various clinical conditions. In addition, it has been demonstrated with low-performance columns that a variety of antibodies can be placed onto the same column for multiresidue analysis. This approach, known as *multi-immunoaffinity chromatography* (*MIAC*), has been used with HPLC for the determination of testosterone, nortestosterone, methyltestosterone, trenbolone, zeranol, estradiol, and diethylstilbestrol in urine [58], as well as in the detection of these compounds in related applications [78,79].

C. Other Methods

Although boronate and immunoaffinity columns have received the most attention in clinical uses of HPAC, there are additional ligands that have also been employed in such a setting. For instance, there have been several reports that have used HPAC and protein A or protein G as the stationary phase. These ligands are cell wall proteins produced by *Staphylococcus aureus* and group G *streptococci* that have the ability to bind to the constant region of many types of immunoglobulins [80–82]. Protein A and protein G have their strongest binding to immunoglobulins at or near neutral pH but readily dissociate from these solutes when placed into a lower pH buffer. These two ligands differ in their ability to bind to antibodies from different species and classes [80,83], but recombinant proteins that blend the activities of these two ligands can also be obtained [84].

The ability of protein A and G to bind to antibodies make these good ligands for the analysis of immunoglobulins in humans. For

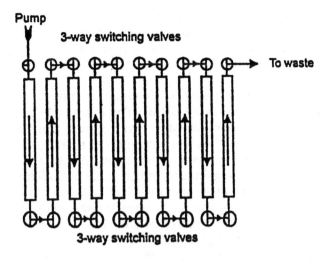

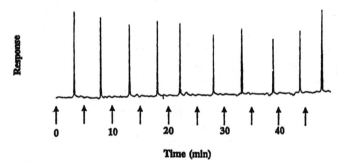

Fig. 5 A multicolumn immunoaffinity system for the determination of (left-to-right) interleukin-1 (IL-1), IL-2, IL-4, IL-5, IL-6, IL-10, IL-12, IL-13, tumor necrosis factor alpha, and gamma interferon. (Adapted with permission from Ref. 77.)

instance, both immobilized protein A and protein G have been used in HPLC systems for the determination of IgG in serum samples (see Fig. 1) [8,85,86]. There has also been work in which a protein A column has been combined with a column containing anti-albumin antibodies for the simultaneous analysis of IgG and albumin in serum [87]. But probably the greatest application of protein A and protein G has been as secondary ligands for the adsorption of antibodies for use in immunoaffinity chromatography. This latter method is especially useful when high antibody activities are needed or if it is necessary to frequently replace the antibodies in the affinity column [41,42].

Lectins are a second class of ligands that have been used for detecting clinical analytes by HPAC. The lectins are non-immune system proteins that have the ability to recognize and bind certain carbohydrate residues [88]. Two examples of such ligands include concanavalin A, which binds to α-D-mannose and α-D-glucose residues, and wheat germ agglutinin, which binds to D-N-acetylglucosamines. The ability of these and other lectins to bind to sugar residues has made them popular in the isolation and study of biological compounds that contain carbohydrate chains, such as polysaccharides, glycoproteins, and glycolipids [2,3].

One reported use of lectin columns in HPAC has been in the separation and analysis of alkaline phosphatase isoenzymes. In this case a high-performance column containing wheat germ agglutinin was used to separate the liver- and bone-derived isoenzymes of alkaline phosphatase in human serum [89]. This method showed improved resolution of the isoenzymes versus a low-performance affinity column [90] and gave good correlation when compared to a solid-phase immunoassay for alkaline phosphatase (see Figure 6) [91].

Another application of lectin columns has been their use to study the structures and properties of various glycoproteins. For instance, columns based on concanavalin A have been used to separate ApoA- and apoB-containing lipoproteins in human plasma [92], to study the microheterogeneity of serum transferrin during alcoholic liver disease [93], and to characterize the carbohydrate structure of follicle-stimulating hormone and luteinizing hormone under various clinical conditions [94]. Other work involving a combination of concanavalin A and wheat germ agglutinin columns has been used to identify changes that occur during prostate cancer in asparagine-linked sugars on human prostatic acid phosphatase [95].

Several additional ligands that have been employed for the direct analysis of clinical analytes. For instance, immobilized heparin has been utilized in affinity columns for the determination of antithrombin III in plasma [96,97]. Another example is S-octylglutathione, which has been used for the separation and analysis of glutathi-

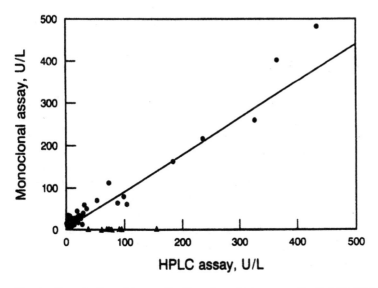

Fig. 6 Correlation of bone alkaline phosphatase results in 139 clinical samples a solid-phase immunoassay versus a method that employed an HPLC lectin affinity column. (Reproduced with permission from Ref. 91.)

one S-transferase isoenzymes in human lung and liver samples [98,99]. And finally, immobilized p-aminobenzamidine has been used for the separation of human plasminogen species, with the addition of an immobilized urokinase column for on-line detection [100].

III. CHIRAL SEPARATIONS

A second group of clinical methods that have employed affinity ligands are those used in chiral separations. This has been a topic of interest for over a decade in the pharmaceutical industry and is beginning to appear more frequently in the field of clinical chemistry, where the ability to quantitate the different chiral forms of a drug or its metabolites is employed in drug metabolism studies or therapeutic monitoring. HPLC methods containing chiral stationary phases make up one set of tools that have been shown to be particularly valuable in quantitating and separating chiral compounds [101]. Because many of the ligands used in affinity chromatography are inherently chiral, this makes them natural choices as stationary phases for such separations.

Table 3 provides a list of chiral affinity ligands that have been used for clinical separations [102–149]. These are usually based on a protein or carbohydrate stationary phase, but several synthetic ligands can be employed as well. Most of the examples in Table 3 were performed by carrying out a routine liquid–liquid or solid-phase extraction of the sample, with the content of this extract later being injected onto the chiral column of interest. However, a chiral column can also be used to first resolve the different forms of a particular solute, followed by fraction collection and on-line or off-line injection onto a second achiral column for further separation and quantitation. Another possibility is to use a standard reversed-phase or normal-phase column to first isolate the compounds of interest from a sample and then apply these on-line or off-line to a chiral column for their stereoselective separation.

A. Protein-Based Stationary Phases

Proteins are one group of affinity ligands that have received some attention for use as HPLC chiral stationary phases. Although all proteins are chiral, only one (α_1-acid glycoprotein) has seen any significant degree of use in the analysis of chiral drugs in clinical samples. α_1-Acid glycoprotein (also known as AGP, AAG or orosomucoid) is a human serum protein that is involved in the transport of many small solutes throughout the body. AGP differs from human serum albumin (another drug-binding protein in serum) in that AGP has a lower isoelectric point and contains carbohydrate residues as part of its structure. The lower isoelectric point makes AGP more useful than serum albumin in binding cationic compounds, while the carbohydrate residues may play a role in determining the stereoselectivity of AGP's binding properties [101]. Table 3 gives some examples of drugs and related solutes that have been examined in human samples by AGP columns. A specific illustration is provided in Figure 7, where an AGP column was used to quantitate the *R*- and *S*-isomers of thiopentone in human plasma [113].

Other proteins that have received some attention in clinical applications of chiral HPLC are ovomucoid [143] and bovine serum albumin (BSA) [119]. Ovomucoid is a glycoprotein obtained from egg whites that has been shown to be useful in the separation of cationic solutes [101]. BSA is a member of the serum albumin family, which

Table 3 Chiral Separations for Clinical Samples Based on Immobilized Affinity Ligands in HPLC

Ligand	Analyte (Sample [Refs.])
α₁-*Acid glycoprotein*	Bunolol (urine [102]); citalopram (plasma [103]); fenoprofen (plasma [104]); flurbiprofen (plasma [115]); ibuprofen (plasma [104,106]); ketamine (plasma [107]); ketoprofen (plasma [104]); methadone (plasma [108,109], serum [110]; norketamine (plasma [107]); norverapamil (plasma [111]); pindolol (serum [112]); thiopentone (plasma [113]); vamicamide (serum, urine [114]); verapamil (plasma [111,115])
Amylose derivatives	Carboxyibuprofen (urine [116]); N-demethyltramadol (urine [117]); hydroxyibuprofen (urine [116]); norverapamil (plasma [118]); tramadol (urine [117]); verapamil (plasma [118])
Bovine serum albumin	Leucovorin (plasma [119])
Cellulose derivatives	Celiprolol (plasma, urine [120]); O-demethyltramadol (urine [117]); N-desmethylzopiclone (urine [121]); disopyramide (plasma, urine [122]); felodipine (plasma [123]); fluoxetine (plasma [124]); 7-hydroxywarfarin (urine [125]); ketamine (serum [126]); mandipine (serum [127]); metoprolol (serum [128,129]); metyrapol (plasma, urine [130]); mono-N-dealkyldisopyramide (plasma, urine [122]); nisoldipine (plasma [131]); norfluoxetine (plasma [124]); N-oxide zopiclone (urine [121]); pindolol (serum, urine [132]); praziquantel (serum [133]); propafenone (serum [134]); propanolol (plasma, urine [135]); remoxipride (plasma [136]); tramadol (plasma [137]); warfarin (urine [125]); zopiclone (plasma, urine [138])
Cyclodextrins	Chlorpheniramine (serum [139]); citalopram (plasma [140]); desmethylcitalopram (plasma [140]); didesmethylcitalopram (plasma [140]); hexobarbital (serum [139]); moguisteine M1 metabolite (plasma, urine [141]); moguisteine M2 metabolite (plasma, urine [141]); propranolol (plasma, urine [142])
Ovomucoid	Pentazocine (serum [143])
Pirkle-type ligands	Albuterol (serum [144]); mepivacaine (serum [145]); phenylpropylamine (plasma [146]); salbutamol (plasma, urine [147])
Miscellaneous ligands	Atenolol (plasma [148]); thioridazine (serum [149])

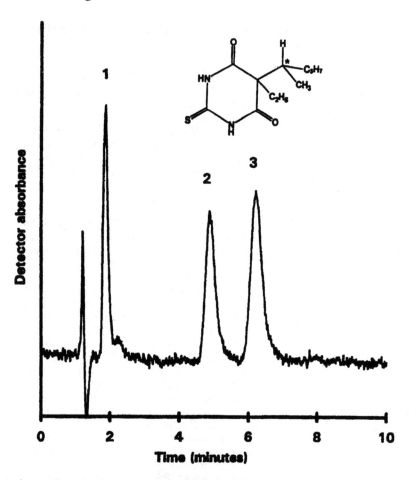

Fig. 7 Separation of R-thiopentone (peak 2) and S-thiopentone (peak 3) in human plasma by an immobilized AGP column. Peak 1 is racemic ketamine, which was used as an internal standard. The structure of thiopentone is shown above the chromatogram, with the asterisk indicating the location of the chiral center. (Reproduced with permission from Ref. 113.)

make up most of the protein content of serum in mammals and are known to be involved in the transport of a wide range of small organic and inorganic compounds throughout the body, including many pharmaceutical agents. BSA, and the related protein human serum albumin, tend to bind best to neutral or anionic compounds, thus making these proteins complementary to AGP and ovomucoid in their applications [101].

B. Carbohydrate-Based Stationary Phases

There are several types of carbohydrates and related derivatives that have been used as ligands for the resolution of chiral compounds in clinical samples. From Table 3 it can be seen that the most popular of these ligands have been those based on cellulose derivatives. Cellulose is a linear polymer of β-1,4-D-glucose and occurs throughout nature as a structural component of plants. There are a wide variety of cellulose derivatives that are available for chiral separations and that have been placed onto silica-based supports for HPLC [101]. The most common derivatives used in clinical testing are those based on cellulose carbamates, especially cellulose *tris*-3,5-dimethylphenyl carbamate. Other cellulose derivatives that have been used in such work include cellulose triphenyl carbamate, cellulose *tris*-(4-methylbenzoate), and microcrystalline cellulose triacetate [126].

Amylose, a polysaccharide that is closely-related to cellulose, has also seen a fair amount of use in clinical applications of chiral HPLC [116–118]. Like cellulose, amylose consists of a linear polymer of D-glucose units but now these units are held together through α-1, 4-linkages. This gives these two carbohydrates different three-dimensional configurations and different binding properties. Amylose is similar to cellulose in that its use in chiral separations of clinical analytes has generally been based on amylose carbamate derivatives. For example, all of the applications listed in Table 3 for amylose have used amylose *tris*-3,5-dimethylphenyl carbamate as the chiral ligand.

Another class of carbohydrates that can be employed as stereoselective ligands in clinical HPLC are the cyclodextrins [139–142]. These are circular polymers of α-1,4-D-glucose that are produced through the degradation of starch by the microorganism *Bacillus macerans*. The most common forms of these polymers are α-, β-, and γ-cyclodextrin, which contain six, seven or eight glucose units, respectively. The cone-shaped structure and hydrophobic interior cavity of cyclodextrins gives them the ability to form inclusion complexes with numerous small, aromatic solutes. Furthermore, the well-defined arrangement of hydroxyl groups about the upper and lower faces of the cyclodextrins give these agents the ability to discriminate between various chiral compounds. The applications for cyclodextrin columns listed in Fig. 3 were based on silica supports that contained either immobilized β-cyclodextrin or acetylated β-cyclodextrin. Commercial columns that contain immobilized α- or γ-cyclodextrin are also available but have not yet seen much use in clinical analyses.

C. Other Chiral Phases

Along with the natural ligands that have already been described, several types of synthetic compounds have been employed as chiral stationary phases for clinical testing. Some examples are the Pirkle brush-type columns based on (S)-tert-leucine or (R)-1-(α-naphthyl)ethylamine, and a Pirkle naphthylurea column. In addition, a closely-related stationary phase based on (R)-(1-naphthyl)ethyl has been successfully with human samples [146]. Other synthetic chiral ligands that have been employed in clinical HPLC methods include a phenylmethylurea-based column [149] and a column that contains (R,R)-diaminocyclohexanedinitrobenzyl as the stationary phase [148].

IV. CHROMATOGRAPHIC IMMUNOASSAYS

Although a large number of analytes can be captured and quantitated directly by HPAC, there are many clinical agents that are too dilute for this approach. One way of handling such analytes is to instead use a scheme that employs a separate labeled compound for the indirect detection of analytes. The analyte is monitored in this case by looking at how its presence in samples affects the retention of the labeled agent to an affinity column. Immobilized antibodies are most often used in such columns, giving rise to a technique which can be considered a type of immunoassay. The resulting approach is known as a *chromatographic* (or *flow-injection*) *immunoassay* and has been the source of increasing interest in recent years as a means of determining trace analytes which, by themselves, may not produce a readily detectable signal when analyzed directly by affinity chromatography [150–157].

Many of the same labels that have been used in traditional immunoassays have also been employed within chromatographic-based immunoassays, as shown by the examples provided in Table 4

Table 4 Chromatographic Immunoassays for Clinical Analytes

Analyte	Label/Monitoring method[a]	Sample [Refs.]
Competitive binding immuno- *assays*		
Human chorionic gonado-tropin	HRP/absorbance	Serum [158]
Human serum albumin	HSA/absorbance	Serum [159]
Immunoglobulin G	GOD/electrochemical detection	Serum [160]
	Fluorescein/fluorescence	Serum [161]
Testosterone	Texas red/fluorescence	Serum [162]
Theophylline	Liposomes/fluorescence	Serum [163–165]
	Fluorescein/fluorescence	Serum [166]
	ALP/electrochemical detection	Serum [167]
Thyroid stimulating hormone	HRP/absorbance	Serum [158]
Thyroxine	Acridinium ester/chemiluminescence	Serum [168]
Transferrin	Lucifer yellow/fluorescence	Serum [169]
Transferrin	Transferrin/absorbance	Serum [159]
Immunometric assays		
Anti-Bovine IgG antibodies	GOD/electrochemical detection	Serum [170]
α-(Difluoromethyl)ornithine	HRP/fluorescence	Plasma [171]
Immunoglobulin G	Acridinium ester/chemiluminescence	Serum [172]
Parathyrin	Acridinium ester/chemiluminescence	Plasma [173,174]
Thyroxine	Acridinium ester/chemiluminescence	Serum [175]

[a] Abbreviations: ALP, alkaline phosphatase; GOD, glucose oxidase; HSA, human serum albumin; HRP, horseradish peroxidase. For ALP, GOD, and HRP, the means of detection was based on the substrate/product systems that were used along with these enzyme labels.

[158–175]. For instance, enzyme labels like horseradish peroxidase, alkaline phosphatase, and glucose oxidase have all been used in such methods. Other labels that have been reported include fluorescent tags like fluorescein, Texas red, or Lucifer yellow; chemiluminescent labels based on acridinium esters; and liposomes impregnated with fluorescent dye molecules. The detection of these labels is generally performed on-line as they elute in the non-retained or retained peaks of the immunoaffinity column; however, fraction collection and off-line detection can also be used.

A. Competitive Binding Immunoassays

There are several different ways of conducting chromatographic immunoassays, but the most common type is one that uses a competitive binding format. The easiest approach for doing this is to mix the sample with a labeled analyte analog (i.e., the "label") and simultaneously inject these onto an immunoaffinity column that contains a relatively small amount of antibody. This method, known as a *simultaneous injection competitive binding immunoassay*, is the most popular approach for conducting a chromatographic immunoassay. An alternative format is to first apply only the sample to the immunoaffinity column, followed later by a separate injection of the label. This method is known as a *sequential injection competitive binding immunoassay*. As shown by the example in Fig. 8, these assays can often be performed in the time range of only a few minutes and provide results comparable to those of standard immunoassays. In addition, such assays can easily be automated and tend to have much better precision than traditional competitive binding immunoassays [158–167,169].

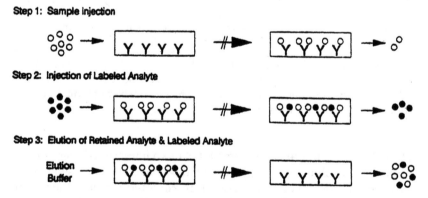

Step 1: Sample Injection

Step 2: Injection of Labeled Analyte

Step 3: Elution of Retained Analyte & Labeled Analyte

Elution Buffer

Fig. 8 General format for a chromatographic immunoassay using the sequential injection competitive binding format. The open circles represent the analyte and the closed circles represent a labeled analog of this analyte. (Reproduced with permission from Ref. 176.)

Both the simultaneous and sequential injection methods can be used for either small or large analytes, as shown in Table 4. One advantage of using sequential over simultaneous injection of the sample and label is that even an unlabeled preparation of analyte can potentially be used as the "label," provided that this species produces a sufficient signal for detection. This is particularly useful for complex samples that contain analytes at moderate-to-high concentrations in complex mixtures. Another advantage of using sequential injection is that there are no matrix interferences present during detection of the label since it is never in contact with the actual sample. However, the simultaneous method is simpler to use because it does not require an additional step for the separate application of label to the immunoaffinity column. These two assay formats also have some differences in terms of how they respond to changes in the injection flow rate and the amount of label that is used within the assay [176,177].

Another type of competitive binding assay that can be performed by HPAC is the *competitive displacement method*. This method has been used to detect such clinical analytes as cortisol and thyroxine [150,168]. This technique is carried out by first applying a labeled analog to an antibody column to occupy some or all of the column's binding sites. When a sample is later passed through this column, the analyte will bind to sites that are momentarily unoccupied by the labeled analog as it undergoes local dissociation and reassociation. The net result of this competition is a displacement of the labeled analog from the column, with the size of its displaced peak being proportional to the amount of analyte in the injected sample. A big advantage of this method is that its response has a direct, rather than inverse, relationship to analyte concentration. Another advantage is that a single application of the labeled analog to the antibody column can often be used for many sample injections, which helps to increase sample throughput. However, caution must be followed in using this technique because it is highly dependent on flow rate and the labeled analog's rate of dissociation. In addition, this approach tends to produce broad peaks when using high affinity antibodies and slowly-dissociating analogs [41].

B. Immunometric Assays

The sandwich immunoassay, or two-site immunometric assay, can also be performed as part of an affinity chromatographic system

[170,172–174]. In this technique, two different types of antibodies are used that each bind to the analyte of interest. The first of these two antibodies is attached to a chromatographic support and is used to extract the analyte from samples. The second antibody contains an easily measured tag and is added in solution to the analyte either before or after sample injection; this second antibody serves to place a label onto the analyte, thus allowing the amount of analyte on the immunoaffinity support to be quantitated as it and the label are eluted from the column. As shown in Fig. 9, one specific application of this method has been the determination of intact parathyrin (PTH) in plasma. In this case the total time per injection was 6.0–6.5 min and the limit of detection was 0.2 pmol/L PTH when using acridinium ester labels and post-column chemiluminescent detection [173,174].

An important advantage of the chromatographic sandwich immunoassay is that it produces a signal for the bound label that is directly proportional to the amount of injected analyte. The fact that two types of antibodies are used in a sandwich immunoassay gives this technique higher selectivity than chromatographic-based competitive binding immunoassays. The main disadvantage of the sandwich immunoassay is that it can only be used for analytes like large peptides or proteins that are big enough to bind simultaneously to two separate antibodies.

A one-site immunometric assay is another format that can be used to perform immunoassays by affinity chromatography. This approach has been utilized in determining such agents as thyroxine [175] and α-(difluoromethyl)ornithine [171]. In this technique, the sample is first incubated with a known excess of labeled antibodies or Fab fragments that are specific for the analyte of interest. After binding between the analyte and antibodies has occurred, this mixture is applied to a column that contains an immobilized analog of the analyte. This column serves to extract any antibodies or Fab fragments that are not bound to the original analyte. Meanwhile, those antibodies or Fab fragments that are bound to analyte from the sample will pass through the column in the non-retained peak, which is then detected and used for analyte quantitation.

Like the chromatographic competitive binding immunoassays, the one-site immunometric method is able to detect both small and large solutes. However, like a sandwich immunoassay, it gives a signal for the non-retained label that is directly proportional to the

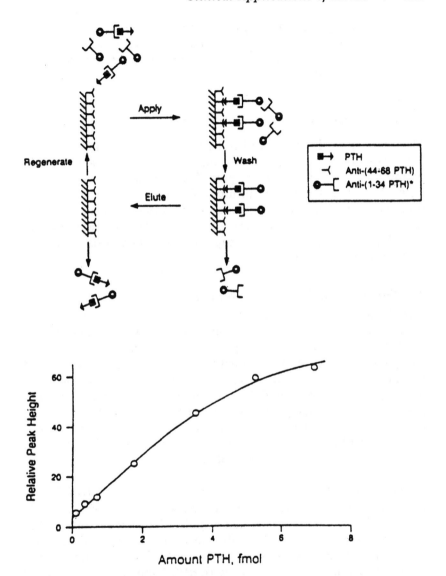

Fig. 9 Format (top) and calibration curve (bottom) for a chromatographic sandwich immunoassay for parathyroid hormone (PTH). (Reproduced with permission from Ref. 173.)

amount of analyte in the original sample. The main disadvantage of this approach is that relatively pure and highly-active labeled antibodies/Fab fragments must be used in order to provide a low background signal.

V. COMBINED ANALYTICAL SYSTEMS

All of the examples discussed to this point have used affinity columns that were operated either alone or as an off-line method in combination with other techniques. But it is also possible to use HPAC or low-performance affinity columns on-line with systems that employ multiple analytical methods. Examples that will be considered in this section will include on-line affinity extraction in HPLC, gas chromatography or capillary electrophoresis, and post-column affinity detection.

A. On-Line Affinity Extraction in HPLC

The direct coupling of affinity extraction with other analytical methods has been the subject of increasing research, with the use of affinity extraction in HPLC being of particular interest [41]. The selectivity and relative ease with which affinity columns can be incorporated into an HPLC system makes this appealing as a means for automating extraction methods and for reducing the time required for sample pretreatment. Also, the relatively high precision of HPLC pumps and injection systems gives on-line affinity extraction better precision than off-line extraction methods, because the on-line approach has more tightly controlled sample application and elution conditions.

Clinical applications of on-line affinity extraction in HPLC have been developed for a large number of analytes, as indicated in Table 5 [178–188]. All of these particular applications used immunoaffinity columns combined with analytical columns for RPLC. For instance, Fig. 10 shows one such method in which an on-line immunoaffinity column was used along with RPLC and mass spectrometry for the separation and analysis of β-agonists in urine samples [189]. In addition, it is possible to perform on-line affinity extraction with other combinations, such as the use of boronate affinity columns [29–38] or the use of size exclusion or ion-exchange chromatography as the second separation step [190–192].

Table 5 Clinical Examples of On-Line Immunoextraction/RPLC

Analyte	Detection method[a]	Samples [Refs.]
α_1-Antitrypsin	UV absorbance	Plasma [178]
Cortisol	UV absorbance	Plasma, urine, milk, saliva [179]
Digoxin	Post-column fluorescence	Serum [180]
Estrogens	UV absorbance	Plasma [181], urine [181,182]
Human epidermal growth factor	Fluorescence	Urine [183]
LSD	UV absorbance, API-MS	Urine [184]
LSD analogs & metabolites	UV absorbance, API-MS/MS	Urine [185]
Phenytoin	UV absorbance	Plasma [186]
Propranolol	UV absorbance, API-MS	Urine [184]
Δ^9-Tetrahydrocannabinol	UV absorbance	Saliva [187]
Transferrin	UV absorbance	Serum [188], plasma [178]

[a] Abbreviations: API-MS, atmospheric pressure ionization-mass spectrometry.

One reason for the large number of reports involving the combination of on-line affinity extraction with RPLC undoubtedly has to do with the popularity of RPLC in routine analytical separations. But another crucial factor is the fact that the elution buffer for most affinity columns is an aqueous solvent that generally contains little or no organic modifier, making this buffer act as a weak mobile phase for RPLC. This means that as a solute elutes from an affinity ligand it will tend to have strong retention as it passes through an on-line reversed-phase column, leading to analyte reconcentration. This effect is valuable in dealing with analytes that have slow desorption from affinity columns, making them impractical to analyze by direct detection techniques [41].

One common format for on-line affinity extraction in RPLC involves injecting the sample onto an affinity column, with the nonretained components being allowed to go to a waste container. The

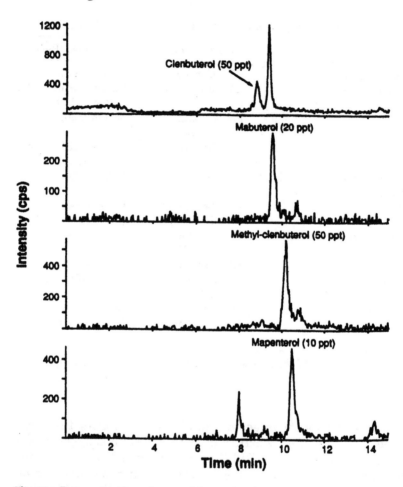

Fig. 10 Determination of several β-agonists in urine by immunoaffinity extraction coupled on-line with reversed-phase HPLC and mass spectrometry. (Reproduced with permission from Ref. 189.)

affinity column is then switched on-line with a RPLC precolumn and an elution buffer is applied to dissociate any retained analyte. As these analytes elute, they are captured and reconcentrated at the head of the RPLC precolumn. After all solutes have left the affinity column, this column is switched back off-line and regenerated by passing through the initial application buffer. Meanwhile, the RPLC precolumn is placed on-line with a larger analytical RPLC column and both are developed with an isocratic or gradient elution scheme involving the application of a solvent with an increased organic modifier content. This causes analytes at the head of the RPLC precolumn to move through the analytical column and to be separated based on their differences in polarity. As these solutes elute, they are monitored and quantitated through the use of a flow-through detector.

B. On-Line Affinity Extraction in GC and CE

Although not as common as in HPLC, there has been some work investigating the use of on-line affinity extraction with GC. This has been used for the determination of β-19-nortestosterone and related steroids in urine [193]. In this case a RPLC precolumn was again used to capture and reconcentrate retained analytes as they eluted from an immunoaffinity extraction column. However, this RPLC precolumn also now served to remove any water from the analytes and to place them into a volatile organic solvent (ethyl acetate, which was used as the elution mobile phase). A portion of the analytes that eluted from the RPLC precolumn were then passed into an injection gap of a GC system. Once the solute/organic solvent plug had entered the GC system, a temperature program was initiated for solute separation. One advantage of this approach is that large volumes of sample can usually be applied to the affinity column, thus allowing the analyte to be concentrated for low detection limits. The main disadvantage of on-line immunoextraction in GC is the greater complexity of this method versus off-line extraction or on-line extraction in HPLC.

Several recent studies have considered the additional possibility of combining on-line immunoextraction with capillary electrophoresis (CE). For example, a capillary packed with a protein G chromatographic support has been used to adsorb antibodies for the extraction and concentration of insulin from serum prior to quantitation by CE [194]. Also, immunoextraction based on immobilized Fab fragments was used to extract and concentrate tear samples for the CE analysis of cyclosporin and its metabolites in samples from corneal transplant patients [195]. In addition, antibodies were covalently immobilized in microcapillary bundles or in laser-drilled glass rods that were then connected to a CE capillary for the on-line immunoextraction and detection of immunoglobulin E in serum [196].

C. Post-Column Affinity Detection

Yet another way in which affinity columns can be employed is for monitoring the elution of specific solutes from other chromatographic columns. This involves the use of a post-column reactor and an affinity column attached to the exit of an analytical HPLC column. As in many other affinity methods, most research in the area of post-column affinity detection has used immobilized antibody (or immobilized antigen) columns. This has given rise to a specific type of detection scheme known as *post-column immunodetection* [41,197].

The direct detection mode of affinity chromatography represents the simplest approach for post-column quantitation of an analyte, provided that the solute is capable of generating a sufficiently strong signal for detection. An example of this approach is a report in which size exclusion chromatography and post-column immunodetection were used for the analysis of acetylcholinesterase (AChE) in amniotic fluid (see Fig. 11) [198]. The method developed in this report used an immunoaffinity column containing anti-AChE antibodies to capture AChE as it eluted from the analytical column. After the AChE was adsorbed to the immunoaffinity column, a substrate solu-

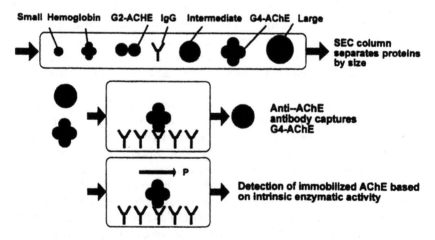

Fig. 11 Post-column affinity detection of a acetylcholinesterase tetramer (G4-AChE) by size exclusion chromatography followed by an anti-AChE immunoaffinity column. (Reproduced with permission from Ref. 198.)

tion for AChE was passed through the column and the resulting colored product was detected by an on-line absorbance detector.

Other formats are also possible for post-column immunodetection, including techniques based on competitive binding immunoassays [197,199] and sandwich immunoassays [200]. But the one-site immunometric assay is the most common format for immunodetection and it is the only additional approach that has previously been used in clinical applications. The basic operation of this format involves taking the eluent from the HPLC analytical column and combining this with a solution of labeled antibodies or Fab fragments that will bind to the analyte of interest. The column eluent and antibody or Fab mixture is then allowed to react in a mixing coil and passed through an immunodetection column that contains an immobilized analog of the analyte. The antibodies or Fab fragments that are bound to the analyte will pass through this column and onto the detector, where they will provide a signal that is proportional to the amount of bound analyte. If desired, the immunodetection column can later be washed with an eluting solvent to dissociate the retained antibodies or Fab fragments; but a sufficiently high binding capacity is generally used so that a reasonably large amount of analytical column eluent can be analyzed before the immunodetection column must be regenerated.

One-site immunometric detection was originally used to quantitate digoxin and digoxigenin as they eluted from a standard RPLC column by employing fluorescein-labeled Fab fragments (raised against digoxigenin) and an immobilized digoxin support in the post-column detection system. This method was then used to successfully monitor both digoxin and its metabolites in plasma and urine [201]. The same general system was later used along with a restricted-access RPLC column to monitor digoxin, digoxigenin, and related metabolites in serum [202].

D. Other Methods

Yet another example of a method in which HPAC has been used with another technique is a clinical method that was developed for the determination of urinary albumin [50]. In this report, a flow injection analysis (FIA) system was attached to the outlet of an anti-albumin antibody column (see Fig. 12). As urine samples were applied to this system, any albumin present was extracted while the remaining nonretained components were passed into the FIA sys-

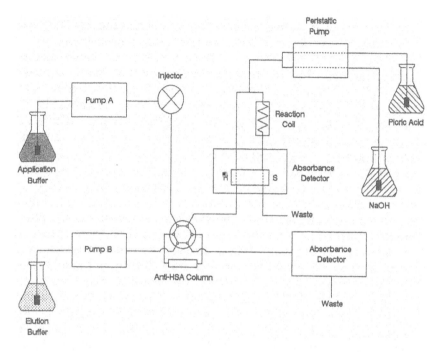

Fig. 12 HPIAC/FIA system for the simultaneous determination of albumin and creatinine in urine samples. The FIA portion of the system is shown in the upper right. The symbols "R" and "S" represent the reference and sample sides of the absorbance detector. (Reproduced with permission from Ref. 50.)

tem. In the FIA system, reagents for the colorimetric Jaffe reaction were used to quantitate the amount of creatinine in the sample. Meanwhile, the sample's albumin concentration was measured by direct UV absorbance detection of the retained HPIAC fraction. The resulting values were then combined to provide the albumin/creatinine ratio, which was used to adjust for variations in sample concentration due to fluctuations in the urine output and dilution.

REFERENCES

1. IUPAC. Nomenclature for chromatography. http://wingate. merck.de/english/services/chromatographie/iupac/ chrnom.htm.

2. D. S. Hage, in E. Katz, R. Eksteen, P. Shoenmakers, N. Miller, (Eds.), *Handbook of HPLC*. Marcel Dekker, New York, 1998, Chap. 13.
3. R. R. Walters, *Anal. Chem.*, *57*: 1099A (1985).
4. P.-O. Larsson, *Methods Enzymol.*, *104*: 212 (1987).
5. J. Turkova, *Affinity Chromatography*, Elsevier, Amsterdam, 1978.
6. I. Parikh and P. Cuatrecasas, *Chem. Eng. News*, *63*: 17 (1985).
7. G. T. Hermanson, A. K. Mallia, and P. K. Smith, *Immobilized Affinity Ligand Techniques*, Academic Press, New York, 1992.
8. S. C. Crowley and R. R. Walters, *J. Chromatogr.*, *266*: 157 (1983).
9. M. Glad, S. Ohlson, L. Hansson, M.-O. Mansson, and K. Mosbach, *J. Chromatogr.*, *200*: 254 (1980).
10. D. Zopf and S. Ohlson, *Nature*, *346*: 87 (1990).
11. A. K. Mallia, G. T. Hermanson, R. I. Krohn, E. K. Fujimoto, and P. K. Smith, *Anal. Lett.*, *14*: 649 (1981).
12. S. Hjerten and J. P. Li, *J. Chromatogr.*, *500*: 543 (1990).
13. R. Fluckiger, T. Woodtli, and W. Berger, *Diabetes*, *33*: 73 (1984).
14. B. J. Gould, P. M. Hall, and J. G. H. Cook, *Clin. Chim. Acta*, *125*: 41 (1982).
15. D. C. Klenk, G. T. Hermanson, R. I. Krohn, E. K. Fujimoto, A. K. Mallia, P. K. Smith, J. D. England, H. M. Wiedmeyer, R. R. Little, and D. E. Goldstein, *Clin. Chem.*, *28*: 2088 (1982).
16. R. N. Johnson and J. R. Baker, *Clin. Chem.*, *34*: 1456 (1988).
17. R. P. Singhal and S. S. M. DeSilva, *Adv. Chromatogr.*, *31*: 293 (1992).
18. F. Q. Nutall, *Diabetes Care*, *21*: 1475 (1998).
19. K. P. Peterson, J. G. Pavolvich, D. Goldstein, R. Little, J. England, and C. M. Peterson, *Clin. Chem.*, *44*: 1951 (1998).
20. A. Duncan, J. D. Hoyer, V. F. Fairbanks, D. J. McCormick, and K. S. Kubik, *Blood*, *92*: 3099 (1998).
21. H. Stirk, K. R. Allen, *Anal. Clin. Biochem.*, *36*: 233 (1999).
22. W. J. Schnedl, R. Krause, G. Halwachs-Baumann, M. Trinker, R. W. Lipp, and G. J. Krejs, *Diabetes Care*, *23*: 339 (2000).
23. L. Bry, P. C. Chen, and D. B. Sacks, *Clin. Chem.*, *47*: 153 (2001).
24. N. Kitagawa and L. G. Treat-Clemens, *Anal. Sci.*, *7*: 195 (1991).

25. A. C. Silver, E. Lamb, W. R. Cattell, and A. B. S. J. Dawnay, *Clin. Chim. Acta, 202*: 11 (1991).
26. M. Panteghini, R. Bonora, and F. Pagani, *Ann. Clin. Biochem., 31*: 544 (1994).
27. K. Shishino, M. Murase, H. Makino, and S. Saheki, *Anal. Clin. Biochem., 37*: 498 (2000).
28. P. Chellan and R. H. Nagaraj, *J. Biol. Chem., 276*: 3895 (2001).
29. P. O. Edlund and D. Westerlund, *J. Pharm. Biomed. Anal., 2*: 315 (1984).
30. P. Ni, F. Guyon, M. Caude, and R. Rosset, *J. Liq. Chromatogr., 12*: 1873 (1989).
31. K. S. Boos, B. Wilmers, R. Sauerbrey, and E. Schlimme, *Chromatographia, 24*: 363 (1987).
32. P. O. Edlund, *J. Pharm. Biomed. Anal., 4*: 625 (1986).
33. L. Hansson, M. Glad, and C. Hansson, *J. Chromatogr., 265*: 37 (1983).
34. C. Hansson, B. Kagedal, and M. Kallberg, *J. Chromatogr., 420*: 146 (1987).
35. B.-M. Eriksson and M. Wikstrom, *J. Chromatogr., 567*: 1 (1991).
36. P.-O. Larsson, M. Glad, L. Hansson, M.-O. Mansson, S. Ohlson, and K. Mosbach, *Adv. Chromatogr., 21*: 41 (1983).
37. E. Hagemeier, K.-S. Boos, E. Schlimme, K. Lechtenboerger, and A. Kettrup, *J. Chromatogr., 268*: 291 (1983).
38. E. Hagemeier, K. Kemper, K.-S. Boos, and E. Schlimme, *J. Chromatogr., 282*: 663 (1983).
39. K. Nakano, K. Shindo, T. Yasaka, and H. Yamamoto, *J. Chromatogr., 332*: 127 (1985).
40. J. Jankowski, W. Potthoff, M. van der Giet, M. Tepel, W. Zidek, and H. Schluter, *Anal. Biochem., 269*: 72 (1999).
41. D. S. Hage, *J. Chromatogr., 715*: 3 (1998).
42. T. M. Phillips, *LC Mag., 3*: 962 (1985).
43. T. M. Phillips, *Clin. Chem., 34*: 1689 (1988).
44. T. M. Phillips and J. V. Babashak, *J. Chromatogr., 512*: 387 (1990).
45. J. P. McConnell and D. J. Anderson, *J. Chromatogr., 615*: 67 (1993).
46. W. T. Wang, J. Kumlien, S. Ohlson, A. Lundblad, and D. Zopf, *Anal. Biochem., 182*: 48 (1989).

47. D. Zopf, S. Ohlson, J. Dakour, W. Wang, and A. Lundblad, *Methods Enzymol.*, *179*: 55 (1989).
48. T. M. Phillips, *J. Chromatogr. B*, *662*: 307 (1994).
49. D. S. Hage, R. R. Walters, *J. Chromatogr.*, *386*: 37 (1987).
50. P. F. Ruhn, J. D. Taylor, D. S. Hage, *Anal. Chem.*, *66*: 4265 (1994).
51. D. M. Zhou, H. F. Zou, J. Y. Ni, L. Yang, L. Y. Jia, Q. Zhang, and Y. K. Zhang, *Anal. Chem.*, *71*: 115 (1998).
52. T. M. Phillips, N. S. More, W. D. Queen, T. V. Holohan, N. C. Kramer, and A. M. Thompson, *J. Chromatogr.*, *31*: 173 (1984).
53. T. M. Phillips, N. S. More, W. D. Queen, and A. M. Thompson, *J. Chromatogr.*, *327*: 205 (1985).
54. T. M. Phillips, *Biomed. Chromatogr.*, *6*: 287 (1992).
55. T. M. Phillips, *Biomed. Chromatogr.*, *11*: 200 (1997).
56. M. Mogi, M. Harada, T. Adachi, K. Kojima, and T. Nagatsu, *J. Chromatogr.*, *496*: 194 (1989).
57. S. Ohlson, B.-M. Gudmundsson, P. Wikstrom, and P.-O. Larsson, *Clin. Chem.*, *34*: 2039 (1988).
58. L. A. van Ginkel, *J. Chromatogr.*, *564*: 363 (1991).
59. S. E. Katz and M. Siewierski, *J. Chromatogr.*, *624*: 403 (1992).
60. S. E. Katz and M. S. Brady, *J. Assoc. Off. Anal. Chem.*, *73*: 557 (1990).
61. N. Kobayashi and J. Goto, *Bunseki Kagaku*, *47*: 537 (1998).
62. H. Ong, A. Adam, S. Perreault, S. Marleau, M. Bellemare, and P. Du Souich, *J. Chromatogr.*, *497*: 213 (1989).
63. K. Shimada, K. Mitamura, and T. Higashi, *J. Chin. Chem. Soc.*, *47*: 285 (2000).
64. C. L. Liu and L. D. Bowers, *J. Chromatogr. B*, *687*: 213 (1996).
65. R. K. Bentsen-Farmen, I. V. Botnen, H. Noto, J. Jacob, and S. Ovrebo, *Int. Arch. Occup. Envir. Health*, *72*: 161 (1999).
66. J. Rohrich, S. Zorntlein, and J. Becker, *Forensic Sci. Int.*, *107*: 181 (2000).
67. A. Koole, J. Bosman, J. P. Franke, and R. A. de Zeeuw, *J. Chromatogr. B*, *726*: 149 (1999).
68. K. Kawatsu, T. Shibata, and E. Hamano, *Toxicon*, *37*: 325 (1999).
69. B. Zimmerli and R. Dick, *J. Chromatogr. B*, *666*: 85 (1995).
70. A. Bachi, E. Zuccato, M. Baraldi, R. Fanelli, and C. Chiabrando, *Free Radical Biol. Med.*, *20*: 619 (1996).

71. G. Mackert, M. Reinke, H. Schweer, H. W. Seyberth, *J. Chromatogr.*, *494*: 13 (1989).
72. C. Chiabrando, V. Pinciroli, A. Campoleoni, A. Benigni, A. Piccinelli, and R. Fanelli, *J. Chromatogr.*, *495*: 1 (1989).
73. M. Ishibashi, K. Watanabe, Y. Ohyama, M. Mizugaki, Y. Hayashi, and W. J. Takasaki, *J. Chromatogr.*, *562*: 613 (1991).
74. V. Prevost, D. E. G. Shuker, M. D. Friesen, G. Eberle, M. F. Rajewsky, and H. Bartsch, *Carcinogenesis*, *14*: 199 (1993).
75. M. D. Friesen, L. Garren, V. Prevost, and D. E. G. Shuker, *Chem. Res. Toxicol.*, *4*: 102 (1991).
76. J. K. Huwe, W. L. Shelver, L. Stanker, D. G. Patterson, and W. E. Turner, *J. Chromatogr. B*, *757*: 285 (2001).
77. T. M. Phillips and J. M. Krum, *J. Chromatogr. B*, *715*: 55 (1998).
78. L. A. van Ginkel, R. W. Stephany, H. J. van Rossum, H. M. Steinbuch, G. Zomer, E. van de Heeft, and A. P. J. M. de Jong, *J. Chromatogr.*, *489*: 11 (1989).
79. M. Dubois, X. Taillieu, Y. Collemonts, B. Lansival, J. De Graeve, and P. Delahaut, *Analyst*, *123*: 2611 (1998).
80. R. Lindmark, C. Biriell, and J. Sjoequist, *Scand. J. Immunol.*, *14*: 409 (1981).
81. P. L. Ey, S. J. Prowse, and C. R. Jenkin, *Immunochemistry*, *15*: 429 (1978).
82. L. Bjorck and G. Kronvall, *J. Immunol.*, *133*: 969 (1984).
83. B. Aakerstrom and L. Bjoerck, *J. Biol. Chem.*, *261*: 10240 (1986).
84. M. Eliasson, A. Olsson, E. Palmcrantz, K. Wibers, M. Inganas, B. Guss, M. Lindberg, and M. Uhlen, *J. Biol. Chem.*, *263*: 4323 (1988).
85. S. Ohlson, in I. M. Chaiken, M. Wilchek, I. Parikh (Eds.), *Affinity Chromatography and Biological Recognition*. Academic Press, New York, 1983, p. 255.
86. P. Cassulis, M. V. Magasic, and V. A. DeBari, *Clin. Chem.*, *37*: 882 (1991).
87. D. S. Hage and R. R. Walters, *J. Chromatogr.*, *386*: 37 (1987).
88. I. E. Liener, N. Sharon, and I. J. Goldstein, *The Lectins: Properties, Functions and Applications in Biology and Medicine*, Academic Press, London, 1986.
89. D. J. Anderson, E. L. Branum, and J. F. O'Brien, *Clin. Chem.*, *36*: 240 (1990).

90. D. G. Gonchoroff, E. L. Branum, and J. F. O'Brien, *Clin. Chem.*, *35*: 29 (1989).
91. D. G. Gonchoroff, E. L. Branum, S. L. Cedel, B. L. Riggs, and J. F. O'Brien, *Clin. Chim. Acta*, *199*: 43 (1991).
92. M. Tavella, P. Alaupovic, C. Knight-Gibson, H. Tournier, G. Schinella, and O. Mercuri, *Prog. Lipid Res.*, *30*: 181 (1991).
93. T. Inoue, M. Yamauchi, G. Toda, and K. Ohkawa, *Alcohol Clin. Exp. Res.*, *20*: 363A (1996).
94. M. J. Papandreou, C. Asteria, K. Pettersson, C. Ronin, and P. Beck-Peccoz, *J. Clin. Endocrin. Metab.*, *76*: 1008 (1993).
95. K. I. Yoshida, M. Honda, K. Arai, Y. Hosoya, H. Moriguchi, S. Sumi, Y. Ueda, and S. Kitahara, *J. Chromatogr. B*, *695*: 439 (1997).
96. A. L. Dawidowicz, T. Rauckyte, and J. Rogalski, *Chromatographia*, *37*: 168 (1993).
97. D. M. Zhou, H. F. Zou, J. Y. Ni, L. Yang, L. Y. Jia, Q. Zhang, and Y. K. Zhang, *Anal. Chem.*, *71*: 115 (1998).
98. A. L. Dawidowicz, T. Rauckyte, and J. Rogalski, *J. Liq. Chromatogr.*, *17*: 817 (1994).
99. J. B. Wheatley, M. K. Kelley, J. A. Montali, C. O. A. Berry, and D. E. Schmidt, Jr., *J. Chromatogr. A*, *663*: 53 (1994).
100. J. B. Wheatley, J. A. Montali, and D. E. Schmidt, Jr., *J. Chromatogr. A*, *65*: 676 (1994).
101. S. Allenmark, *Chromatographic Enantioseparation: Methods and Applications*, Ellis Horwood, New York, 1991, Chap. 7.
102. F. Li, S. F. Cooper, M. Cote, and C. Ayotte, *J. Chromatogr.*, *660*: 327 (1994).
103. D. Haupt, *J. Chromatogr. B*, *685*: 299 (1996).
104. S. Menzel-Soglowek, G. Geisslinger, and K. Brune, *J. Chromatogr.*, *532*: 295 (1990).
105. G. Geisslinger, S. Menzel-Soglowek, O. Schuster, and K. Brune, *J. Chromatogr.*, *573*: 163 (1992).
106. K.-J. Pettersson and A. Olsson, *J. Chromatogr.*, *563*: 414 (1991).
107. G. Geisslinger, S. Menzel-Soglowek, H.-D. Kamp, and K. Brune, *J. Chromatogr.*, *568*: 165 (1991).
108. N. Schmidt, K. Brune, and G. Geisslinger, *J. Chromatogr.*, *583*: 195 (1992).
109. O. Beck, L. O. Boreus, P. LaFolie, and G. Jacobsson, *J. Chromatogr.*, *570*: 198 (1991).

110. K. Kristensen, H. R. Angelo, and T. Blemmer, *J. Chromatogr. A, 666:* 283 (1994).
111. Y.-Q. Chu and I. W. Wainer, *J. Chromatogr., 497:* 191 (1989).
112. F. Mangani, G. Luck, C. Fraudeau, and E. Verette, *J. Chromatogr. A, 762:* 235 (1997).
113. D. J. Jones, K. T. Nguyen, M. J. McLeish, D. P. Crankshaw, and D. J. Morgan, *J. Chromatogr., 675:* 174 (1996).
114. A. Suzuki, S. Takagaki, H. Suzuki, and K. Noda, *J. Chromatogr., 617:* 279 (1993).
115. H. Fieger and G. Blaschke, *J. Chromatogr., 575:* 255 (1992).
116. S. C. Tan, S. H. D. Jackson, C. G. Swift, and A. J. Hutt, *J. Chromatogr. B, 701:* 53 (1997).
117. B. Elsing and G. Blaschke, *J. Chromatogr., 612:* 223 (1993).
118. A. Shibukawa and I. W. Wainer, *J. Chromatogr., 574:* 85 (1992).
119. L. Silan, P. Jadaud, L. R. Whitfield, and I. W. Wainer, *J. Chromatogr., 532:* 227 (1990).
120. C. Hartmann, D. Krauss, H. Spahn, and E. Mutschler, *J. Chromatogr., 496:* 387 (1989).
121. C. Fernandez, F. Gimenez, B. Baune, V. Maradeix, A. Thuillier, and R. Farinoitti, *J. Chromatogr., 617:* 271 (1993).
122. H. Echizen, K. Ochiai, Y. Kato, K. Chiba, and T. Ishizake, *Clin. Chem., 36:* 1300 (1990).
123. P. A. Soons, M. C. M. Roosemalen, and D. D. Breimer, *J. Chromatogr., 528:* 343 (1990).
124. S. Pichini, R. Pacifici, I. Altieri, M. Pellegrini, and P. Zuccaro, *J. Liq. Chrom. Rel. Technol., 19:* 1927 (1996).
125. H. Takahashi, T. Kashima, S. Kimura, N. Muramoto, H. Nakahata, S. Kubo, Y. Shimoyama, M. Kajiwara, and H. Echizen, *J. Chromatogr. B, 701:* 71 (1997).
126. H. Y. Aboul-Enein and M. R. Islam, *J. Liq. Chromatogr., 15:* 3285 (1992).
127. M. Yamaguchi, K. Yamashita, I. Aoki, T. Tabata, S.-I. Hirai, T. Yashiki, *J. Chromatogr., 575:* 123 (1992).
128. D. R. Rutledge and C. Garrick, *J. Chromatogr., 497:* 181 (1989).
129. R. Straka, K. A. Johnson, P. S. Marshall, and R. P. Remmel, *J. Chromatogr., 530:* 89 (1990).
130. J. A. Chiarotto and I. W. Wainer, *J. Chromatogr. B, 665:* 147 (1995).

131. R. Heinig, V. Muschalek, and G. Ahr, *J. Chromatogr. B*, *655*: 286 (1994).
132. H. Zhang, J. T. Stewart, and M. Ujhelyi, *J. Chromatogr. B*, *668*: 309 (1995).
133. J. Liu and J. T. Stewart, *J. Chromatogr. B*, *692*: 141 (1997).
134. H. Y. Aboul-Enein and S. A. Bakr, *Biomed. Chromatogr.*, *7*: 38 (1993).
135. H. Takahashi, S. Kanno, H. Ogata, K. Kashiwada, M. Ohira, and K. Someya, *J. Pharm. Sci.*, *77*: 993 (1988).
136. M. E. DePuy, J. L. Demetriades, D. G. Musson, and J. D. Rogers, *J. Chromatogr. B*, *700*: 165 (1997).
137. A. Ceccato, P. Chiap, P. Hubert, and J. Crommen, *J. Chromatogr. B*, *698*: 161 (1997).
138. C. Fernandez, B. Baune, F. Gimenez, A. Thuiller, and R. Farinotti, *J. Chromatogr.*, *572*: 195 (1991).
139. J. Haginaka and J. Wakai, *Anal. Chem.*, *63*: 997 (1990).
140. B. Rochat, M. Amey, and P. Baumann, *Ther. Drug Monit.*, *17*: 273 (1995).
141. D. Castoldi, A. Oggioni, M. I. Renoldi, E. Ratti, S. DiGiovine, and A. Bernareggi, *J. Chromatogr. B*, *655*: 243 (1994).
142. C. Pham-Huy, B. Radenen, A. Sahui-Gnassi, and J. R. Claude, *J. Chromatogr. B*, *665*: 125 (1995).
143. J. W. Kelly, J. T. Stewart, and C. D. Blanton, *Biomed. Chromatogr.*, *8*: 255 (1994).
144. A. G. Adams and J. T. Stewart, *J. Liq. Chromatogr.*, *16*: 3863 (1993).
145. M. Siluveru and J. T. Stewart, *J. Chromatogr. B*, *690*: 359 (1997).
146. T. D. Doyle, C. A. Brunner, and J. A. Vick, *Biomed. Chromatogr.*, *5*: 43 (1991).
147. D. W. Boulton and J. P. Fawcett, *J. Chromatogr. B*, *672*: 103 (1995).
148. C. Egginger, W. Lindner, S. Kahr, and K. Stoschitzky, *Chirality*, *5*: 505 (1993).
149. S. A. Jortani and A. Poklis, *J. Anal. Tox.*, *17*: 374 (1993).
150. D. S. Hage, M. A. Nelson, *Anal. Chem.*, *73*: 198A (2001).
151. Y. Fintschenko and G. S. Wilson, *Mikrochim. Acta*, *129*: 7 (1998).
152. K. Shahdeo and H. T. Karnes, *Mikrochim. Acta*, *129*: 19 (1998).

153. M. de Frutos and F. E. Regnier, *Anal. Chem.*, *65*: 17A (1993).
154. G. Gubitz and C. Shellum, *Anal. Chim. Acta*, *283*: 421 (1993).
155. C. H. Pollema, J. Ruzicka, A. Lernmark, and J. D. Christian, *Microchem. J.*, *45*: 121 (1992).
156. B. Mattiasson, M. Nilsson, P. Berden, and H. Hakanson, *Trends Anal. Chem.*, *9*: 317 (1990).
157. R. Puchades, A. Maquieira, and J. Atienza, *Crit. Rev. Anal. Chem.*, *23*: 301 (1992).
158. M. A. Johns, L. K. Rosengarten, M. Jackson, and F. E. Regnier, *J. Chromatogr. A*, *743*: 195 (1996).
159. S. A. Cassidy, L. J. Janis, and F. E. Regnier, *Anal. Chem.*, *64*: 1973 (1992).
160. U. De Alwis, G. S. Wilson, *Anal. Chem.*, *59*: 2786 (1987).
161. M. J. Valencia-Gonzalez and M. E. Diaz-Garcia, *Ciencia*, *4*: 29 (1996).
162. D. A. Palmer, M. Evans, J. N. Miller, and M. T. French, *Analyst*, *119*: 943 (1994).
163. R. A. Durst, L. Locascio-Brown, A. L. Plant, in R. D. Schmid (Ed.), Flow Injection Analysis (FIA) Based on Enzymes or Antibodies, VCH, New York, 1991, p. 181.
164. L. Locascio-Brown, A. L. Plant, R. Chesler, M. Kroll, M. Ruddel, and R. A. Durst, *Clin. Chem.*, *39*: 386 (1993).
165. W. T. Yap, L. Locascio-Brown, A. L. Plant, S. J. Choquette, V. Horvath, and R. A. Durst, *Anal. Chem.*, *63*: 2007 (1991).
166. C. M. Rico, M. Del Pilar Fernandez, A. M. Guiterrez, M. C. P. Conde, and C. Camara, *Analyst*, *120*: 2589 (1995).
167. D. A. Palmer, T. E. Edmonds, and N. J. Seare, *Anal. Lett.*, *26*: 1425 (1993).
168. W. A. Clarke, Development of Sandwich Microcolumns for use in Rapid Chromatographic Immunoassays, Ph.D. Dissertation, University of Nebraska-Lincoln, 2000.
169. D. A. Palmer, R. Xuezhen, P. Fernandez-Hernando, and J. N. Miller, *Anal. Lett.*, *26*: 2543 (1993).
170. W. U. De Alwis and G. S. Wilson, *Anal. Chem.*, *57*: 2754 (1985).
171. P. C. Gunaratna and G. S. Wilson, *Anal. Chem.*, *65*: 1152 (1993).
172. A. Hacker, M. Hinterleitner, C. Shellum, and G. Gubitz, *Frenz. J. Anal. Chem.*, *352*: 793 (1995).
173. D. S. Hage and P. C. Kao, *Anal. Chem.*, *63*: 586 (1991).

174. D. S. Hage, B. Taylor, and P. C. Kao, *Clin. Chem.*, *38*: 1494 (1992).
175. D. S. Hage, *J. Clin. Ligand Assay*, *20*: 293 (1997).
176. D. S. Hage, D. H. Thomas, and M. S. Beck, *Anal. Chem.*, *65*: 1622 (1993).
177. D. S. Hage, D. H. Thomas, A. Roy Chowdhuri, and W. Clarke, *Anal. Chem.*, *71*: 2965 (1999).
178. C. L. Flurer and M. Novotny, *Anal. Chem.*, *65*: 817 (1993).
179. B. Nilsson, *J. Chromatogr.*, *276*: 413 (1983).
180. E. Reh, *J. Chromatogr.*, *433*: 119 (1988).
181. A. Farjam, A. E. Brugman, H. Lingeman, and U. A. T. Brinkman, *Analyst*, *116*: 891 (1991).
182. A. Farjam, A. E. Brugman, A. Soldaat, P. Timmerman, H. Lingeman, G. J. de Jong, R. W. Frei, and U. A. T. Brinkman, *Chromatographia*, *31*: 469 (1991).
183. T. Hayashi, S. Sakamoto, I. Wada, and H. Yoshida, *Chromatographia*, *27*: 574 (1989).
184. G. S. Rule and J. D. Henion, *J. Chromatogr.*, *582*: 103 (1992).
185. J. Cai and J. Henion, *Anal. Chem.*, *68*: 72 (1996).
186. B. Johansson, *J. Chromatogr.*, *381*: 107 (1986).
187. V. Kircher and H. Parlar, *J. Chromatogr. B*, *677*: 245 (1996).
188. L. J. Janis and F. E. Regnier, *Anal. Chem.*, *61*: 1901 (1989).
189. J. Cai and J. Henion, *J. Chromatogr. B*, *691*: 357 (1997).
190. A. Riggin, J. R. Sportsman, and F. E. Regnier, *J. Chromatogr.*, *632*: 37 (1993).
191. L. J. Janis and F. E. Regnier, *J. Chromatogr.*, *1*: 444 (1988).
192. L. J. Janis, A. Grott, F. E. Regnier, and S. J. Smith-Gill, *J. Chromatogr.*, *476*: 235 (1989).
193. A. Farjam, J. J. Vreuls, W. J. G. M. Cuppen, U. A. T. Brinkman, and G. J. de Jong, *Anal. Chem.*, *63*: 2481 (1991).
194. L. J. Cole and R. T. Kennedy, *Electrophoresis*, *16*: 549 (1995).
195. T. M. Phillips and J. J. Chmielinska, *Biomed. Chromatogr.*, *8*: 242 (1994).
196. N. A. Guzman, *J. Liq. Chromatogr.*, *18*: 3751 (1995).
197. H. Irth, A. J. Oosterkamp, U. R. Tjaden, and J. van der Greef, *Trends Anal. Chem.*, *14*: 355 (1995).
198. M. Vanderlaan, R. Lotti, G. Siek, D. King, and M. Goldstein, *J. Chromatogr. A*, *711*: 23 (1995).
199. A. J. Oosterkamp, H. Irth, U. R. Tjaden, and J. van der Greef, *Anal. Chem.*, *66*: 4295 (1994).

200. B. Y. Cho, H. Zou, R. Strong, D. H. Fisher, J. Nappier, and I. S. Krull, *J. Chromatogr. A, 743*: 181 (1996).
201. H. Irth, A. J. Oosterkamp, W. van der Welle, U. R. Tjaden, and J. van der Greef, *J. Chromatogr., 633*: 65 (1993).
202. A. J. Oosterkamp, H. Irth, M. Beth, K. K. Unger, U. R. Tjaden, and J. van der Greef, *J. Chromatogr. B, 653*: 55 (1994).

Index